DIGITAL REMOTE SENSING

About the Authors

Dr. Prithvish Nag (*b*. 1951) is M.Sc. (Gold Medalist) and Ph.D. from Banaras Hindu University. He was the Chairman of the Commission on Population Cartography of the International Cartographic Association (ICA) for the period 1987-91. He was the first Indian to become chairman of an ICA body. He was also Full Member of the Commission on Population Geography of the International Geographical Union (IGU) for two terms. He has been closely associated with the International activities related to geography and cartography, such as a Member of the International Working Group on the Thematic Maps of the World, and a Member of the Editorial Board of the International Journal of the Geographical Information Systems. He has been a Senior Visiting Member of the Staff of the School of Oriental and African Studies, University of London.

Dr. Nag is author and editor of several books and about 75 research papers. He is holder of the Commonwealth Bursary awarded by the Royal Geographical Society on its 150th anniversary. He has been UN Consultant in Cartography in the Sultanate of Oman recently.

Dr. Nag is now the Director of the National Atlas and Thematic Mapping Organisation (NATMO) of the Ministry of Science and Technology, Government of India; and President of the Indian National Cartographic Association and the Indian Association of Special Libraries and Information System.

Dr. M. Kudrat (*b*. 1953) is a M.Sc. and Ph.D. in Agriculture Chemistry from the University of Calcutta. In 1979 he joined the All India Soil and Land Use Survey as Senior Soil Surveyor. In 1986 he was appointed as Scientist/Engineer in the Regional Remote Sensing Service Centre of the Indian Space Research Organisation in Dehra Dun.

Dr. Kudrat has published about 32 papers on remote sensing applications including pedology. He has handled national projects like (a) Land Use Mapping, (b) Waste Land Mapping, (c) Integrated Mission for Sustainable Development, (d) Development of Methodology for Watershed Management Using Remote Sensing (Digital Image Processing) and GIS.

Dr. Kudrat is also involved in teaching in different courses on remote sensing conducted by the RRSSC and the Indian Institute of Remote Sensing, Dehra Dun.

DIGITAL REMOTE SENSING

Dr. P. NAG
Director
National Atlas and Thematic Mapping Organisation
Calcutta

Dr. M. KUDRAT
Scientist/Engineer
Indian Space Research Organisation
Dehradun

Foreword by

PROFESSOR B.L. DEEKSHATULU
Former Director, NRSA
Director, Centre for Space Science & Technology Education (CSSTE-AP)
(Affiliated to UN), Dehradun

CONCEPT PUBLISHING COMPANY, NEW DELHI - 110059

ISBN 81-7022-725-9 (P.B.)

First Published 1998

Published and Printed by
Ashok Kumar Mittal
Concept Publishing Company
A/15-16, Commercial Block, Mohan Garden,
New Delhi - 110059 (India)
Phones: 5648039, 5649024
Fax No. 091-11-5648053

FOREWORD

One of the very first thoughts that crossed my mind even as I turned the pages of this book was "why another book on digital remote sensing? Therefore, the fundamental question I asked myself as I read through the book was "Is it just another book on the subject"? With such a critical bias, I set out an extraordinary expectation and to my pleasure, I discovered that the book is well entrenched in a justified context.

The book is written in a simple and lucid English that makes it easier for inexperienced and beginners alike, to follow the intricate details of digital remote sensing. This character is of topical interest to us who are faced with a seemingly simple but inherently complex problem of teaching scientific workers across the Asia-Pacific region on a common language base. The book uses not so vast a vocabulary and the approach to explanation of scientific details is direct and organised. This raises the hope of effectively transferring the technology in a region of diverse linguistic background.

As a corollary to the simplicity in approach, the book appeals to me as a potential document that may be used as a text book in national and international training programmes. Its wide coverage gives it the analogous benefits of a comprehensive study parlour and I find it suitable to a broad range of syllabi. In addition to all these, it has this rare quality of absorbing the reader into its many splendoured corridors, which is an essential quality of text books directed towards students who might not voluntarily involve themselves into study and may need to be involuntarily drawn in.

Finally, the book is characterised by its composite fabric made of umpteen yarns of examples and case studies drawn from the Indian sub-continent served by the several Indian satellites that have adorned the space orbit in the recent times. Notwithstanding the other factors, this character alone gives a justifiable contemporary context to the book. It is perhaps for the first time the world is treated with such an exemplary collection of literature from this region. In this context, it is to be noted that India has made significant forays into almost every discipline of earth sciences, with the support of remote sensing technology. The self-sufficiency attained by India in making remote sensing satellites, data reception, training and applications have provided sustenance to National Natural Resources Management System (NNRMS). Under this mechanism and with the spirit of applied experimentation, several national level projects have been taken up in diverse resource themes. Moreover, the Indian Remote Sensing Satellites are an important element of the NNRMS and serve a national goal in terms of providing continuous and operational remote sensing data services for the management of natural resources. Further IRS data is also being distributed world over by EOSAT company of USA and many countries have set up earth stations to receive IRS data. The developments are placing Indian endeavour in a pioneering position of eminence in the world and as such it is often felt very appropriate to have books and documents written to herald the benefits of these efforts. Against this background, it is encouraging to see that a compilation of this nature exists in this book on digital remote sensing.

With all these observations, I opine that the book provides material for compulsive reading to the remote sensing community at large. I also hope that the book will play the role of a cultural ambassador to the development and progress that the Indian minds have attained and professed in the field.

B.L. DEEKSHATULU
Former Director, NRSA
Director, CSSTE-AP
(Affiliated to UN)
Dehradun

PREFACE

REMOTE SENSING has provided a new impetus for the earth, resource and environmental scientists. This gift from the space technology has to be fully harnessed for tackling the problems of the country. Increasing population and diminishing resources have compelled our planners, leaders and scientists to consider better ways for the management of natural resources. Plea for sustainable development has added new dimensions to resource and environmental management. Synoptic view and quick decision-making process have become essential to meet the challenges appearing over the national scene. At this juncture, remote sensing technology has come to the rescue.

The Indian space programme has provided opportunities for the scientists and technologists to work out strategies in the management of resources. The application of technology obviously required trained manpower to take benefit of the opportunities at their disposal. The thematic scientists have to be exposed in their own application areas. Five satellites, specifically for remote sensing (IRS-1A/1B/P2/1C/1D), have been launched. Regional remote sensing centres, state remote sensing centres; national and regional institutions, and universities have facilities for handling remote sensing data. There are institutions like the National Remote Sensing Agency, Indian Institute of Remote Sensing, Institute of Remote Sensing of Anna University, Space Applications Centre, National Atlas & Thematic Mapping Organisation which provide regular courses related to remote sensing applications. Further, the five Regional Remote Sensing Service Centres also run courses on digital remote sensing. Such efforts have increased the awareness about the potentials of the remote sensing technology among the scientists and planners. But, it is the responsibility of the application scientists to work out methodologies for integrating this latest technology with the traditional and conventional methods which is being practiced in the country over decades. This is perhaps the only way to make remote sensing most acceptable among the decision-makers in their own set-up.

This book has been written with an objective to introduce the remote sensing technology to the application scientists in India. The language has been kept as simple as possible. In such a technology based book, jargons, acronyms, abbreviations are perhaps inevitable. Nevertheless, we have tried to make it easy by repeating the meanings. Still it may be of interest for some users to refer to books on mathematics or statistics or even physics for more details as the scope of this book has its own limitations. The idea is to provide a comprehensive picture of the scenario of remote sensing in India.

The intention here is not to bring out a text book on remote sensing, but institutions, teachers and students may find that it covers their course contents in one way or other. Attempts have been made to reduce the publication time as much as possible as the related technology is changing very fast. In due course of time such changes may effect the conceptual aspects of the book as well.

We have been extremely fortunate in getting the blessings and guidance of Dr. K. Kasturirangan, Chairman, Indian Space Research Organisation and Secretary, Department of Space, in organising this volume. This book has become valuable to us because Professor B.L. Deekshatulu, Director, Centre for Space Science and Technology Education (CSSTE-AP), Dehra Dun has kindly agreed to write the foreword. He has been a source of inspiration for all remote sensing related activities in India, including writing of this book. We are grateful to both of them.

We are also grateful to Professor V.S. Ramamurthy, Secretary, Department of Science & Technology, Government of India for encouraging us in promoting the application of science and technology in the country.

A volume of this nature cannot be completed in isolation. We have to take help and suggestions from our friends and colleagues all over the country. Our association with the Regional Remote Sensing Service Centres in Bangalore and Dehra Dun; and the national institutions like the National Atlas and Thematic Mapping Organisation and the All India Soil and Land Use Survey Organisation gave ample opportunities to keep in touch with the latest technology as well as the scientists working in this field. Our associations with the RRSSC Bangalore brought us closer to the headquarters of the Indian Space Research Organisation, particularly Dr. Mukund Rao, Dr. B. Manikiam and Mr J. Krishnamurthy of the Central Management Office of the National Natural Resources Management System (NNRMS) and Dr. S. Adiga, Head, Regional Remote Sensing Service Centre, Bangalore. Similarly, RRSSC Dehra Dun gave us opportunity to interact with the Indian Institute of Remote Sensing. We are grateful to Prof. R. Mishra, Prof. S.K. Bhan, Prof. N.D. Sharma, Dr. R.C. Lakhera, Mr. I.J. Dhingra, Dr. S.K. Saha, Mr. S.K. Mittal and Mr. S.C. Gupta of the IIRS; and Dr. M.L. Manchanda, Mr. K.V. Ravindran, Dr. R.K. Sharma, Dr. A.K. Tiwari of the RRSSC Dehra Dun. We have been in close touch with the application scientists of the NRSA, particularly Dr. D.P. Rao, Dr. N.C. Gautam, Mr. G. Behera, Dr. A. Perumal, Mr. S.K. Subramanian, and Dr. C.B.S. Dutt. Our discussions about possible remote sensing applications with Dr. A.K.S. Gopalan and Mr. A.R. Dasgupta of the SAC, Ahmedabad were very useful. From time to time we have been interacting with our colleagues and friends and were benefited by their valuable suggestions. To mention a few Mr. Y.S. Rajan and Dr. R.K. Midha of the Department of Science and Technology, New Delhi, Mr. Dilip Pal of RRSSC Kharagpur, Dr. R.

Ramachandran of the Advanced Data Processing Research Institute (ADRIN), Hyderabad, Mr. K. Krishnan Unni and Mr. E.V.R. Parthasarthi of the Geological Survey of India, Hyderabad and Prof. A.K. Bagchi of the Sultan Qabos University, Muscat. In fact we have been persuaded by some of our geographer-friends to bring this book out soon as remote sensing has been included in the post-graduate courses in geography all over the country. Some of them are Dr. G.K. Dutt, former Director of NATMO, Prof. N.K. De of Burdwan University, Prof. D.C. Barai of Bangalore University, Dr. V.K. Kumra, Dr. M.B. Singh and Dr. J. Singh of BHU, Dr. V.C. Jha of Visva Bharati, Dr. P. Venketachalam and Dr. Smita Sengupta of IIT Mumbai, Dr. Surya Kant of Panjab University, Chandigarh, Prof. T. Natrajan of the Institute of Remote Sensing of the Anna University, Madras; Mr. M.P. Singh of DAV College, Dehra Dun, Dr. Haushila Prasad of Kenyatta University, Nairobi, Dr. S.K. Singh of Kumaun University, Almora, Dr. S. Chattopadhyay of the Centre for Earth Science Studies, Trivandrum, and Prof. H.R. Betal of Calcutta University. Our special thanks are for our colleagues in NATMO, Calcutta. They are Mr. G.N. Saha, Mr. A.K. Malik, Dr. R. Prasad, Dr. R.K. Shukla, Dr. R.K. Dhabai, Mrs. Baisakhi Sarkar, Smt. Alpana Chatterjee, Smt. Subhra Choudhury and others. For preparing the manuscript we had to consult several libraries including that of NATMO and IIRS. We are thankful to the staff of both the libraries. The manuscript could not have been completed without the help of Mr. S.K. Banerjee, Ms. Sangeeta, Mrs. Indira Choudhury, Ms. Shashi Garg and Mrs. Swarnlata. Needless to say, Shri Ashok Kumar of the Concept Publishing Company, New Delhi gladly accepted our proposal to bring out this book. We are thankful to him as well.

DR. PRITHVISH NAG
DR. M. KUDRAT

CONTENTS

LIST OF FIGURES

LIST OF COLOURED PLATES

ONE

INTRODUCTION

REMOTE SENSING in India has developed within last twenty-five years. India is considered to have entered the space age in 1972 with the establishment of the Department of Space. Since then, there has been a systematic growth of this programme resulting a well integrated system having space, ground and application segments. After the success of the first Indian satellite, Aryabhatta, in April 1975, a programme was initiated for launching experimental remote sensing satellites Bhaskara I and II (June 7, 1979 and November 20, 1981 respectively). The experience gained through these two satellites and Apple, the experimental phase was over. The situation was ripe for launching operational remote sensing satellite. Simultaneously, data reception station was established in Hyderabad. Joint experimental projects (JEPs) were conducted with the user agencies in order to harness this technology for the development in the priority areas.

Why Remote Sensing ?

The concept of National Natural resources Management System or NNRMS was mooted in order to find out possible solutions of the pressing problems in India concerning the increasing population and diminishing resources. Such a step was necessary because of phenomenal growth of population from 350m to 850m since independence. Even after Green Revolution, when the food production rose from 55m tonnes to 172m tonnes, the crop yield in India remained one of the lowest in the world (1.7 tonnes/ha). Obviously there was pressure on forest land which got reduced. The productive utilization of wastelands was considered as a possible option to meet the demand of the teeming population. In order to bring more land under production, it was necessary to quickly assess the status of different natural resources such as water, land, soils, vegetation and the like. At this juncture, remote sensing became an essential tool for management, development, and planning. The synoptic view available at regular interval was a welcome change in the decision-support or decision-making system.

With the Bruntland Commission report pleading for sustainable development and the resolutions of the Earth Summit held in Rio-de-Janeiro (Agenda 21), along with resources, the development and environmental parameters have become important. Here again remote sensing has very important applications. As the resolution becomes better, more details could be monitored by the remote sensing technology.

Remote Sensing Activities in India

NNRMS was aimed at optimal utilization of country's natural resources by a proper and systematic inventory of resource availability using remote sensing data in conjunction with conventional techniques (Kasturirangan, 1995). User involvement and participation was very much desired. For this purpose, the Regional Remote Sensing Service Centres or RRSSCs were established, and the support was extended to the different state remote sensing centres. The biggest boost was the successful launching of the Indian Remote Sensing or IRS-1A satellite in March 1988. Following this, the IRS-1B and IRS-P2 satellites were also successfully launched in August 1991 and October 1994 respectively. The latter satellite was launched from the indigenously developed Polar Satellite Launch Vehicle (PSLV). The next satellite in this series is the IRS-1C which has been launched in December 1995.

Even before launching of the IRS satellites, the user community within the Department of Space and outside were using the Landsat data of the US and SPOT data of France. In fact the Landsat false colour composites (FCCs) were very popular with the scientists involved in landuse, soil, water resources and forest mapping. Till today, apart from IRS, Landsat and SPOT data are widely used. Due to the availability of digital data processing facilities now available all over the country, digital remote sensing of the data from these three satellites is becoming more and more practiced. Hence in this book, the application of the Landsat, SPOT and IRS data have been discussed at great length.

Apart from the DOS sponsored institutions and setup, different ministries of the Central and State governments have established remote sensing cells which have at least visual interpretation facilities. In addition, the university departments dealing with natural resources, and planning and development activities have similar facilities. A new brand of scientists and engineers are required to avail this technology. Since, remote sensing technology has several dimensions, such trained manpower are from various disciplines, such as geography, soil science, geology, forestry, statistics, physics and computers. In order to train them, courses have been planned in the national level institutions like the National Remote Sensing Agency (NRSA), Hyderabad, Indian Institute of Remote Sensing (IIRS), Dehra Dun, Space Applications Centre (SAC), Ahmedabad and the like. Universities have also opened regular post-graduate level (M. Tech.) courses in remote sensing, such as Institute of Remote Sensing of Anna University and the Roorkee University. In some universities remote sensing or image processing has also become a full course paper at the post-graduate level in subjects like geography and geology. Hence, there is an awareness of this technology in the scientific community.

Due to the initiatives taken by the Indian Space Research Organisation (ISRO) and the Department of Space (DOS), there has been a close co-operation between them and the private sector. NRSA has been transferring patents for manufacturing remote sensing instruments under their guidance. These arrangements have encouraged private investments in remote sensing. In fact there is a confederation of remote sensing industries in the country. On the other hand, the Indian Remote Sensing Satellites IRS-1B and IRS-P2 data are now being received at Norman Ground Station, Oklahoma in the US and distributed by EOSAT. This indicates the wider acceptability of the Indian data across its boundaries. The Polar Satellite Launch Vehicle (PSLV) has put India at par with the most advanced nations in the world.

Need for a Remote Sensing Book

Though a good number of scientists have been trained in remote sensing, there has been a serious problem of getting a book on this subject. There are different reports and newsletters from the Indian Space Research Organisation or the Department of Space which are being very generously distributed, but they have limited circulation and may not reach the students who are pursuing post-graduate courses. Further some edited books have been brought out including one by us (Nag, 1992). But they hardly meet the requirements of the students who would like to have the details more systematically. Hence, this book has been planned with an aim to fill this void.

We have been working on this book since 1991. Both the authors met several times to work out the details. The remote sensing scenario, computers and software capabilities are changing so fast that we have to keep on revising frequently. We have done our best to include the latest developments, but there may still be some gaps. Perhaps such problems cannot be avoided in a book like this.

Outline of the Book

Keeping in view of the above objectives, this book was planned basically for the Indian scientists and engineers involved in using this gift from space technology. There are 13 chapters in this volume. Some of the chapters are directly dealing with the title of this publication, i.e. digital remote sensing. But there are some chapters which do not exactly fit in the scheme of things. For example, the visual interpretation chapter was thought essential because after digital analysis, one has to interpret the features visually. The details of the visual interpretation instruments do provide different options and support system as well as the process of the development of the instruments from purely visual to digital. Further, not exactly for the above reason, the chapter on the Geographical Information System (GIS) has been included because the available technologies and methodologies have brought them closer. Both are getting integrated. Software companies are claiming live links between the both. Other associated technologies like Land Information System (LIS) and Global Positioning System (GPS), have also been included for almost similar reasons. The chapter on "applications" is meant to give an idea of the research and application of the remote sensing technology in India. It cannot be claimed comprehensive, but indicative only. There are several remote sensing applications centres at national, regional and university or institution levels. Their research work are either mentioned in their newsletters or reports. In addition, there are Indian journals which are dealing with remote sensing directly or along with other works. The most important one if the *Photonirvachaka* of the Indian Society of Remote Sensing which is published from Dehra Dun. The journals of geography, geology, forestry and land or earth based multi-disciplinary subjects do publish articles on remote sensing. In addition, several papers of applications by the Indian authors are published abroad. It may not be possible to keep track of all such journals. However, *NNRMS Bulletin* of ISRO, *Space News* from Space Applications Centre, Ahmedabad and *Interface* of the National Remote Sensing Agency, Hyderabad provide valuable information related to remote sensing activities in India. Further, in order to make the book more comprehensive, several annexure have been included. They deal with essential and relevant information.

The Chapters

Including the two chapters mentioned above, this book has been divided into following chapters:

— Electromagnetic radiation
— Imaging systems
— Visual interpretation
— Hardware and software options
— Preprocessing and rectification
— Enhancement techniques
— Spatial filtering
— Band combination
— Classification techniques
— Geographical information system
— Application trends
— Conclusion

The first three chapters are essential for any application scientists using either visual interpretation and/or digital remote sensing techniques. The chapter on hardware and software is basically introduction to digital processing of data including geometric corrections. The following chapters on preprocessing and rectification, enhancements, filtering techniques, band combination and classification are essential for any book on digital remote sensing. The chapter on GIS is meant to integrate the remote sensing data with conventional information sets. Such an aspect has become important in the Indian situation because a lot of data have been collected traditionally which is important as well. For the acceptability of remote sensing data such integration has become pivotal.

In the last chapter on applications, the versatility of remote sensing data has been highlighted. On the other hand, attempts have been made to work out hybrid models consisting of all sorts of data. Such models have internal mechanism for checking and corrections. Hence, step by step integration is possible.

Foregleam

India was not very late in catching up with the other countries of the world in space technology. In fact in the beginning of the 1970s when the US and Russia took initiatives in sending their satellites, India was planning a forward leap in this field. Within a period of a quarter of century, India is one of the foremost countries in the world. It has got a well chalked out plan for next 25 years. This country has a bright future in harnessing the fruit of the space technology.

REFERENCES

Baliga, B.B. (1995), Remote sensing and development, *Science and Culture*, vol. 61, No. 1-3, pp. 1-10.

Chaurasia, R., P.K. Sharma, D.C. Loshali and Minakshi (1995), The science of remote sensing and its application in natural resource management, *Science and Culture*, vol. 61, No. 1-3, pp. 11-15.

Kasturirangan, K. (1995), Remote sensing in India: Present scenario and future thrusts, *Photonirvachak*, vol. 23, No.1, pp. 1-6.

Nag, P. (ed) (1992), *Thematic Cartography and Remote Sensing*, Concept Publishing Company, New Delhi.

Pisharoty, P.R. (1993), Remote sensing applications for resource development, *Photonirvachak*, vol. 21, No. 4, pp. 185-90.

Rao, U.R. (1987), India Reiterates Commitment to Peaceful Uses of Outer Space, Address to the 24th Session of the Scientific and Technical Sub-committee of the United Nations Committee on the Peaceful Uses of the Outer Space (UNCOPUOS), New York, *Space India*, vol. 1987/2, pp. 18-9.

TWO

ELECTROMAGNETIC RADIATION

ELECTROMAGNETIC energy or radiation is the medium for transmitting information from an object to sensor. The information is propagated by electromagnetic radiation at the velocity of light from the source (earth surface) directly through free space or indirectly by reflection, scattering and radiation to the sensor. The interaction of electromagnetic radiation with natural surfaces and the atmosphere is strongly dependent on the frequency or wavelength of the radiations. Electromagnetic spectrum in different wavelength region or bands tends to excite different mechanism namely electronic, molecular or conductive mechanisms. This is dealt separately in this chapter.

Electromagnetic Spectrum

Electromagnetic spectrum or energy have oscillating electric and magnetic fields. They travel in a harmonic sinusoidal fashion at the velocity of light (3×10^8 m/sec). Electromagnetic spectrum propagate in the form of wave. The distance between one wave peak and the next wave peak is called wave length (λ) and the number of peaks passing a fixed point in space per unit time is called frequency (ν) of the wave. The electric field (E) and the magnetic field (H) of electromagnetic wave are perpendicular to each other and also perpendicular to the direction of propagation of the wave (Slater, 1980; Deekshatulu and Bajpai, 1982). Maxwell showed that radiation is described as rapid variations in the electromagnetic field surrounding a charged particle and the variation in field being caused by the vibration of the particle. Since velocity of light is constant (3×10^8 m/sec), the wave length (λ) of electromagnetic radiation and the frequency (ν) are inversely related to each other. The relationship is expressed in equation 1.

$$\nu = \frac{C}{\lambda} \qquad \ldots \qquad (1)$$

The wavelength of electromagnetic radiation depends upon how long a time the charged particle is accelerated, and the frequency (ν) of the radiation depends on the frequency of vibration of the particle.

Classification of Electromagnetic Spectrum

The electromagnetic spectrum includes the entire spectrum starting from cosmic rays (λ 10^{-11} cm to 10^{-8} cm) at the short wavelength to audio wave at the longest end (λ = 10 km). This is shown in Fig. 2.1 and Table 2.1. The radio band covers the region of wave length longer than 10 cm (30 cm to 3 km or frequency less than 3 GH_Z). This region is used by the active radio/sensors such as imaging radar, altimeters, and sounders and to a lesser extent passive radiometers.

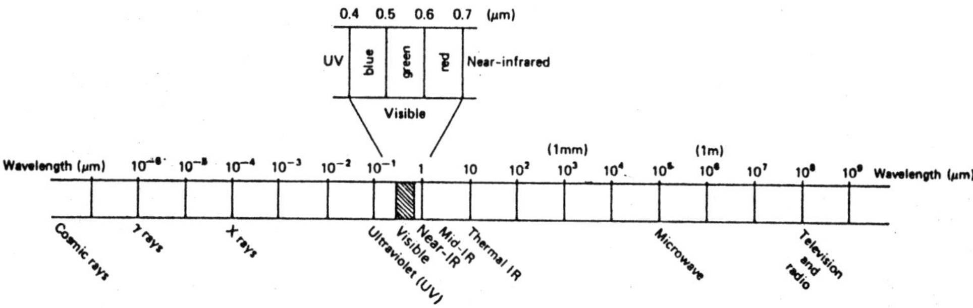

Fig. 2.1: Classification of electromagnetic radiation

TABLE 2.1: Different electromagnetic radiations and their wavelength ranges

Sl. No.	Name/class	Wavelength range (λ)
1	Cosmic rays	< 10^{-11} cm
2	Gama rays	10^{-11} – 10^{-8} cm
3	X-rays	0.3×10^{-8} cm — 3×10^{-6} cm
4	Ultraviolet rays	0.3μ — 0.4μ
5	Visible rays	0.4μ — 0.7μ
(i)	Blue	0.4μ — 0.5μ
(ii)	Green	0.5μ — 0.6μ
(iii)	Red	0.6μ — 0.7μ
6	Infrared rays	0.7μ — 1μ
(i)	Near infrared	0.7μ — 1.3μ
(ii)	Middle infrared	1.3μ — 3.0μ
(iii)	Far infrared or Thermal infrared	3.0μ — 0.3mm
7	Micro wave	1mm — 1m
8	Radio wave	30 cm — 3 km
9	Audio wave	> 3 km

The micro wave bands have the wavelength range 1mm to 1m. In this region most of the interactions with matters are governed by molecular rotation. This region is mostly used by microwave radiometers or spectrometers and radar system. This region is also useful for earth science studies.

The infrared band covers the spectral region from 0.7μm to 1μm. This region is further subdivided into near infrared (0.7μm to 1mm), middle infrared (3mm to 30mm) and thermal infrared (30μm to 0.3μm). Molecular rotation and vibration take place within the region. This region is very important in remote sensing for vegetation, hydro thermal alteration of rocks and soil moisture.

The visible band covers the spectral region from 0.4 to 0.7μm. The interaction mechanisms are similar to that of infrared region. In addition, electronic transition or excitation may also takes place during interaction with the matter.

The next neighbouring region is ultraviolet band. It covers the spectral region 0.03μm to 0.4μm. Electronic energy level play the main role in wave-matter interaction. This region is mainly used to sense the planetary atmospheres or surfaces with no atmospheres because of the optical density of gases. X-rays (0.03°A to 0.3A°), gamma rays and cosmic rays less than (0.03 A°) have been used to a lesser extent because of atmospheric absorptions capacity. Their use has been limited to planetary surfaces. Cosmic rays is radioactive and of limited use in remote sensing. In developed country gamma ray was being used for soil moisture study.

Nature of Electromagnetic Radiation

The electromagnetic radiation shows dual nature. Under identical same circumstances, electromagnetic radiation behaves as wave, while in other situation, it functions as particle or packet of energy called quantum. Newton first postulated the corpascular nature of light emitted by shining substance. The corpascular nature of light was derived from the observation that in a uniform medium light appears to travel in a straight line. Newton also had the idea about the wave nature of light from his experiment with colour interference effect in soap bubbles and between the glass plates. This colour interference is referred as Newton's ring (Hunt, 1980; Slater, 1980). From the beginning of seventeenth century, many workers described light as a wave motion spreading out from source in all direction. Huyghens (1629-1695) and Young (1802) contended that each point on a wave front may be considered to be a new source of spherical waves, and the envelop of those wave constitutes wave front at a later time. Maxwell first formulated the wave theory of electromagnetic radiation. He successfully characterized the magnetic and electric field and their relation to charge and current. He expressed these relationship in a set of partial differential equation, now popularly known as Maxwell's equations of electromagnetic waves. According to Maxwell, it was possible to have wave like configuration of electric and magnetic fields which are entirely disconnected from their source, and are propagated throughout the space in all possible directions with a constant speed (velocity of light). Maxwell theory of electromagnetic wave fails to explain the interaction of electromagnetic energy with matter on an atomic and molecular level, though it successfully explains the propagation of light, reflection, refraction and interference of electromagnetic radiation.

Max Plank in 1900 proposed a revolutionary hypothesis on quantum theory of electromagnetic radiation. He stated that an oscillator emits or takes up energy continuously and energy changes occur in the form of packet or quanta. Max Plank introduced three assumptions:

(i) The energy of oscillator cannot vary continuously but only by discrete amounts. Hence each oscillator has definite energy level,
(ii) The oscillator emitting an energy of frequency (ν) can only radiate in units or quanta of magnitude hy, where 'h' is Plank constant,
(iii) The energy levels of oscillator can only be integral multiple of quantum.

Interaction of EMR and the Atmosphere

The sun is the source of electromagnetic radiation (EMR) in passive remote sensing. The atmosphere comes in the way of the radiation when it travel from the sun to target as well as when reflected or emitted radiation travel from the earth surface (target) to the sensor (Slater, 1980; Swain & Davis, 1978). The radiation flux either from the sun or from the target interact with the atmosphere in two ways. These are either absorption and or scattering. Remote sensing mainly concerned with the radiant flux (exitence from the earth surface). It includes the reflectance and emitance. But the remote sensing sensors detect the radiant exitance of the earth atmosphere as well as earth surface. For earth resource remote sensing detectors, atmosphere acts as a villain. Absorption and scattering, both the phenomena affect the original signature of the target and both the phenomena are referred to attenuation. Unless the atmospheric effect is constant throughout the scene, corrections for the effect may have a significant impact on the final analysis of the scene (Parsons & Jurica, 1975). Strong absorption of EMR in some specific wavelength region by the constituents of the atmosphere restricts the availability of the entire electromagnetic spectrum for passive remote sensing. Only some portion of the EMR is available. The electromagnetic spectra of specific wavelength regions are transparent to the atmosphere and are termed as atmospheric window (Fig. 2.2). Only window region are being used. The atmospheric absorption while irradiance incident on the earth surface can be overcome by choosing specific atmospheric window region. The main problem is the modification of exitence of EMR from the earth either by absorption or scattering. Both scattering and absorption reduces the scene contrast (Slater, 1980).

The detail discussion on the mechanism and fundamentals of scattering is

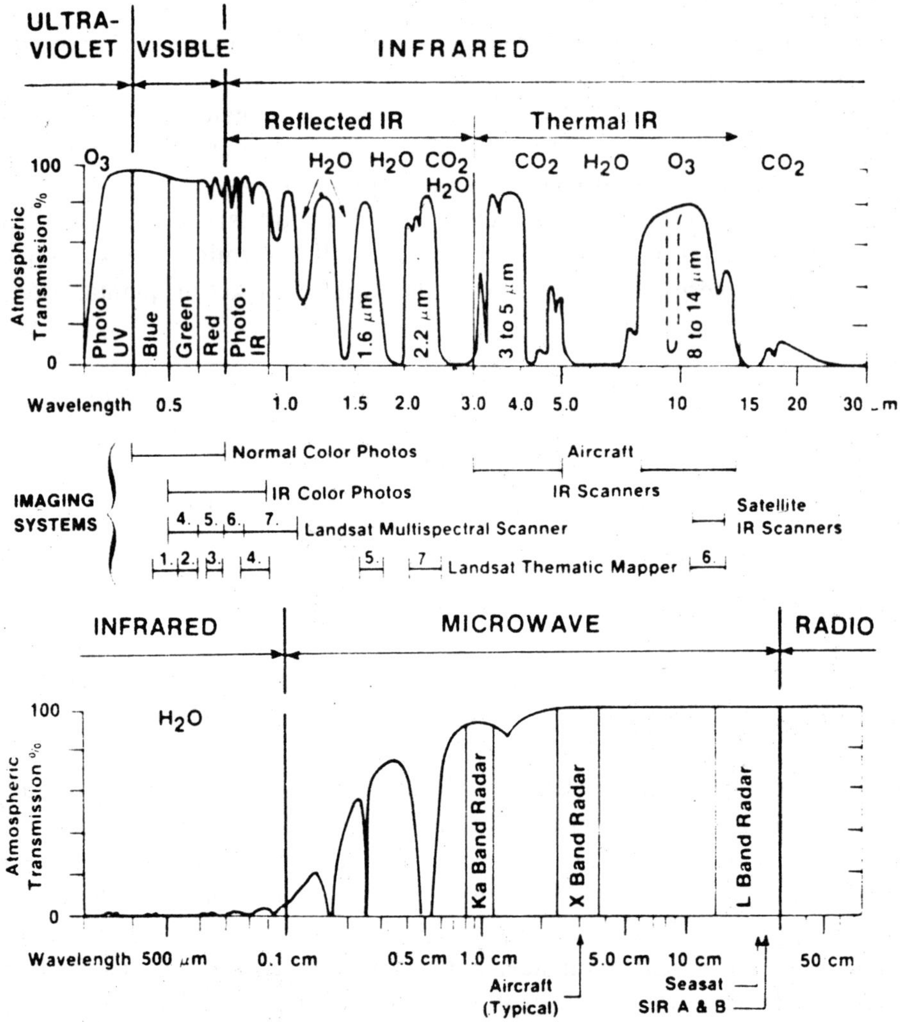

Fig. 2.2: Atmospheric windows

Expanded diagrams of the visible and infrared regions (upper) and the microwave region (lower) showing atmospheric windows. Wavelength bands of commonly used remote sensing systems are indicated. Gases responsible for atmospheric absorption are shown.

beyond the scope of this volume. Here the authors have highlighted the basic idea and familiarization about absorption and scattering.

Absorption

Absorption is a very simple phenomenon and is wavelength dependent and composition of the interacting medium (here atmosphere). Absorption is a thermodynamically transformation of radiant energy into heat energy. This implies that with the absorption of EMR, each and every body in the universe will be heated up. In the visible and photographic infrared region (0.4 to 0.9μm) absorption in the clear atmosphere is negligible. Ozone show a strong absorption in the ultraviolet region of the spectrum below the wavelength region 0.29 to 0.39μm. Water vapour and carbon di-oxide, the constituent of the atmosphere absorb a number of bands in the infrared region as shown in Fig. 2.3.

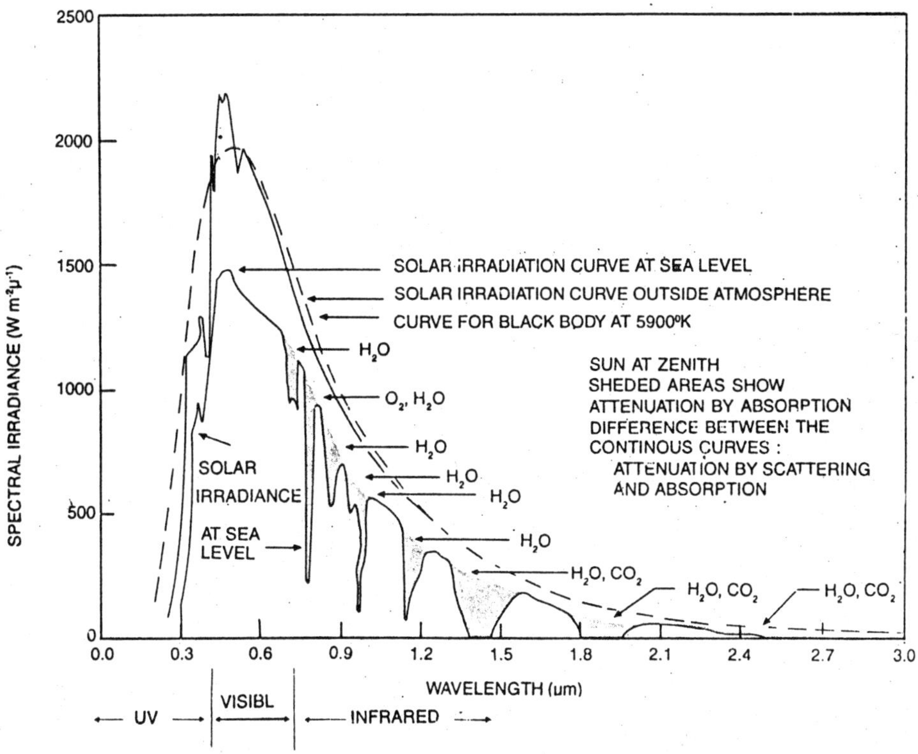

Fig. 2.3 : Spectral irradiance of the earth surface

Scattering

Pure scattering generally occurs in the absence of any absorption. During scattering there is no loss of energy. Scattering is only a redirection of energy. If the field of view of the sensor is narrow, then scattering causes some of the energy to be redirected outside the field of view of the sensors. On the other hand, some scattered radiation may reach the sensor. So in the former case scattering causes an apparent attenuation or dimming of the image, whereas in the latter case increases the intensity of the signal being received by the sensor (Swain & Davis, 1978). With long experience in handling remote sensing data it will be apparent that in Landsat TM or IRS-1A, band 1 shows higher digital counts than that expected. So in either way it modifies the signal of the target and modifies the scene contrast. Scattering may be defined as the process by which a small particle suspended in a medium of different refractive index, diffuse a portion of light in all direction. The reradiation or redirected radiant flux incident upon a rough surface is due to scattering. The scattering process depends on the size distribution of the scattering elements, their composition and concentration and wavelength of radiant flux incident on them (Slater, 1980).

In the context of remote sensing, scattering can be divided into three different categories depending on the size of the particle and wavelength of scatter/flux (Barrett & Curtis, 1992). Lord Rayleigh explained the scattering based on the following four assumptions:

(a) Scattering elements are distributed randomly in the scattering medium;
(b) The scattering elements are well separated from each other so that the scattering action, if any, is dependent of that its neighbouring elements;
(c) The elements are neither metallic nor absorbing; and
(d) Scattering is independent of shape and anistrophy of the scattering elements.

Rayleigh Scattering

This type of scattering involves molecules and other tiny particles with a diameter less than the effective wavelength of radiant flux (Coulson *et al.*, 1960). Based on the above assumption atmospheric scattering can be described in terms of Rayleigh's scattering coefficient as follows :

$$\beta_{\theta\lambda} = \frac{2\pi}{N\lambda^4} [n(\lambda) - 1]^2 (1 + \mathrm{Cos}^2\theta) \qquad \ldots\ldots \quad (1)$$

where $\beta_{\theta\lambda}$ = Rayleigh's scattering coefficient
$n(\lambda)$ = Refractive index of the molecules
N = Number of scattering molecules per unit volume in the atmosphere
θ = Angle between the incident and scattered flux
λ = Effective wavelength

The spectral intensity (I_λ) of incident unpolarized flux is directly related with the scattering coefficient and is presented in the following equation:

$$I_{\theta\lambda} = \beta_{\theta\lambda} I_\lambda \qquad \ldots\ldots \quad (2)$$

where $I_{\theta\lambda}$ = Scattered flux per unit volume
I_λ = Spectral intensity of incident unpolarized flux

From the above two equations, it can be noted that the scattered flux is inversely proportional to the fourth power of the effective wavelength of the incident flux and to the number of scattering molecules per unit volume present in the atmosphere. Fig 2.4 shows the scattered flux is distributed symmetrically around the scattered centre, the slope of the forward scattering is as pronounced as that of backward scattering. This also explain the blue colour of the sky since wavelength of blue colour is less than that of red. In the evenings as well as in the mornings, the sun is located near the horizon and radiant flux has to travel more path or more grazing. During grazing time most of the short waves scattered leaving higher wavelength. This is the reason why sky becomes red or orange at the time of sun set or sun rise. Scattering in the atmosphere is actually a multiple process in which the flux scattered by one element can be scattered again by other elements.

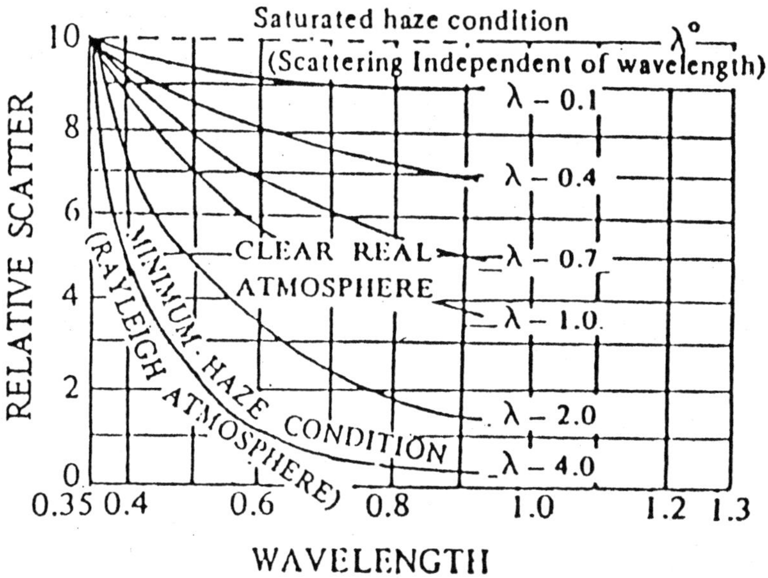

Fig. 2.4: Relative scatter as a function of wavelength for various magnitudes of atmospheric haze

The scattering element size is a very important parameter to decide which types of scattering is dominant. Size parameter of scattering elements (ρ) can be defined as in the following equation :

$$\rho = 2\pi r/\lambda \qquad .... \qquad (3)$$

where r = radius of the scattering elements
λ = effective wavelength of incident flux

This is a very useful equation because it allows for the convenient separation of the different types of scattering. For a non-absorbing elements, whose size is such that $\rho < 1$, the scattering process is identical to Rayleigh scattering.

Mie Scattering

This occurs when the atmosphere contains essentially spherical particles whose size, i.e. diameters approximate to the effective wavelength of the incident radiant flux. This means the ρ in equations should lie between 2 to 6. This scattering is due to water vapour and atmospheric aerosol. In the Mie regime, scattering is not symmetrical and the amount of the lope forward scattering is higher than that of the backward scattering.

Non-selective Scattering

Non-selective scattering takes place when size of the diameter of the scattering elements are several times more than the effective wavelength of radiant flux in question. Here ρ must be above 6. For example water droplets having a diameter ranging from 5 to 100μm scatter all wavelengths of visible light (0.4 to 7μm) with equal efficiency. Cloud and fog are the examples of non-selective scattering because all the colour in rainbow (VIBGYOR) with equal intensity (approximately) produces white colour.

Interaction of EMR with matter on the Surface of the Earth

There is a possibility of three fundamental interactions with electromagnetic energy and the matter. These reactions are reflection, absorption and transition. This means spectral radiant energy incident on the surface of the earth is partially reflected at the surface and the remaining part is being transmitted with or without absorption by the material on the earth surface (Lillesand & Keifer, 1990; Slater, 1980; Swain & Davis, 1978). According to the law of conservation of energy, we can write the interrelationship between the above reactions as follows :

$$E_{i(\lambda)} = E_{R(\lambda)} + E_{A(\lambda)} + E_{T(\lambda)} \qquad \dots \quad (4)$$

where $E_{i(\lambda)}$, $E_{R(\lambda)}$, $E_{A(\lambda)}$ and $E_{T(\lambda)}$ refer to the incident, reflected, transmitted and absorbed radiant flux respectively. All the energy components are function of wave length. By $E_{i(\lambda)}$, the equation 4 becomes :

$$\rho_\lambda + \tau_\lambda + \alpha_\lambda = 1 \qquad \dots \quad (5)$$

where ρ_λ, τ_λ and α_λ denote the spectral reflectance, spectral transmittance and absorptance respectively. This is explained by Fig. 2.5. The interaction of EMR with matter on the earth surface and their interrelationship is a function of the nature of the matter and wavelength of interacting radiations. This phenomenon leads the analyst to indentify different features on an image in specific wavelength regions.

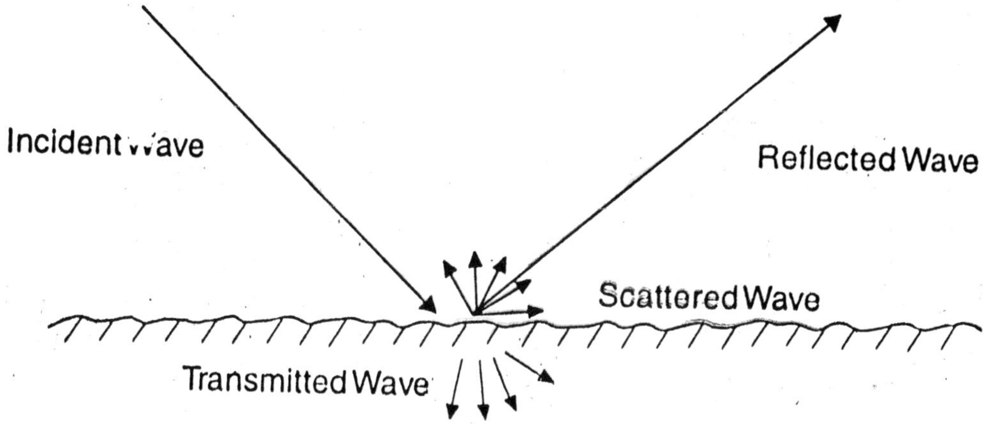

Fig. 2.5 : Wave interaction with an interface

In remote sensing, the spectral exitance is being measured. This is in general a combination of reflected flux and thermally emitted flux. Thus total outcoming influx form a target (matter) is given by the following equation :

$$E_{\lambda T} = E_{\lambda r} + E_{\lambda e} \quad \ldots \quad (6)$$

where $E_{\lambda T}$, $E_{\lambda r}$ and $E_{\lambda e}$ are total outcoming influx, reflected flux and emitted flux respectively. The remote sensing community is more interested in separate component ($E_{\lambda r}$ and $E_{\lambda e}$). Considering sun is a black body radiator, the maximum incident radiant flux is being received by the earth is as 0.5μm, which is well removed from the maximum for thermal exitence at the earth at about 10μm (Slater, 1980). Since the temperature of the earth is assumed to be about 300°K and behave like black body.

Fig. 2.6 shows the comparison between radiant exitance and solar incident radiant flux (left hand ordinates) and radiance (right hand ordinates). Based on Kirchhoff's law, curves plotted for various values of surface diffuse reflectance ρ and emissivity. From the Fig. 2.7 it is noticed that the point 'A', the intersection between the curve 4 and 5 represents the approximate wavelength below which solar reflected flux exceeds the emitted flux. But the point 'B', the intersection between the curve 2 and 6 represents the approximate wavelength above which thermal emission becomes dominant component. Point 'A' at about 3.0μm represents a reliable estimate of lower wavelength limit because the reflectance value 0.01 are not often found in nature. The point 'B' at about 4.5μm is less certain because the slope of the curves 2 and 6 are less and because higher values of reflectance

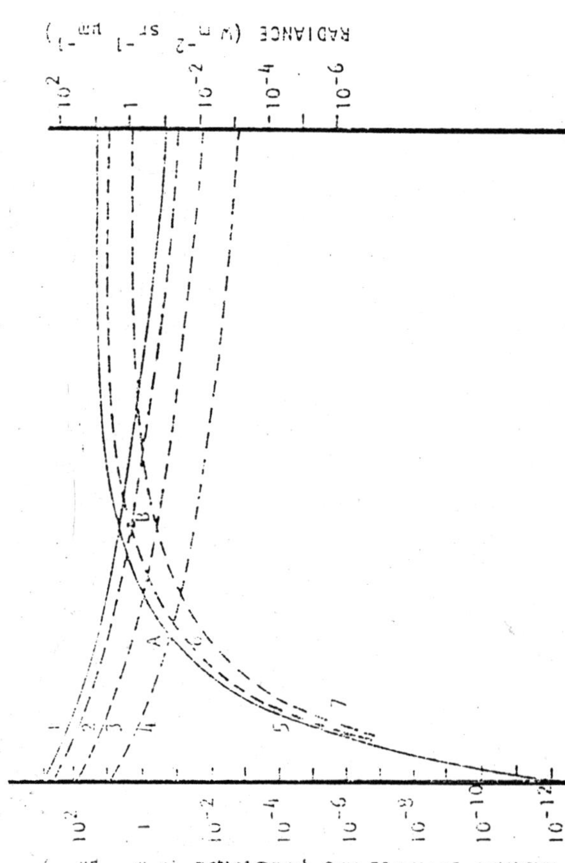

Fig. 2.6(a): Comparison between radiant exitences and irradiances (left-hand ordinate) and irradiances (right hand ordinate) due to a 6000-K blackbody sun and a 300-K Lambertian earth surface, ignoring atmospheric absorption.

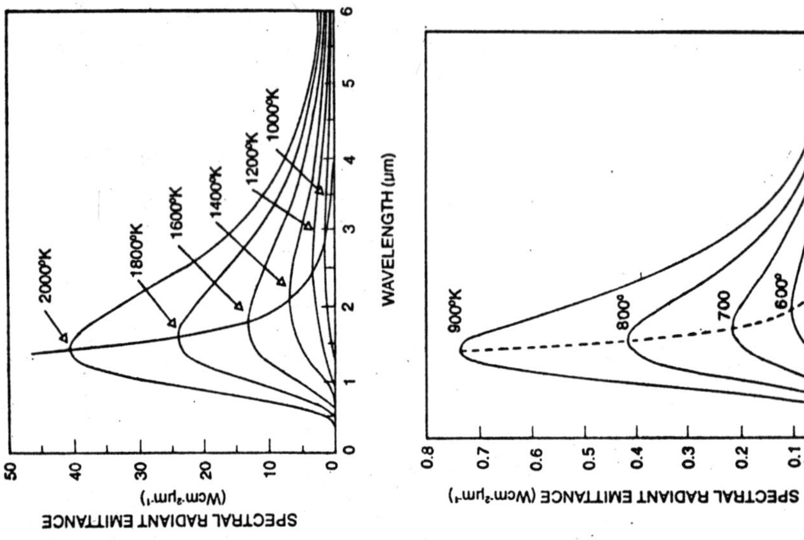

Fig. 2.6(b): Solar irradiance at earth's surface.

ρ (ρ > 0.5), are sometimes encountered. In practice spectral transmittance changes by the atmosphere in the range of 2.5 to 5.5μm because CO_2 shows strong atmospheric absorption at about 4.3μm. Thus for remote sensing, we refer to the radiant flux in the region below 2.5μm as solar reflected flux and that in the region above 6μm as thermally emitted flux. Because of the above phenomenon popular sensor sensed between the region 0.4 to 2.5μm and 8-10μm for remote sensing.

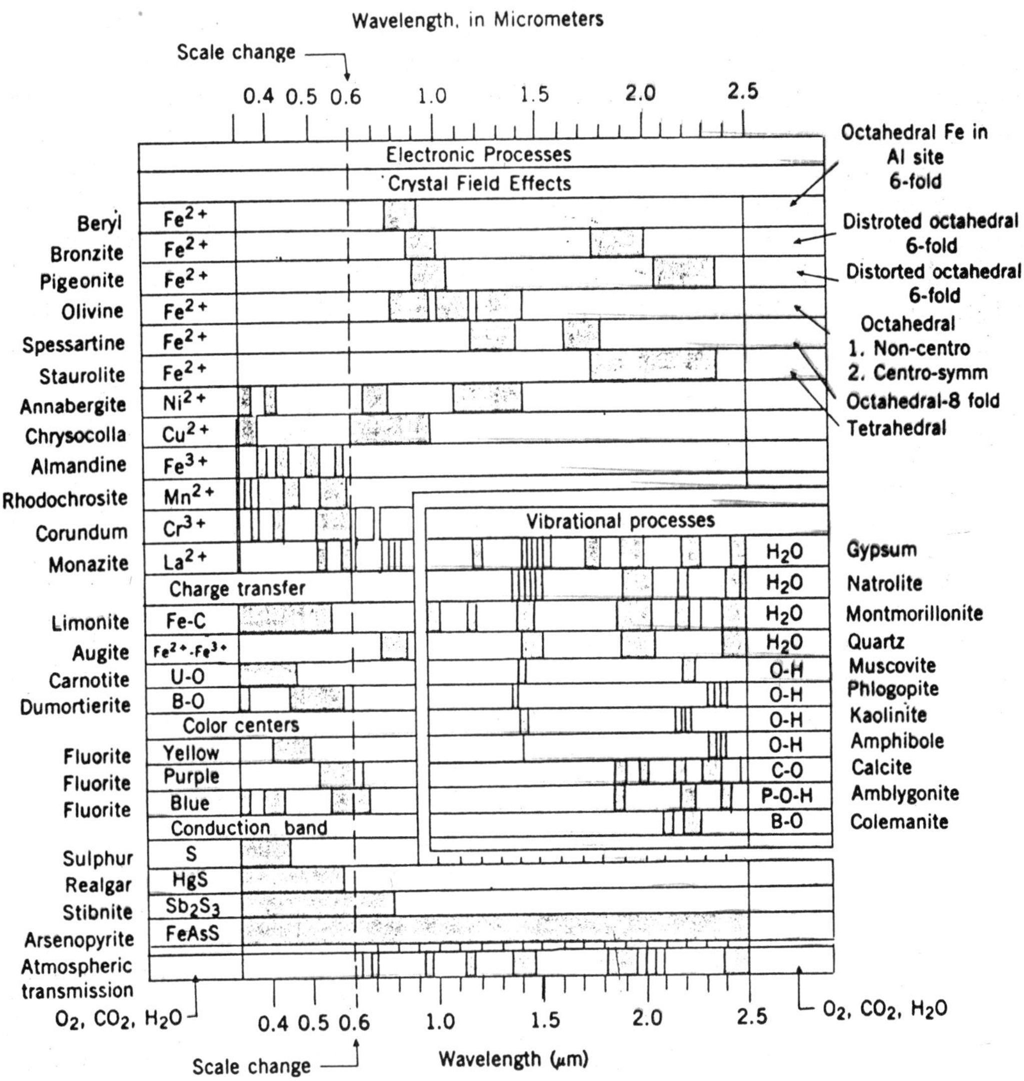

Fig. 2.7(a): Spectral signature diagram of a variety of geologic materials. (Based on Hunt, 1997)

The main interest in remote sensing is the study of the interaction of energy with solids which in particular, may be characterized by refractive index and extinction coefficient. These are known as optical constants. These optical constants are function of wavelength and nature of the material. The magnitude of refractive index indicate the speed with which electromagnetic radiation passes through the material. The extinction coefficient, on the other hand, refers to an exponential damping to the electromagnetic radiation (absorption) as it travels the materials (Hunt, 1980).

The quantum mechanical approach determines the stationary states of a system to be determined. The stationary states are determined by specific energy level. This is constant for specific material once a system or material occupies a particular stationary state or energy level, it will remain stationary for an indefinite period until and unless something causes it to change, thus causing a transition from stationary state to another energy level known as stationary state.

The interaction of electromagnetic energy with matter (e.g. molecular and atomic structure) takes place in a number of ways. These are dependent on the frequency or wavelength of EMR (i.e. its quantum energy) and the energy levels of the structure of the matter. As the EMR interact with certain material, be it a gas, or nuclei are put into motion. The motion may be rotational, vibrational or

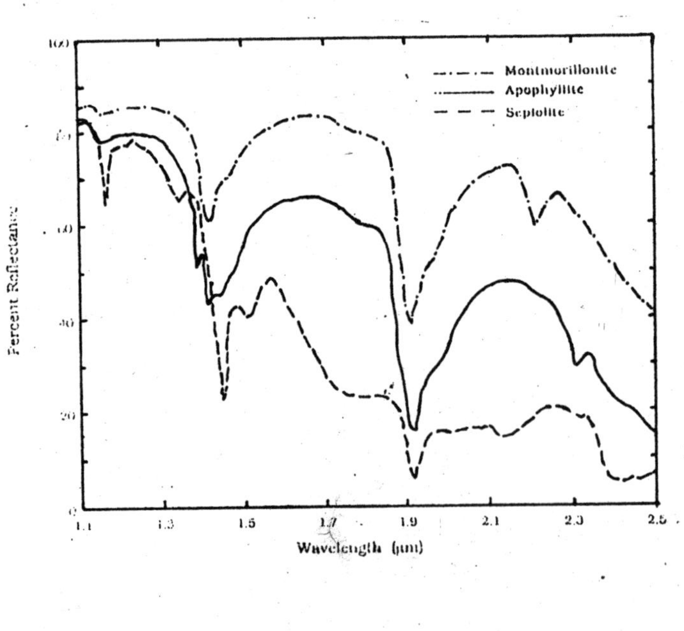

Fig. 2.7(b): HO-bearing minerals showing diagnostic absorption features at 1.4 and 1.9μm.

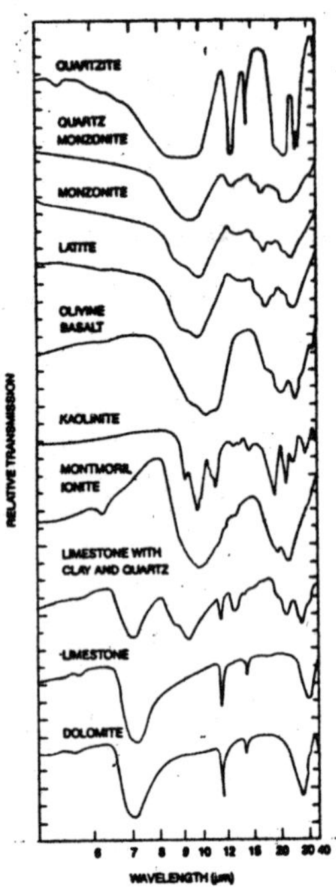

Fig. 2.7(c): Transmission spectra of common silicate minerals

displacement. All these phenomenon lead to exchange energy between EMR and the matter. The interaction mechanism or process are molecular rotation, vibrations, electronic processes and florescence. The electronic processes, sometimes referred to transitions of electrons from stationary state to transition state. All the above mechanism takes place due to the absorption of EMR by the matter. All the reactions are confined to geologic material (i.e. minerals, soils, etc.). The mechanism of biological material (plant) is same.

Molecular Rotations

Rotational movement is restricted to gas molecules. The rotation of molecule can be described in terms of angular momentum. Rotation may take place in maximum along the axis (X, Y, Z) whose origin is the centre of gravity of the system and the moment of inertia about these axis.

Molecules can be classified into four categories, depending upon the relationship between the molecules of its three principal moment of inertia (X-axis, Y-axis, Z-axis). The molecules are said to be linear if the moment of inertia on Z-axis is zero and the moment of inertia along X-axis and Y-axis are equal. The molecules are spherical top when all three moment are equal and are symmetric top when the moment on X-axis and Y-axis are equal but that on X-axis is different. For this type of movement, energy requirement is less. This occurs due to the absorption of EMR of higher wavelength (microwave and far infrared) region (King *et al.*, 1943 & 1949). This is important for the meteorology and atmospheric studies.

Molecular Vibrations

Molecular vibrational process corresponds to small displacement of atoms from their equilibrium positions. Each molecule may have three possible mode of vibrations, because atom has three degree of freedom. If we consider a molecule that contains N atoms, there are possibilities of maximum 3N mode. Of motion of these mode of motion, three constitute translations, and three constitute rotation of the molecule as a whole (only exception is linear molecule, where there are only two rotations) and there are thus 3N-6 (or 3N-5 for linear molecules) possible independent types of vibrations (Hunt, 1980). For a molecule with many classical frequencies (say n_1, n_2......n_{3N-6}) the energy is given by the following equation

$$E = (n_1+1/2)h_{n1} + (n_2+1/2)h_{n2} + \dots\dots\dots(3N\text{-}6+1/2)h_{3n-6} \quad \dots \quad (7)$$

where $(n_1+1/2)h_{ni}$ are the energy level of a linear harmonic oscillator and 'ni' are the vibrational quantum number (i.e. ni=0, 1, 2,......). The number and value of energy level for a material are thus determined by molecular structure and number of atoms within the molecule (Elachi, 1987). The band that occur in a spectrum as a result of transition between vibrational energy levels are referred to fundamental overtone or combination tones. The transition between the ground state (ni=0) to a state where ni=1, are termed as fundamental tones. An overtone occurs when the transition between ground state to higher energy level in which ni=2 or some multiple integer. Combination tones occur when the transition from ground level (ni=0) to a level (ni≠0), whose energy is determined by the sum of two or more fundamentals or overtone vibrations. The energy for combination tones can be determined by equation given below :

$$Ec = (n_1+1/2)h_{n1} + (n_2+1/2)h_{n2} \quad \ldots \quad (8)$$

Fundamental tones usually occur in the far to mid infrared region (λ = > 3μm), where as overtones and combination tones occur between 1μm to 5μm region of electromagnetic radiations. The simpler molecules is water that is very much sensitive to this phenomenon. Water molecule consists of three atoms (2 hydrogen and 1 oxygen) has three classical frequencies n_1, n_2 and n_3 which correspond to the three wavelengths.

λ_1 = 3.106μm— corresponding to the symmetric OH stretch
λ_2 = 6.08μm — corresponding to HOH bending
λ_3 = 2.903μm— corresponding to OH stretch

The wavelengths are fundamentals. The lowest order overtones corres-ponding to the frequencies $2n_1$, $2n_2$, $2n_3$ and to wavelength $\lambda_{1/2}$, $\lambda_{2/2}$ and $\lambda_{3/2}$. The combination tone can be $n = n_3+n_2$ corresponding to the wave length

$$\frac{1}{\lambda} = \left(\frac{1}{\lambda_3} + \frac{1}{\lambda_2}\right) \text{ i.e. } \lambda = 1.87\mu m. \quad \ldots\ldots \quad (9)$$

All the mineral that contains water molecules shows two absorption peak—one at near 1.45μm and other at about 1.9μm. The presence of above water absorption peaks is usually the diagnostic of the presence of the molecule. The sharpness of these peaks indicates that water molecule are located in well defined ordered sites. The broaden peak shows the presence of H_2O molecules in unordered or several unequivalent sites. The exact location and appearance of spectral bands give specific information about the way in which H_2O is associated with any inorganic material. Fig 2.8 illustrate the spectral behaviour of various mineral that are present in soil as well as in rocks, and that contains H_2O molecules.

The Si-O, Mg-O and Al-O bonding show fundamental vibration mode near 10μm or longer wavelength. The first order overtone near 5μm and higher order overtone in the near infrared region is not normally clear. Any feature that have

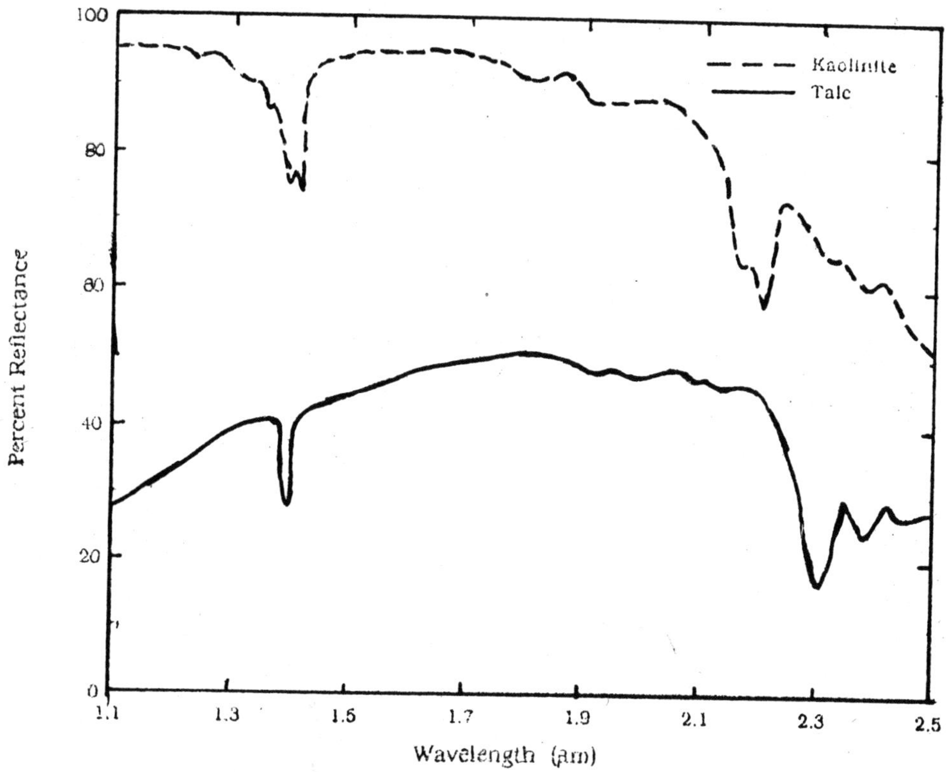

Fig. 2.8: OH-bearing minerals showing a feature at 1.4 μm, but no water feature at 1.9 μm.

very high fundamental frequency shows the overtone and combination in the near infrared regions.

The hydroxyl ion very frequently occur in organic solids. It has one stretching fundamental which occur near 2.77μm but its exact location depends on the site at which OH ion is located and the atom attached to it. In some situation, the spectral feature is doubled, indicating that OH ion is in two slightly different locations. Al-OH and Mg-OH bonding have fundamental wavelength at 2.2 and 2.3μm respectively. The first overtone of OH at about 1.4μm is the best common feature present in near infrared region of EMR. Fig. 2.9 shows example of spectral behaviour of hydroxyl ion (OH) in different clay minerals.

Electronic Process

Electronic processes are associated with electronic energy levels. Electrons in an atom can be characterized by specific quantised orbit with specific energy levels. The most common elements present in the earth crust are silicon, aluminum and

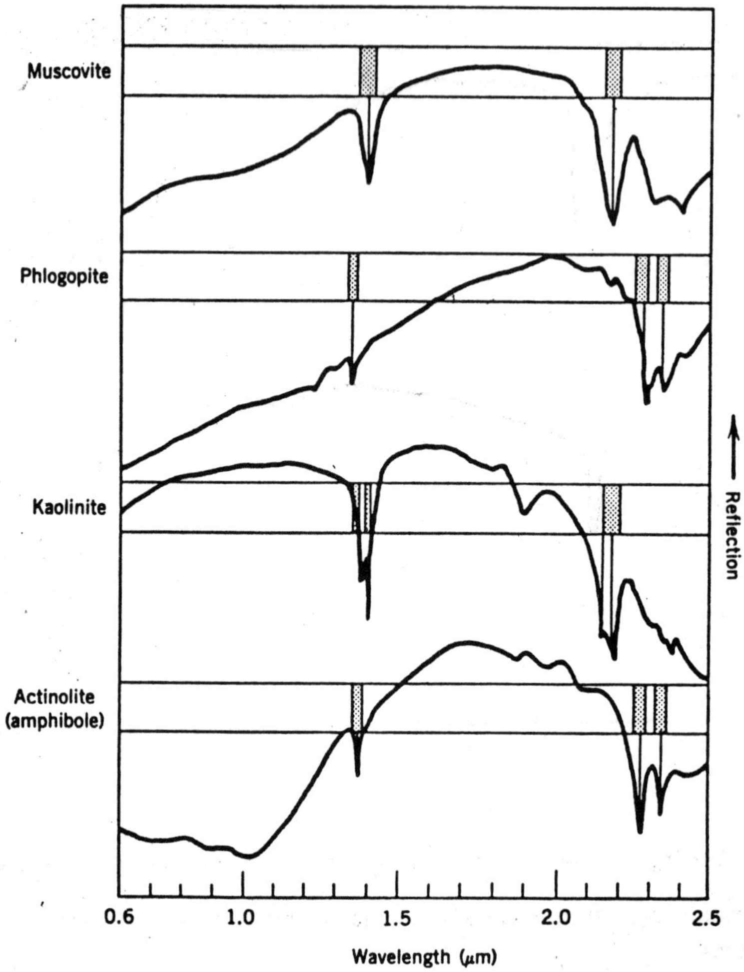

Fig. 2.9: Spectra displaying the hydroxyl group tones: overtone near 1.4 μm and combination tones near 2.3 μm. (From Hunt, 1977)

oxygen. They do not show any transition energy level to show spectral features in the visible or near infrared region. As a result, no direct information is available on the bulk composition of the earth crust (geologic materials or soils). However, indirect information is available as a consequence of crystal or molecular structure imposing its effect upon the energy levels of specific ions. The most important effect in the context of remote sensing are crystal field effect. charge transfer, conjugate bands and transitions in materials.

Crystal Field Effect

The atoms are bound together to form a molecule by valence electrons. In isolated atom, the valence electrons are unpaired and are the primary cause of colour. In molecule, the valence electrons of adjacent atoms form pair which constitute the chemical bonding that holds atoms together. As a consequence of this pair formation the absorption bands of the valence electrons are usually displaced to ultraviolet region. But in the case of transition metals namely iron, chromium, copper, nickel, cobalt and manganese, the atoms have inner shells (d, f) that remains only partially filled. These unfilled inner shells [(n-1)d] which have excited states that often fall in the visible spectrum of electromagnetic radiations (EMR). These states are strongly affected by electrostatic field surrounding the atoms. The field is determined by the surrounding crystalline structure. The different arrangement of the energy levels for different crystal field leads to the different spectra for the same ion. However, not all possible transitions may occur equally strong. The allowed transitions are defined by the selection rule is the context of remote sensing, the most important rule is the one related to the electron spins of the state involved. This rule allows the transition between the status of identical spin only. Typical example of crystal field effects are ruby and emerald.

The basic material of ruby is an oxide of aluminum Al_2O_3 (corundum). In the lattice of ruby a few aluminum ions are replaced by chromium ions. Each chromium atom has three unpaired electrons whose lowest possible energy is the ground state and a complicated spectrum of the excited state. The exact position of the excited state is determined by the crystal electric field in which chromium ion is immersed. The symmetry and strength of the field is in turn determined by the nature of the surrounding chromium ion and their arrangement in the lattice. There are three possible excited states with energy in the visible region.

Selection rule does not allow a direct transition from ground state to excited state but allow transition to both transition states (Fig. 2.10). The energy associated with these transition corresponds to wavelength in the violet, yellow-green region of the spectrum. That is why, it appears as deep red (depleted of violet, blue and green).

Owing to the above selection rule, electron can return from transition states to ground state only through the intermediate transition (excited state). The transition from the transition state to excited state releases infrared radiation, but transition from excited state to ground state gives rise to strong red light emission. Crystal field effect can arise whenever there are ions bearing unpaired electron in solids or crystals.

One of the most important cases in remote sensing is the presence of ferrous ion (Fe^{2+}). Perfectly octahedral ferrous ion only has one spin-allowed transition which is in the near infrared region. However, if the octahedral site is distorted, the resulting field may lead to splitting of the energy level, thus allowing additional transitions. If the ferrous ions are in different nonequivalent sites, as is the case in olivine, additional characteristics transition may occur. Thus important information on the bulk structure of the mineral can be derived indirectly by using ferrous ion

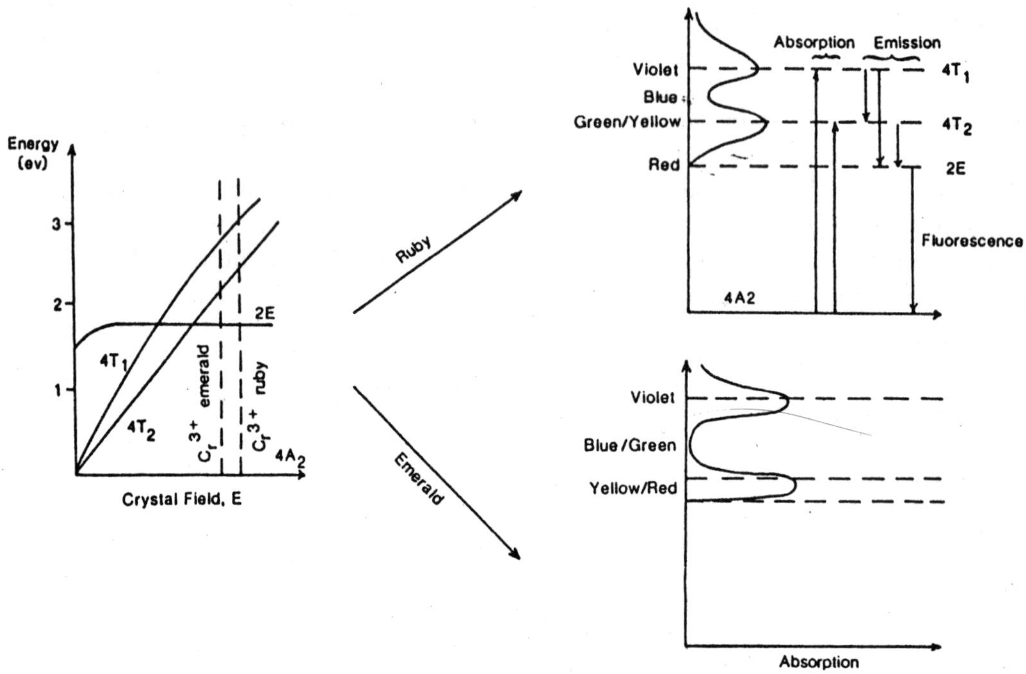

Fig. 2.10: Basis for the color characteristics of emerald and ruby.

spectral feature as a probe. Fig. 2.11 shows the characteristics reflection spectra of different minerals, containing ferrous ions. The crystal field effect is not only limited to electrons in transition metal ions, but also is an electron trapped at a structural defect, such as a missing ion or an impurity, similar transition energy level occurs. These anomalies are called colour centres. The colour centre effect is responsible for the purple colour of Fluorine (CaF_2) in which calcium is surrounded by eight fluorine ions. A colour centre forms when fluorine is missing from its usual position. In order to keep electrical neutrality, an electron is trapped in the fluorine location and is bound to the location by crystal field of the surrounding ions. Within this field it can occur a ground state and various excited states.

Conjugate Bond

In biological pigments and many organic substances, carbon and nitrogen atoms are linked by alternate double bonds and single bands (chlorophyll, butane, phenol, haemoglobin etc.). The alternate double bonds and single bonds are called conjugate bonds. According to molecular orbital theory, each double bond represents two pair

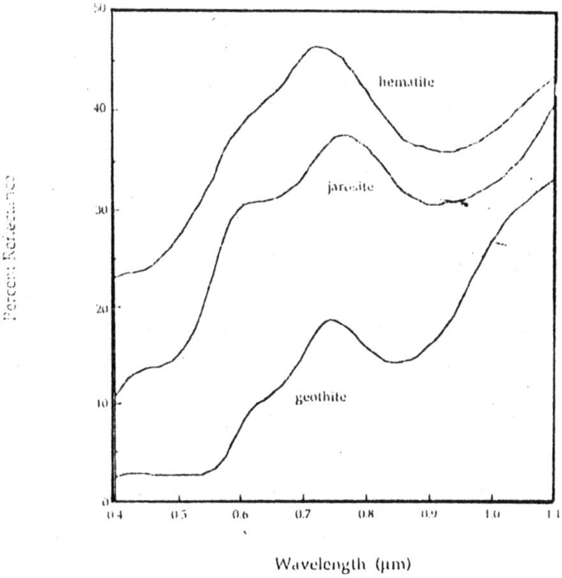

Fig. 2.11: Reflectance spectra of iron bearing minerals

by shared electrons: σ bonding forming and π bond forming electrons. The latter bond are mobile and can move from a double bond atom to adjacent single bond atom. Thus moving a pair of π electrons from double bonds to adjacent single bonds reverse the entire sequence of bonds leading to equivalent structure (e.g. benzene, chlorophyll etc.).

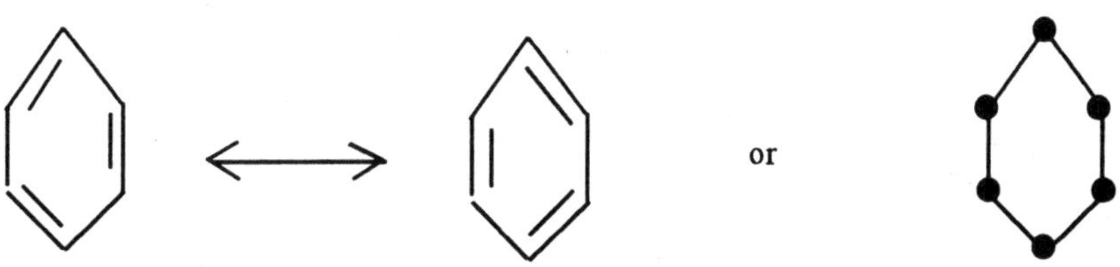

From valence bond concept the actual best representation of the structure shows all the atoms connected by a single bonds with remaining pair of bonding π electrons distributed over the entire structure in molecular orbitals. The presence of orbitals in a system of conjugate bonds tends to diminish the excitation of energy of the electron pair allowing absorption of EMR in the visible region.

Molecular orbital transition play an important role in spectral response by biological pigments like chlorophyll and haemoglobin showed strong absorption in the visible region of EMR.

Spectral Signature

For effective and efficient utilization of remote sensing data for identification and delineation of different objects on the earth surface, it is important to have a thorough knowledge and understanding of spectral characteristics of various earth surface objects or features and the factors influence these spectral characteristics. In the context of remote sensing various objects are categorized as geologic materials that includes soil, minerals and rocks, and biological materials. The biological materials are vegetation. Rocks are the assemblage of minerals and hence the spectrum of rocks is a composite of individual spectra of its constituent minerals. Soil is also a heterogeneous mixture of minerals, considerable amounts of organic matter, and often water. All of them contribute their spectra to the overall spectrum of the soil. Depending on the composition and nature of the materials different objects showed different spectral characteristics or signature. This particular interaction of EMR with earth object is a diagnostic characteristics of earth objects. The spectral characteristics will lead us to understand the spectral signature of earth object. According to Slater (1980), a spectral signature of a ground object may be defined as a set of measurement values for reflectance or radiance of the earth objects, each value corresponding to specific well defined wavelength interval. In general the set of reflectance or radiance value is dependent on irradiance (incoming solar flux), viewing geometry of the sensor and the environmental factors. Here the authors have emphasized on the spectral characteristics of vegetation, soils and water, and snow (Fig. 2.12).

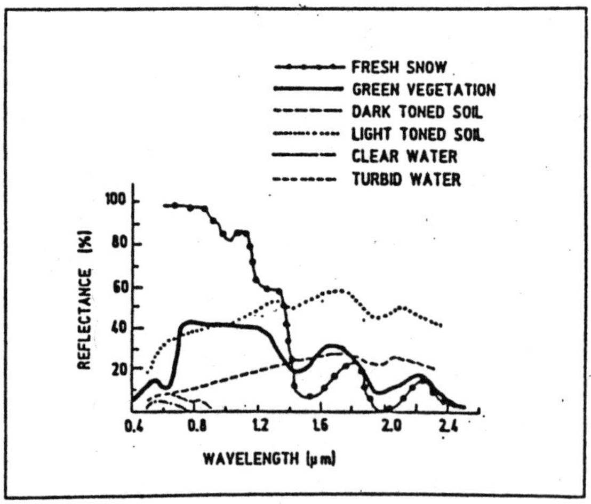

Fig. 2.12: Comparison of reflectance spectra of a few land covers

Spectral Signature of Vegetation

The spectral signature or reflectance of green vegetation is very distinctive and quite variable with wavelength. Fig. 2.13 represents a typical spectral signature of green vegetation and depicts the spectral response region of immense significance. In the visible region, strong absorptions are prominent at about 0.45μm and 0.66μm. The strong depressions or absorptions in the blue and red regions of the visible spectrum corresponds to the chlorophyll absorption (chlorophyll-a and chlorophyll-b) bands, because the chlorophyll in the leaf absorbs most of the incident energy in these wavelength regions for photo synthesis. Therefore, in the visible bands the incident radiation (EMR) which is not absorbed, will be largely reflected back. A relatively poor absorption in the wavelength region between the absorption bands (0.45µm and 0.65µm) shows a reflectance peak at about 0.54µm

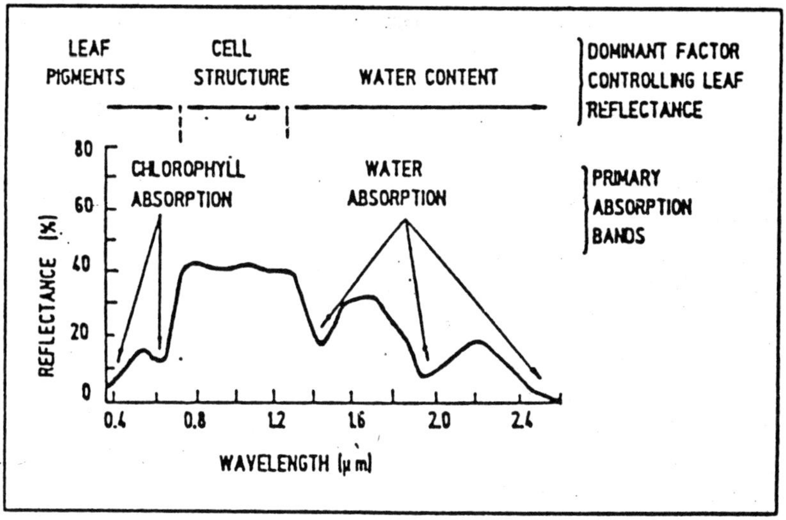

Fig. 2.13(a): Typical vegetation reflectance spectrum

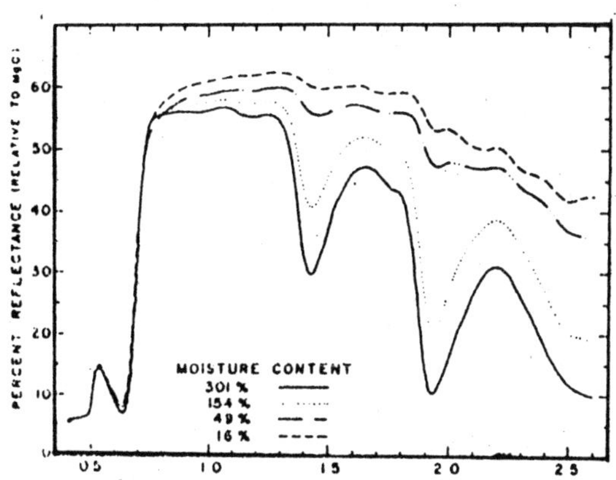

Fig. 2.13(b): Spectral reflectance of moisture stress vegetation

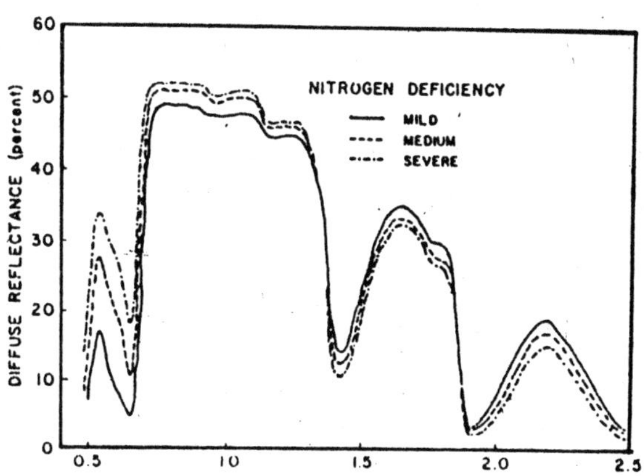

Fig. 2.13(c): Impact of a nitrogen deficiency plant

which is the green wavelength region. That is why, relatively low absorption in the green wavelengths and higher reflection in this region, causes normal, healthy foliage to appear as green (Fig. 2.13a&b) (Hoffer, *et al.* 1969). Further, when a plant is under stress which may be due to deficiency of water nutrients etc., chlorophyll content is decreased. Thus the lack of leaf pigments (chlorophyll) causes plant to absorb less energy in the chlorophyll absorption bands (0.45µm to 0.65µm). Such plants show much higher reflectance particularly in the red region and the foliage appears as yellowish or chlorotic.

Other interested pigments are carotenes, xanthophyll and anthocyanin. Former two are yellow in colour and the latter is red in colour. Carotenes and xanthophyll are frequently present in green leaves and have as absorption band only in the visible region of electromagnetic radiation (nearly at 0.45µm). Since chlorophyll is the dominant pigment in healthy green leaves, it masks these yellow and red pigments. However, as the plant undergoes senescence, the chlorophyll frequently disappears, allows the carotenes and xanthophylls to be dominant. Due to this reason, senescenced leaves appears yellowish. During leaf fall season, chlorophyll production decreases, and some tree species producing anthocyanin in great quantities, giving rise to the leaves. Spectral behaviour of different pigments and its effect are presented in Fig. 2.14 (a). Fig. 2.14 (b) shows the striking differences in the pigment characteristics. Though due to difference in pigment content leaf shows difference in reflectance in the visible region of EMR, there is little difference in reflectance in the near infrared and mid-infrared region.

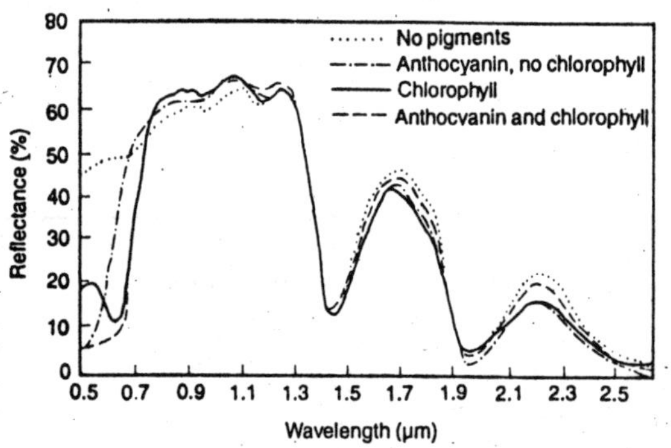

Fig. 2.14(a): The effect of pigmentation on leaf reflectance of Coleus leaf

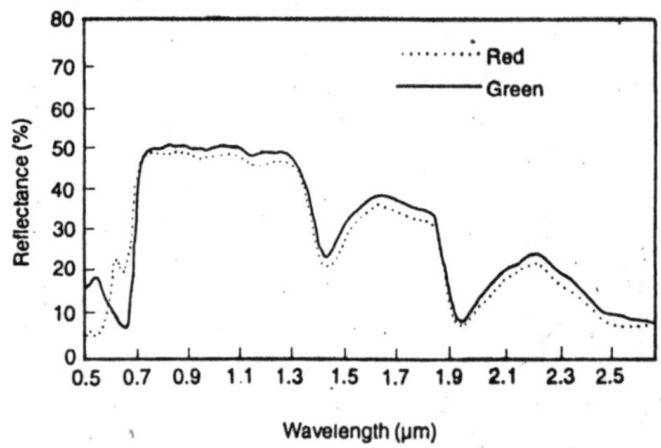

Fig. 2.14(b): The effect of pigmentation on leaf reflectance of Maple leaves

Now we are drawing attention in the near infrared region of spectrum and its interaction with the green leaves. Fig. 2.15 shows that there is a very marked increase in reflectance in passing from visible to near infrared region at approximately 0.7µm. In this region, healthy green vegetation is characterized by high reflectance, very high transmittance and low absorption, as compared to the visible wavelength region. For majority of vegetation, it is observed approximately 45 to 50 per cent reflectance and 45 to 50 per cent transmittance and less than 5 per

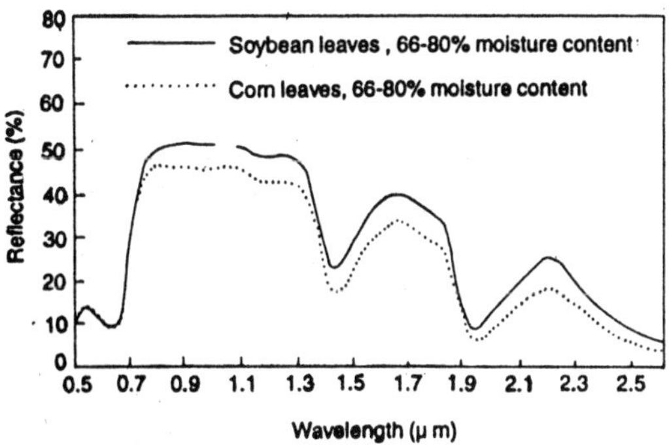

Fig. 2.15: Spectral reflectance of different crops (soybean and corn) with comparable leaves of moisture content

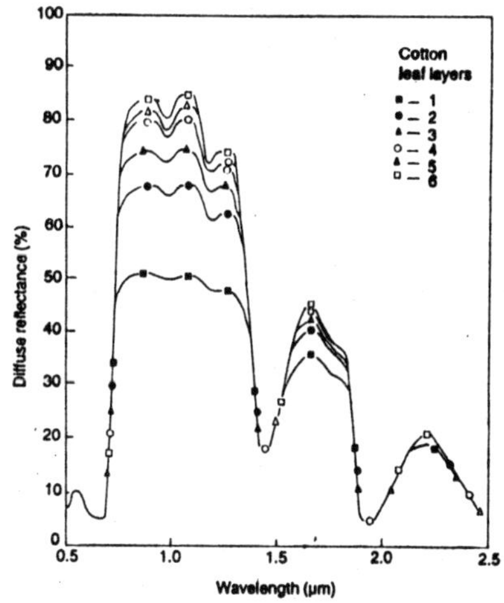

Fig. 2.16: Reflectance from combinations of cotton leaves stacked on one another

cent absorption in the near infrared region (Hoffer, 1978).

The internal structure of plant leaves, which is mainly mesophilic tissues is very complex, and it is this internal structure that largely controls the reflectance of near infrared region (Gates *et al.*, 1965). The internal structure of different species are different and this leads to various amount of reflectance in NIR. Fig.2.15 shows the reflectance curve of soybean and corn. It is observed that there is hardly any difference in the visible region. Only difference is observed in the NIR and mid-infrared region. Many other plant species also behave in the way and showed difference in the entire region of NIR and mid-infrared region and leads to discrimination in the multispectral imagery. Approximately 50 to 60 per cent of irradiance in the near infrared region is being reflected and 50 per cent of it transmitted through leaves. for an area where vegetation is multilayer, the resultant reflectance will also be higher. This phenomenon will lead to identify and distinguish multilayer from single story forest. The impact of multilayer leaves of raw cotton has been discussed by Myer (1970) (Fig. 2.16).

The spectral response of green vegetation in the middle infrared region is mainly due to water absorption bands. The absorption peaks or depressions are nearly at 1.4, 1.9 and 2.7μm (Fig.2.17). The absorption band at 2.7μm is a major one and is due to fundamental vibrational water absorption band, which is discussed earlier. This band is actually a combination of two fundamental vibrational water absorption bands located at 2.66 and 2.73μm. There is another fundamental vibration of water-absorption band at 6.27μm. But the energy regions at 1.9, 1.4, 1.1 and 0.9μm are all overtone and combination bands and are successfully weaker absorption bands than the fundamental water absorption bands (Slater, 1980;

Elachi, 1987 & Hunt, 1980). The absorption bands at 1.4 and 1.9μm are dominant over the spectral reflectance of leaves in the middle infrared wavelength region. However, even a minor water absorption bands at 1.4 and 0.9μm have a significant impact on reflectance of vegetation (Fig. 2.16), particularly for multilayered leaves (Myer, 1970).

According to Hoffer and Johanson (1969), the reflectance of leave in the middle infrared region is approximately inversely proportional to the absorption of a layer of water approximately 1mm in depth. The degree to which incident solar energy in the middle infrared portion of the spectrum is absorbed by vegetation, is a function of the total water content and thickness of the leaf (Fig. 2.17a). With the decrease in water content of the leaves, the reflectance in the middle infrared region increases remarkably. Fig. 2.17b demonstrate the effect of middle infrared reflectance of corn leaves at various moisture content. But it is depicted in this figure that there are substantial changes in reflectance in the wavelength bands between the water absorption bands. This increased reflectance in the wavelengths between the water absorption bands are generally referred as "carry over effect", whereby the increased reflectance in the water absorption bands influences or carry over to the wavelength between the water absorption bands. Consequently there is an increased reflectance throughout the middle infrared region. This carry over effect and absorption regions throughout electromagnetic wave, generally available in the atmospheric window region will provide insight to determine and monitor the health or stress of green vegetation (Knipling, 1967; Sinclair *et al.*, 1971).

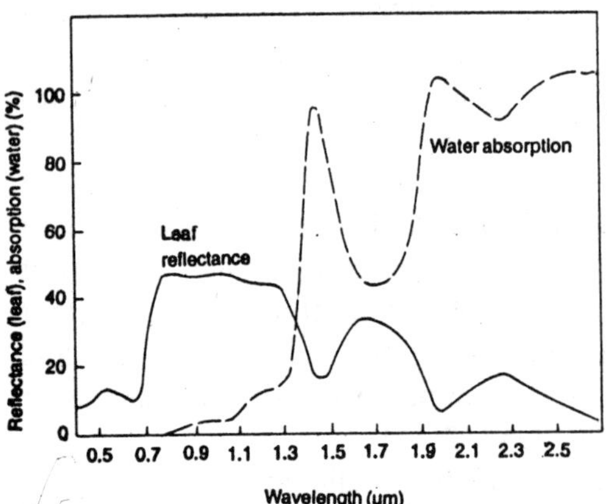

Fig. 2.17(a): The inverse relationship between leaf reflectance and water absorption

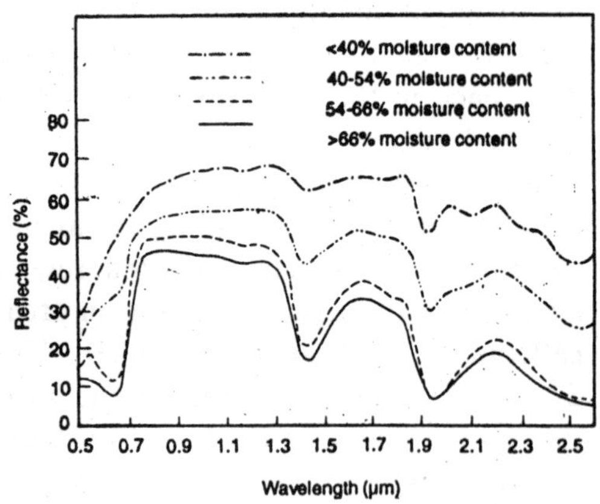

Fig. 2.17(b): Effects of moisture content on reflectance of corn leaf

Spectral Signature of Geologic Materials

The geologic material includes rocks, minerals and soils. The spectral signature of geological material in the visible and infrared region is mainly confined to electronic and vibrational processes. Rocks are the assemblage of minerals so spectrum of a rock is a composite of the individual spectra of its constituent minerals. Soils, on the other hand, is a complex heterogeneous mixture of geologic materials (minerals, sand, silt and clay) along with organic material, air and water having various physical and chemical properties which affect the energy matter interaction (Hunt, 1980).

Since rotational processes are precluded in solids, the spectral signature of minerals and soils are produced as a consequence of either electronic or vibrational process (Elachi, 1987). Energy requirement for electronic process and vibration process are different and they show their occurrences in different spectral region. Electronic process requires more energy than vibrational process, and the former appears primarily in the visible region, whereas the latter show the evidence in the infrared region. Usually this lies between the 0.7μm to 2.5μm region of EMR. Here the authors have discussed the spectral characteristics first.

Soils

In comparison to vegetation, spectral signature of soil materials is much simpler, because all the incoming radiation will either be reflected or absorbed. There is little transmittance. The overall trend and nature of soil signature is similar but the soil properties viz. organic matter, texture, moisture content, salt contents, minerals etc influence the amplitude of the reflectance curve.

The most important soil properties that influence the spectral characteristics are the soil texture, moisture content, organic matter content, colour, presence of sesquioxides, dominant clay minerals etc. Texture refers to relative proportions of sands, silt and clay. These are classified according to the size. Clay is the finer particle having less than 0.002mm size, sand is the coarser particle having a diameter from 0.02 to 2mm, and silt is the intermediate particle of diameter 0.02 to 0.002mm. Clay particles enable to be packed very closely, leaving very minute space between the particles. On the other hand, relatively large sand particles allow large space between the soil particles, thereby enabling more air and water movement freely through the soil as compared to clayey soils. In presence of water each soil particle is covered by a very thin film of water and water will occupy some space between the particles. Because of the finer particle size and air space and large surface area, clayey soil can hold more water. Consequently the finer capillary space restricts the movement of H_2O through underground drainage and evaporation as compared to coarser textured soils. The texture and moisture content relationship has tremendous impact on soil spectral response (Figs. 2.18 to 2.20).

Fig.2.19 shows typical reflectance curve for a sandy soil at three different moisture levels. Sandy soil having relatively large particle size and relatively larger pore space, allow most of the moisture to be removed from the soil when it is air dried. There is no significant decrease in reflectance in the water absorption bands

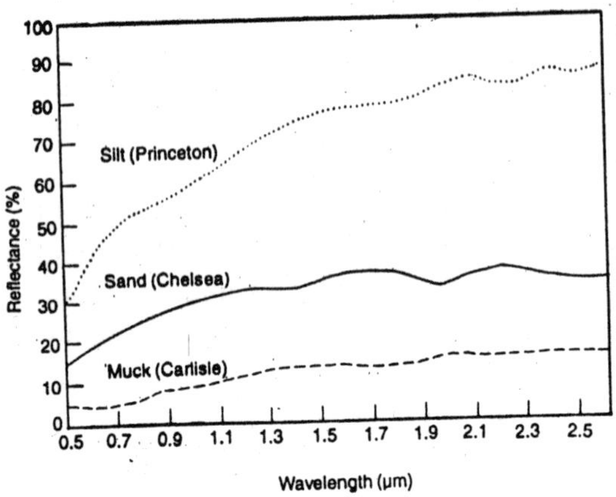

Fig. 2.18: Spectral reflectance curve for three soil types at low moisture contents

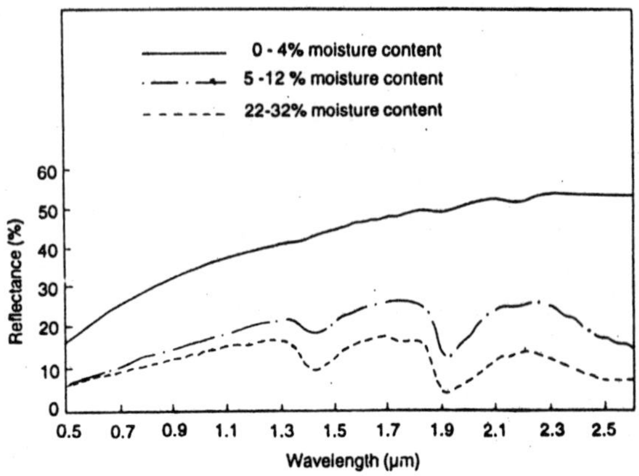

Fig. 2.19: Spectral reflectance for Chelsea sand in three moisture content groupings

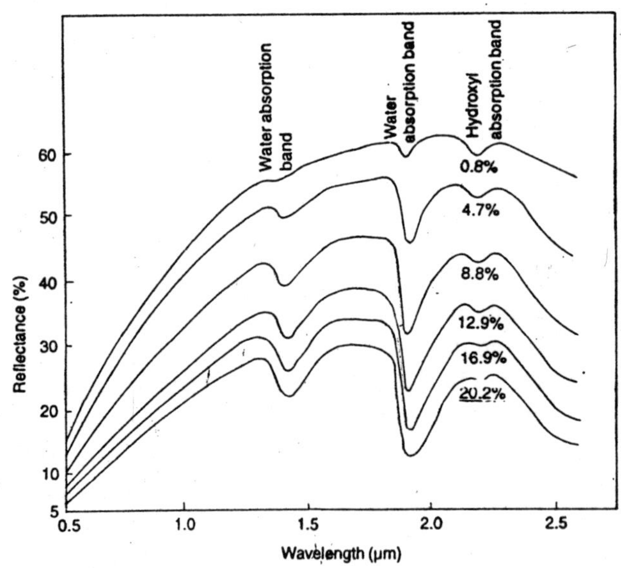

Fig. 2.20: Spectral reflectance curve for Newtonia silt loam at various moisture contents

of air dried sandy soil. However, the sandy soil that contain significant amounts of water have distinct decrease in reflectance in the water absorption bands at approximately 1.4, 1.9 and 2.7μm, which are same water absorption bands that were discussed earlier. To some extent, the trends of reflectance is similar to that of water absorption bands of green vegetation. The overall reflectance decreases as the moisture content increases. The possible reason for this of course is that the incoming radiation is strongly absorbed by water at these particular wavelength, regardless of whether the material containing water is soil or vegetation.

In the visible portion of the spectrum, there is also a distinct decrease in reflectance as moisture content increases. The reason is that in presence of moisture, the soil colour becomes more darker causing less reflectance. After certain limit, if we increase the moisture content, it shows no significant change (Fig. 2.20 and Fig. 2.21).

Soil texture influence the spectral reflectance of soil in two ways because it influence moisture holding capacity and because of the size of the particles. According to Bower and Hanks (1965) with increase in particle size (from 0.22 to 2.65mm) increases the absorption by 14 per cent. Montgomery (1976) claimed that silt content has maximum influence in soil spectral reflectance. Since silts are relatively smaller in size, the relationship was directly proportional. This implies that an increase in silt content increases the level of reflectance.

The organic matter content is an another important soil property that significantly influence the reflectance characteristics of soils. The organic matter content of a soil is extremely important characteristics of soil to an agriculturist because its controls physical and chemical properties of soil as well as soil fertility. General trend of organic matter content of soil varies from 0.5 to 5.0 per cent. Soil with higher organic matter appears as a dark colour soils (mollisols, histosol etc.) whereas the soil with lower organic matter content appears as light brown to grayish in colour. The degree of decomposition of soil organic matter will also influence the colour, but the well decomposed and more fertile organic soil produced from peat is black or nearly so (U.S.D.A., 1951). The relationship between organic matter of the soil and reflectance characteristics throughout the visible region has been shown to be curvilinear (Fig. 2.22). The reader should always keep in mind that soil may not show same relationship because soil is a heterogeneous system and spectral response of soil is the resultant response of all soil components. This may vary due to the variation of climate as well as drainage conditions. For example, same amount of organic matter in colder region will be darker while a soil of the tropical region appears brownish. Also it has been found in tropical region like India, even a dark clayey soil seldom contains as much as 3 per cent organic matter; but they are darkest solid in the world (black cotton soils of India). For interpretating the organic and spectral reflectance, other soil forming conditions should be kept in mind. The lower reflectance of organic matter in rich soil may be due to aromaticity and conjugate bonding.

Iron oxide too has a significant role in spectral characteristics of the soil. In tropics, laterite and lateritic soils are generally characterized by high free iron oxides and they appear reddish in colour. The reddish colour may be due to anhydrated iron oxide, sometimes partially hydrated iron oxide and manganese dioxide can also cause red colouration (U.S.D.A., 1975). There is no inverse relationship between spectral reflectance and iron oxide content especially in the visible wavelength regions (Fig. 2.23). Fig. 2.24 clearly demonstrate that with the removal of iron oxide from the soil will cause a marked increase in reflectance throughout the optical remote sensing region (0.5μm to 1.1μm), but the reflectance above 1.1μm is not significantly affected. Similar effect has been claimed in the case of organic matter by Mathews (1972).

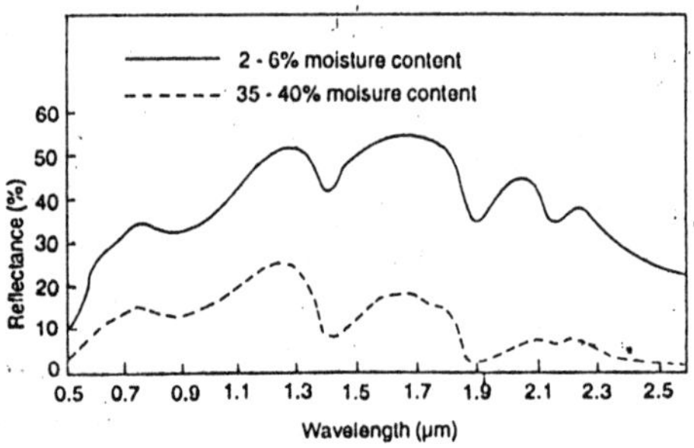

Fig. 2.21: Spectral reflectance for a typical clay soil at two moisture contents

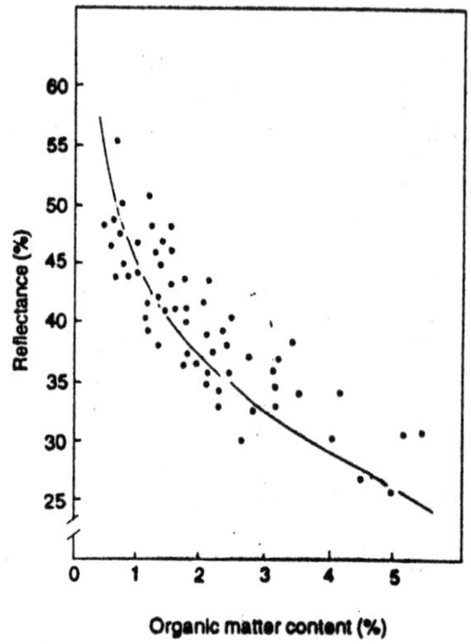

Fig. 2.22: Relationship between organic matter contents and soil reflectance

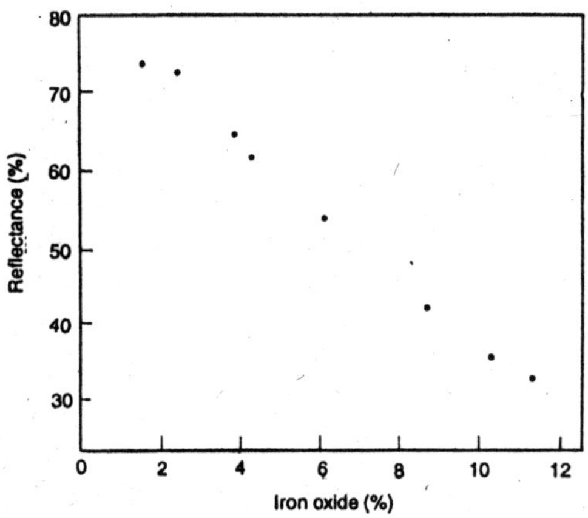

Fig. 2.23: Relationship between iron oxide contents and soil reflectance in the 0.50 to 0.64 microns wavelength band

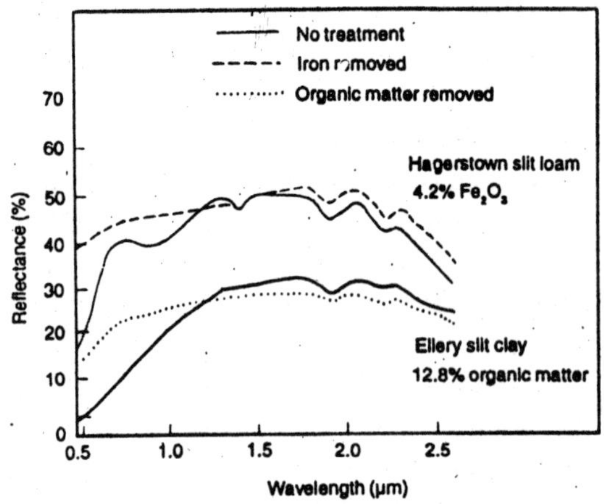

Fig. 2.24: Effect of removal of iron oxide and organic matter on soil spectral reflectance

To summarise the discussion of soil reflectance, it has been observed that moisture contents decrease the reflectance of the soil throughout the reflective portion of the spectrum that texture of the soil will cause an increased reflectance with decreased particle size; that decrease in surface roughness will increase the level of reflectance; that an increase in organic matter will cause a decrease in reflectance and that an increase in iron oxide can cause a decrease in reflectance. However, many of the above parameters are interrelated. Montgomery (1976) mentioned that the relationship between increasing organic matter content and decreasing reflectance exists primarily for soils with more 1.5 per cent organic matter content show decrease in reflectance with increase in organic matter. With less than 1.5 per cent organic matter content, iron oxide content of the soil exerts significant effect on the level of reflectance. Furthermore, for those soils having over 1.5 per cent organic matter, the relationship of decreasing reflectance with increasing organic matter held true only for soils in which the percentage of silt content is approximately constant. Conversely, if two soils having same organic matter content but different silt content, the soil having more silt content have the highest level of reflectance.

The interrelationship between iron oxide and organic matter and their influence on the reflectance characteristics of soils are important in interpretating the remote sensing data from different geographic locations. The relationship between climate and soil development is well recognized and must be taken into consideration at time of interpretation and must be taken into consideration at time of interpretation. For example laterite and lateritic soils of India are poor in organic matter and rich in iron oxide appears red or reddish brown in colour. Similarly soils of Himalayas where temperature is low, soils are rich in organic matter, appears dark in colour, we have realized that spectral reflectance of soil can be dominated by several different but often closely interrelated soil properties. Soil texture, moisture content, surface roughness characteristics, organic matter, and iron oxide are among the most important soil parameters involved. Both the reflective and emissive portions of the spectrum can provide useful information concerning soil features, and data from both of these spectral regions must be utilized to effectively interpret the soil characteristics. The interpreter should keep in mind that there are significant differences in amplitude of reflectance among various soils and soil conditions. These differences are relatively consistent throughout the various wavelength range. But in the case of vegetation, the relative differences in reflectance is a function of wavelength.

Spectral Characteristics of Water and Snow

While dealing with the spectral characteristics of vegetation and soil, we came to know that water body showed spectral absorption in near and mid infrared region. Fig. 2.25 shows the combined spectral signature of vegetation, soil and water. The latter shows always lower spectral reflectance in the entire region of EMR in comparison with soils and vegetation. Usually water body absorbs all the incident energy both in near and mid infrared region, even when water is shallow. So there is very little energy available to be reflected.

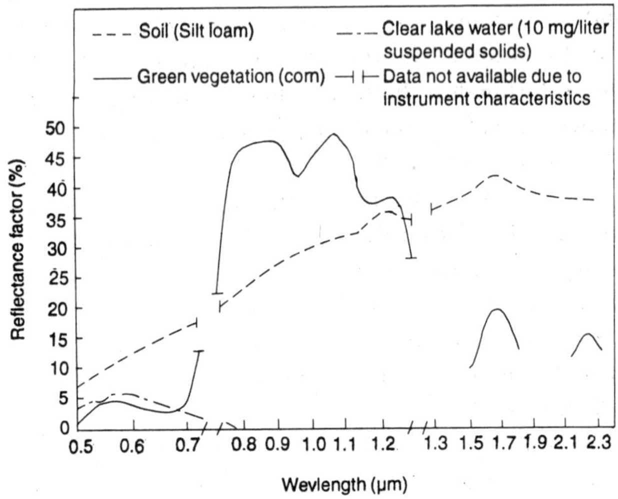

Fig. 2.25: Spectral reflectance of green vegetation, soil and clear water

In the visible portion of the spectrum, energy-matter interaction of water bodies is highly complex. The reflectance of water is dependent on the water quality, bottom material suspended and soluble materials. The reflection of water may be from the surface water, from the bottom material and also from the suspended sediments. Thus the absorption and transmission properties are not only function of water per se but also significantly influences by the various types and size of the suspended material—both organic and inorganic origin.

In nature, water bodies are usually not pure like clear distilled water. It always contains a variety of organic and inorganic material, some of which are in suspension. These suspended material cause scattering and absorption of incident energy and consequently cause significant variation in the transmission of energy through water (Fig. 2.26a, b, c, d). Thus, the interactions between scattering, absorption and transmission control the spectral reflectance can be estimated by remote sensing techniques.

Snow

During the era of Landsat-1 system, people had an idea that snow cannot be separated from cloud spectrally because both indicate higher reflectance within optical region, i.e. 0.4 to 1.1μm.

Further, typical spectral signature of fresh snow and two day old snow has been

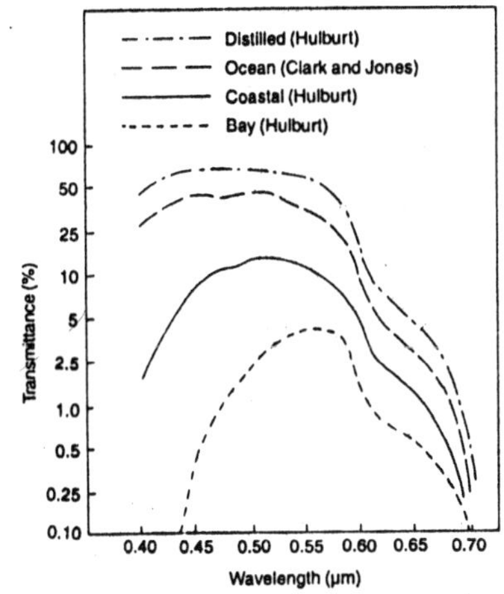

Fig. 2.26(a): Spectral transmitance through 10m of water of various types

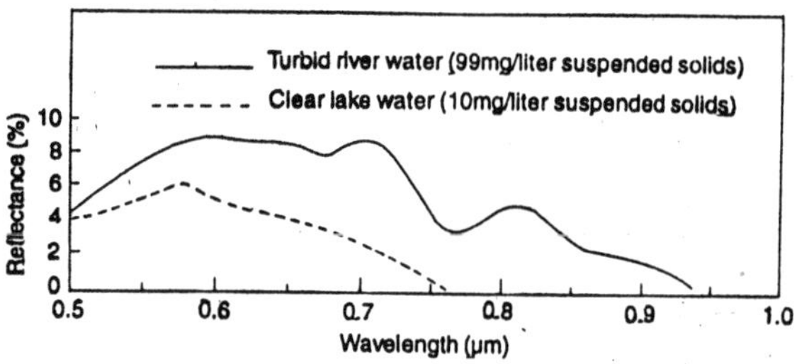

Fig. 2.26(b): Spectral reflectance characteristics of turbid and clear

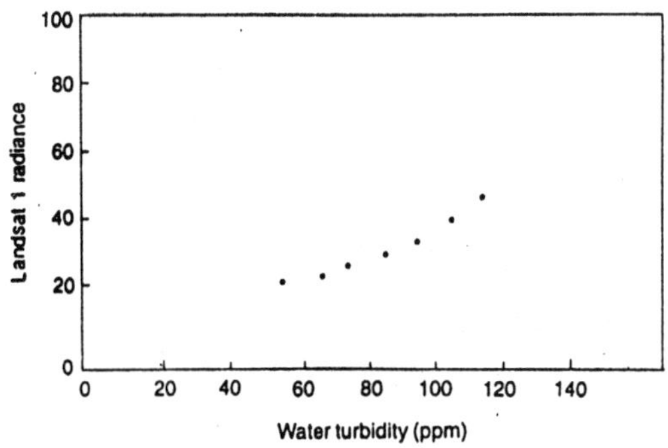

Fig. 2.26(c): Reflectance in the 0.6 - 0.7 micrometer band in relation to difference of water turbidity

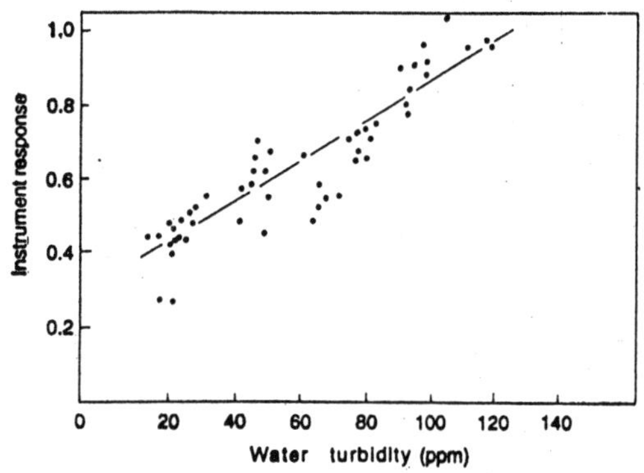

Fig. 2.26(d): Landsat 1 radiance level

studied by O'Brien and Munis (1975) and presented in Fig. 2.27. The spectral reflectance of snow decreases abruptly from 100 per cent at wavelength 0.8μm to almost zero per cent at about 1.5μm. The strong absorption peaks are similar to water absorption peak. Aged snow always shows lower reflectance from that of fresh snow because of its dryness in the near infrared and middle infrared region. In the visible region, there is hardly any difference in reflectance. Sunlight is non selectively scattered by water droplets present in the clouds, so there is a high uniform spectral response throughout the reflective portion of the spectrum (0.3 to 3.0μm) (Bartolucci *et al.*, 1977). That is why in the middle infrared region, the clouds have high reflectance in contrast to the low reflectance of snow. But in the thermal infrared wavelength region, the cloud top and snow often have the same temperature range, so spectral separation in this wavelength region is not reliable.

According to O'Brien and Munis (1975), the reflectance of snow generally decreases throughout the reflective infrared portion of the spectrum.

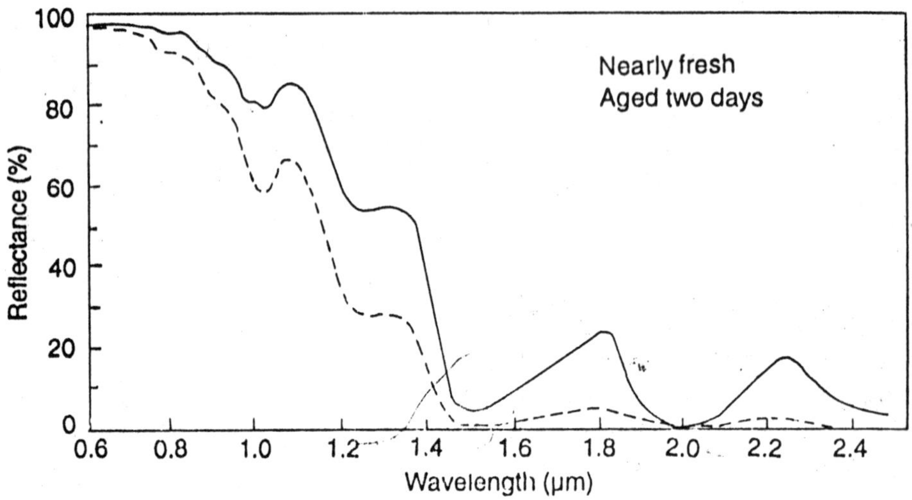

Fig. 2.27: Spectral reflectance characteristics of snow

REFERENCES

Barrett, E.C. and L.F. Curtis (1992), *Introduction to Environmental Remote Sensing*, Chapman & Hall, NY.

Bartolucci, L.A., B.P. Robinson and L.P. Silva (1977), Field measurements of spectral response of natural water, *Photogrammetric Engineering and Remote Sensing*, vol. XLIII, pp. 595-8.

Bower, S.A. and R.J. Ranks (1965), Reflection of radiant energy from soil, *Soil Science*, vol. 100, pp. 130-8.

Coulson, K.L., J.V. Dave and Z. Sekera (1960), *Tables related to Radiation Emerging from a Planetry Atmosphere with Rayleigh Scattering*, University of California Press, Berkeley.

Deekshatulu, B.L. and O.P. Bajpai, (1982), Physics of Remote Sensing, *Current Science*, vol. 51, No. 24, pp. 1132-43.

Elachi, C. (1987), *Introduction to the Physics and Techniques of Remote Sensing*, Wiley, N.Y., pp. 22-113.

Gates, D.M., H.J. Keegan, J.C. Schleter and Weedner (1965), Spectral properties of plants, *Applied Optics*, vol. 4, No. 1, pp. 11-22.

Hoffer, R.M. and P.L. Johanson (1969), Ecological potentials in spectral signature analysis, in P.L. Johanson (ed), *Remote Sensing in Ecology*, University of Georgia Press, Athens.

Hoffer, Roger M. (1978), Biological and Physical considerations in applying computer-aided analysis techniques to remote sensing data in P.H. Swain and S.M. Davis (eds), *Remote Sensing: The Quantitative Approach*, McGraw Hill, Inc.

Hunt, G.R. (1980), Electromagnetic radiation : The communication link in remote sensing, in Seigal & Gillepsie (eds), *Remote Sensing for Geology*, Wiley, N.Y., Chapter 2, pp. 5-45.

King, G.W., R.M. Haimer and P.C. Cross (1943), The asymmetric rotor: 1. Calculation and symmetric classification of energy levels, *Journal of Chemical Physics*, vol. 11, pp. 27-42.

King, G.W., R.M. Haimer and P.C. Cross (1949), The asymmetric rotor: VII, Extension of calculation of energy levels, *Journal of Chemical Physics*, vol. 17, pp. 826-36.

Knipling, E.B. (1967), Physical and physiological basis for differences in reflectance of healthy and diseased plants, in the *Proceedings of the Workshop on Infrared Colour Photography in the Plant Sciences*, Florida Department of Agriculture, Winter Haven.

Lillisand, T.M. & R.W. Keifer (eds) (1987), *Remote Sensing and Image Interpretation*, John Willey & Sons, New York (2nd ed. 1990).

Mathews, H.L. (1972), *Application of Multispectral Reflectance Patterns to Soil Survey Research*, Ph.D. Thesis, Pennsylvania State University, College Station.

Montgomary, O.L. (1976), An Investigation of the Relationship between Spectral Reflectance and the Chemical, Physical and Genetic Characteristics of Soils, Ph.D. desertation, Department of Agronomy, Purdue University, West Lafayette.

Myers, V.I. (1970), *Remote Sensing with Special Reference to Agriculture and Forestry*, National Academy of Sciences, Washington, D.C., pp. 253-97.

Myers, V.I. (1975), Crop and soils, in R.G. Reeves (ed), *Manual of Remote Sensing*, American Society of Photogrammetry, Falls Church, Virginia (2nd ed.), pp. 1715-813.

O'Brien, H.W. and R.H. Munis (1975), Red and Near Infrared Spectral Reflectance of Snow, in *NASA Special Publication* SP-391, Washington.

Parsons, C.L. Jr., and Jurica, G.M. (1975), *Correction of the Earth Resources Technology Satellite Multispectral Scanner Data for the Effect of Atmosphere*, LARS Information Note 061875, Laboratory for Applications of Remote Sensing, Purdue University, West Lafayette.

Satterwhite, M.B. and J. Ehlen (1980), *Vegetation and Terrain Relationship in South Central New Mexico and Western Texas*, Report No. ETL00284, U.S. Army Engineers, Topographic Laboratory, Fort Bdvoir.

Sinclair, T.R., R.M. Hoffer and Schreiber (1971), Reflectance and internal structure of leaves from several crops during a growing season, *Agronomy Journal*, vol. 63, pp. 864-8.

Slater, P.N. (1980), *Remote Sensing: Options and Optical Systems*, Addison-Wesley Reading, Massachusetts & London.

Swain, P.H. and S.M. Davis (eds) (1978), *Remote Sensing: The Quantitative Approach*, McGraw Hill, London & New York.

U.S.D.A. (1951), *Soil Survey Manual*, Handbook No. 18, Washington D.C.

U.S.D.A. (1975), *Soil Taxonomy*, Agricultural Handbook 436, U.S. Department of Agriculture, Soil Conservation service, U.S. Government Printing Office, Washington D.C.

THREE

IMAGING SYSTEMS

Background

REMOTE SENSING technology has given an additional option to assess the environment and natural resources of the earth. This application is becoming more and more important due to attention now being given to the related issues, such as sustainable development. Further, the availability of the information in digital form has given additional dimension and much flexibility for different applications. In order to use such a facility, it is important to understand the imaging system of the different sensors from which the remote sensing data is available. It is interesting to compare different imaging system among themselves, particularly with the Indian Remote Sensing Satellite-IRS.

In last two decades, the scenario related to remote sensing has witnessed rapid changes. The data provided by each satellite and their respective sensors have been better, particularly considering the resolution and repeatability. The pixel size in the LANDSAT 1, 2 and 3 was 79m × 79m which reduced to 10m × 10m in SPOT panchromatic. However, LANDSAT Thematic Mapper (TM) data is considered to be more versatile because it is provided in seven bands, including thermal band. The IRS-1A and 1B have two sensors with the pixel size of 72.5m × 72.5m in LISS I, and 36.25m × 36.25m in LISS II which are close to LANDSAT 5 MSS and TM. Further, the high resolution Russian satellite digital data from KVR 1000 is available having resolution of 2m. One frame covers an area of about 40 km × 40 km in panchromatic spectral range. Such data may be slightly old, but considering the resolution it offers, mapping upto 1:10,000 is possible. Hence, it has a lot of applications in area and urban form, can also be in the form of paper print, film positive and film negative. Since, the digital data is only having single band, a lot of image processing techniques cannot be carried out as is possible with the MSS data. Nevertheless, it can be used with other remote sensing data, if required after doing resampling.

There are several other satellites and sensors which do provide digital remote sensing data such as National Oceanic and Atmospheric Administration or NOAA (1 km × 1 km), Seasat (25m × 25m), Geostationary Operational Environmental Satellite or GOES (1 km × 1 km), Television Infrared Observation Satellite or TIROS, Heat Capacity Mapping Mission Satellite or HCMM, SKYLAB, and Space Shuttle. However, we will concentrate on the LANDSAT, SPOT and IRS because they are primarily used by the application scientists and resource managers for assessing the environment and the resources. Table 3.1 provides the comparative features of these sun synchronous satellites.

TABLE 3.1: Features of different remote sensing satellites

Features	LANDSAT 1, 2, 3		LANDSAT 4, 5		SPOT		IRS 1A, 1B	
				ORBIT DETAILS				
Altitude (km)	918		705		813		904	
Inclination	99°		99°		98°6'		99°	
Repetition cycle (days)	18		16		26		22	
Orbits	252		233		--		307	
Time passing equator	9.42 am		11 am		10.30 am		10 am	
			IMAGING SYSTEM					
Sensor	RBV	MSS	MSS	TM	HRV1	HRV2	LISS-I	LISS-II
Scanning	Video		Oscillating mirror			Push broom		Scanning
FOV (scan angle)	--	11.56°	14.9°	14.9°	4.13°	4.13°	9.4°	4.70°
GRE/IFOV (m sq.)	40	79	79	30	20	10	72.5	36.25
Swath (km)	185	185	185	185	60	60	148	74°
Frame area (km)	185 × 185	185 × 185	185 × 185	185 × 170	60 × 60	60 × 60	148 × 148	74 × 74°
Spectral bands	3	4	4	7	3	1	4	4

Landsat satellites carry three types of sensors: RBV, MSS and TM. High resolution visible camera (HRV) is used in SPOT. IRS use the LISS-I and LISS-II sensors. The details of the above sensors will be discussed here.

Landsat Series of Satellites

Landsat of the United States of America fall under the category of first generation unmanned earth resources satellites. After the success gained in the manned space missions, the National Aeronautics and Space Administration (NASA) and the US Department of Interior mooted an idea of sending unmanned experimental earth resources satellite. It was launched in July 1972 and was known as Earth Resources Technology Satellite-1 or ERTS-1; which was renamed in January 1975 as Landsat 1. This satellite demonstrate and proved the potential use of the remote sensing imagery which were readily available for any part of the world without any restrictions, and fairly low cost. Following this, four Landsat satellites were launched, details of which are given in Table 3.2.

From the Tables 3.1 and 3.2 it is apparent that all the Landsat satellites have multiple sensors: multispectral bands (MSS) (Fig. 3.1), scanner or return beam

TABLE 3.2: Landsat Satellites

Satellite Altitude	Sensors	Period of Operation
Landsat 1 917 km	MSS, RBV, DCP	July 1972--January 1978
Landsat 2 917 km	MSS, RBV, DCP	January 1975--July 1983
Landsat 3 915 km	MSS, RBV, DCP	March 1978--March 1983
Landsat 4 705 km	MSS, TM, DCP	July 1982--October 1983
Landsat 5 705 km	MSS, TM, DCP	March 1984--Operational

vidicon (RBV) (Fig. 3.2) television cameras and Thematic Mapper (TM) (Fig. 3.3). Further, similar to SPOT and IRS, Landsat satellites are also sun synchronous and except the polar region (>82° N and S) they cover the whole earth. In 103 minutes they cover one full revolution of the earth. In a day 14 such orbits can be covered. The sensors on board covers swath of 185 km. In the Landsat 1, 2 and 3 the entire earth used to be completed in 18 days. But in the successive Landsat satellites, it takes on 16 days. The RBV sensor payload for Landsat 1 and 2 (Fig. 3.4) was having three cameras with spectral bands of (a) 0.475-0.575 µm or green, (b) 0.580-0.690µm or red, and (c) 0.690-0.830µm or infrared. However, Landsat 3 was having three RBV payload for only one pair of camera for range of 0.505-0.750µm and with a resolution of 40m. With the help of two cameras, each covering 98.5 km × 98.5 km, it covers the entire visible and near infrared spectral range. The details of the RBV sensors is shown in Table 3.3.

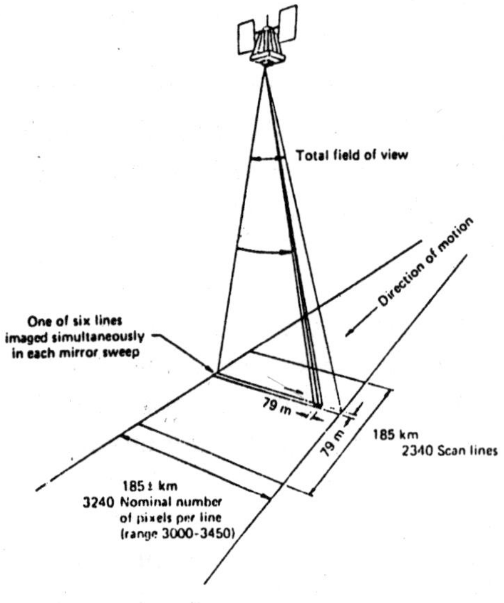

Fig. 3.1: Landsat MSS operating configuration

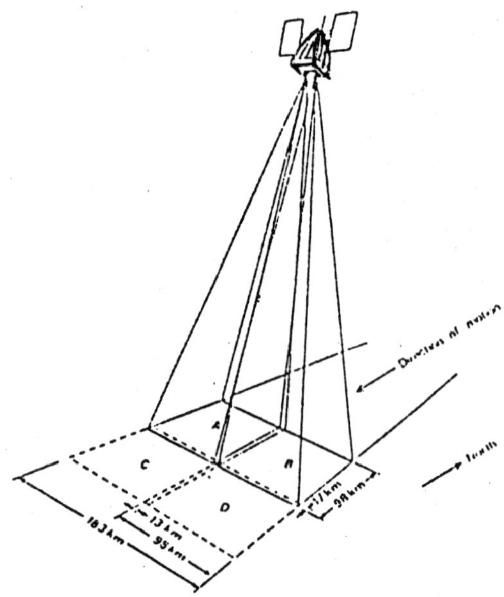

Fig. 3.2: Landsat RBV operating mode (Landsat MSS comprising 2,340 scan lines and 3,240 pixels per line—a total of about 7.5 million pixels per channel and 30 million pixels per scene of 4 channel)

TABLE 3.3: RBV sensors in Landsat 1, 2 and 3

Satellite	Band range (μm)	Resolution
Landsat 1 & 2	0.475-0.575 (green)	80m
Landsat 1 & 2	0.589-0.680 (red)	80m
Landsat 1 & 2	0.690-0.830 (infrared)	80m
Landsat 3	0.505-0.750 (visible)	40m

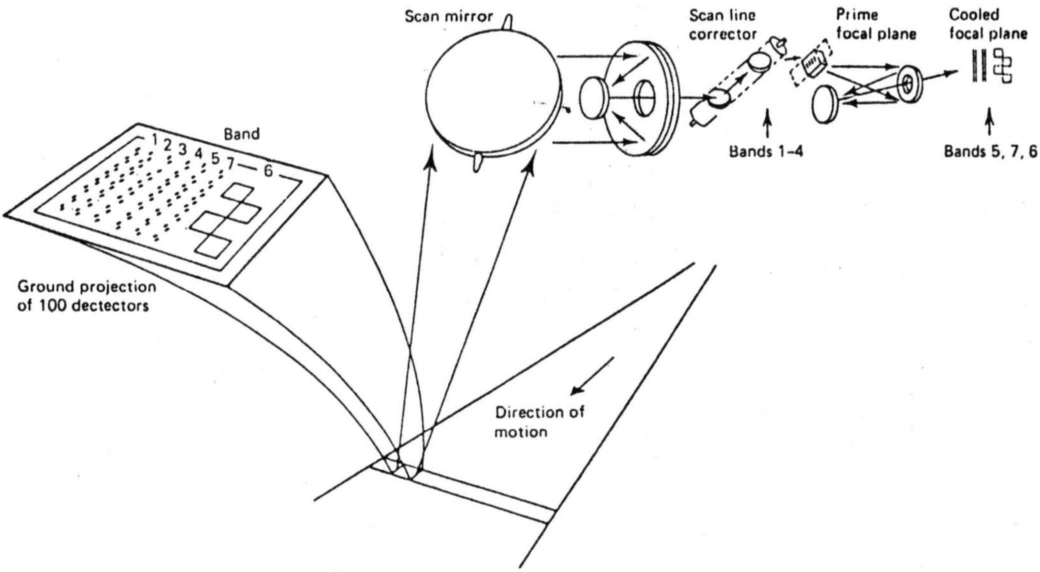

Fig. 3.3: TM optical path and projection of detector IFOVs on earth surface. (Adapted from NASA diagram)

The imaging system for RBV is different from MSS and they do not contain films. Their images are captured by a shutter device on a photosensitive surface within the cameras. This surface is then scanned as a raster file and video signals are produced as in TV cameras. It has better geometric accuracy. Unfortunately, Landsat 1 gave very limited frames (1690 frames from 23-7-1972 to 5-8-1972), and Landsat 2 was used occasionally for this purpose. Landsat 3 was having only one spectral band. The following Landsat satellites discontinued this sensor. The MSS sensor which is in all Landsat satellites have four bands as indicated in Table 3.4.

The above mentioned bands are in visual and infrared ranges. In Landsat 3 (Fig. 3.4) attempt was made to introduce thermal band (10.4-12.6 μm) but it failed to work. As mentioned in Table 3.1, the GRE or IFOV for the MSS data is 79m. The field of view is 11.56°. In a way, the RBV was replaced by Thematic Mapper in Landsat 4 and 5 (Figs. 3.5 a & b) having some similarity with MSS

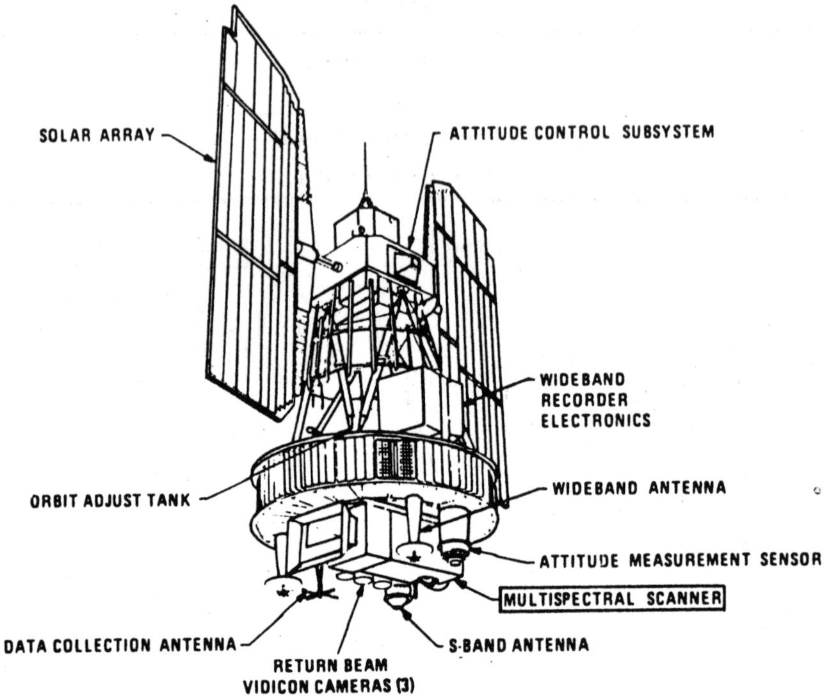

Fig. 3.4: Diagram of Landsat 1 and 2. The multispectral scanner has produced almost all the imagery from these satellites. From NASA (1976, Fig. 2.2).

TABLE 3.4: Landsat MSS bands

Sensors Landsat	Wavelength (μm)	Band Name	Sensitive to
4	0.5-0.6	Green	Movement of sediment, laden water and delineation of shallow water areas.
5	0.6-0.7	Red	Cultural features, discrimination of vegetation types
6	0.7-0.8	Near IR	Vegetation, boundary between land and water, landform
7	0.8-1.1	Near IR	Penetration of atmospheric haze, vegetation, boundary between land and water and landform

payload but having seven spectral bands (Table 3.5).

The resolution of the TM data is 30m, excepting for the thermal band (i.e.

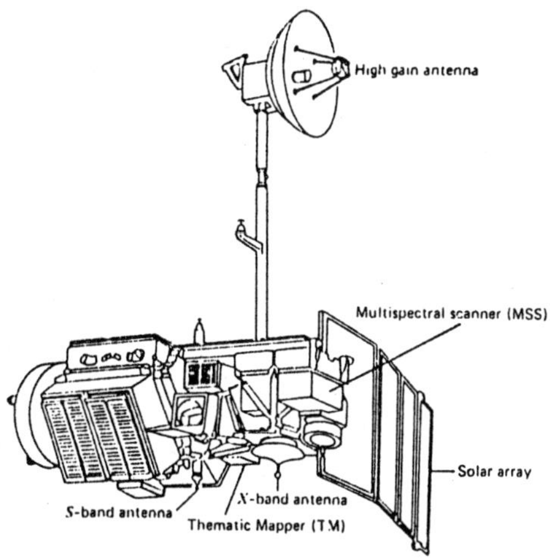

Fig. 3.5(a): Landsat-4 and 5 observatory configuration. (Adapted in NASA diagram)

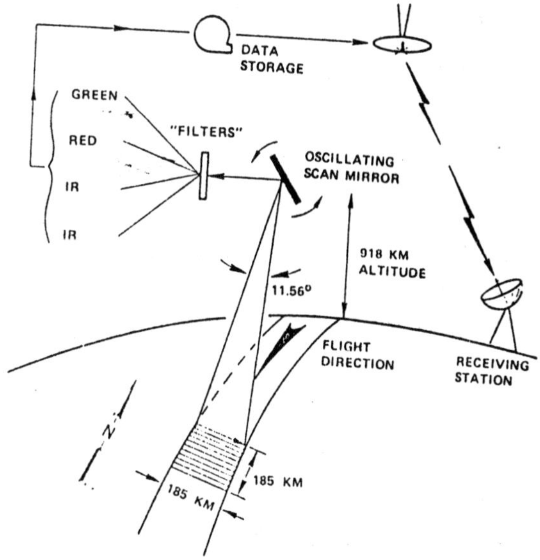

Fig. 3.5(b): Landsat multispectral scanner. For each terrain scene, four images are transmitted to a receiving station. From NASA

TABLE 3.5: Spectral Bands of Landsat Thematic Mapper (TM)

Band	Spectral range (μm)	Applications
TM 1	0.45—0.52	Designed for water body penetration useful in coastal bathymetry. Also useful for differentiation of soil from vegetation, and deciduous from coniferous flora.
TM 2	0.52—0.60	Measurement of visible green reflectance peak of vegetation for vigor assessment.
TM 3	0.63—0.69	Chlorophyll absorption band, important for vegetation discrimination.
TM 4	0.76—0.90	Determination of biomass content and surface water mapping.
TM 5	1.55—1.75	Indicative of vegetation moisture content and soil moisture. Also useful for separating snow from clouds.
TM 6	10.40—12.50	Thermal channels useful in vegetation stress analysis, soil moisture discrimination and thermal mapping.
TM 7	2.08—2.35	Discrimination of rock types and for hydro-thermal mapping.

TM 6) which is having 120m.

SPOT

SPOT or *System Probeteir d'observation de la Terra* is a French remote sensing satellite which was launched on 22nd February 1986 by Ariane-1 Launcher of the European Space Agency (ESA), and was declared operational on 6th May in the same year. It was planned to serve several years. The whole system was designed under the supervision of Centre National d'Etudes Spatiales (CNES) in collaboration with the Institute Geographique National (IGN) and several other French, Belgian and Swedish companies. Till recently it produced remote sensing data having the best resolution of 10m and 20m (Table 3.1).

SPOT satellite consists of two identical high resolution HRV imaging systems and a package compromising of two magnetic tape data recorders and a telemetry transmitter. Since it has better resolution, the data can be used for detailed studies and cartographic applications as well. It can be used for updating maps at about 1:50,000 and 1:25,000 scales as well.

High Resolution Visible (HRV) imaging system of SPOT has push-broom scanner used on board. There are two identical system called HRV-1 and HRV-2. These sensors can operate in two modes: (a) panchromatic i.e. black/white with 10m resolution, and (b) multispectral (MSS), i.e. coloured having resolution 20m resolution in the visible and near infrared range of the spectrum. The former mode has 6,000 detectors while the latter has 3,000 detectors for each band in MSS mode. Each sensor can be pointed so that two adjacent strips of the earth surface are imaged with an overlapping of 3 km, having a total swath width of 117 km. Further, any of the sensor can be tilted up to ± 27° in order to produce stereoscopic images having the frame size of 60m × 60m within a 950 km wide strip centered on the satellite ground tracks (Fig. 3.6). However, at the extremes of the ground track the width of the swath will be about 80 km.

The in-built facility of off-nadir viewing has increased the possibility for revisiting the same area with better frequency even with repeat cycle of 26 days. For example, there will be 7 revisits at the equator within one repeat cycle. This will increase to 11 revisits at the latitude of 45°. Such revisiting facilities has given a very important capability of generating stereoscopic pairs of the same frame or scene. The stereo-pairs are close to photogrammetry in terms of various applications in topographical mapping, geology, geomorphology and hydrology.

The technical characteristics of SPOT images can be gathered from Table 3.6. Here location accuracy is defined for any point on the ground imaged if the SPOT frame as the error between the points estimated position relative to a given mode of cartographic projection and its actual position. For any two points on the ground separated by at least 500m, the scale distortion is defined as difference between the said points identified on the image again is the chosen projection and the actual distance measured on the ground. Further, the superimposition of the

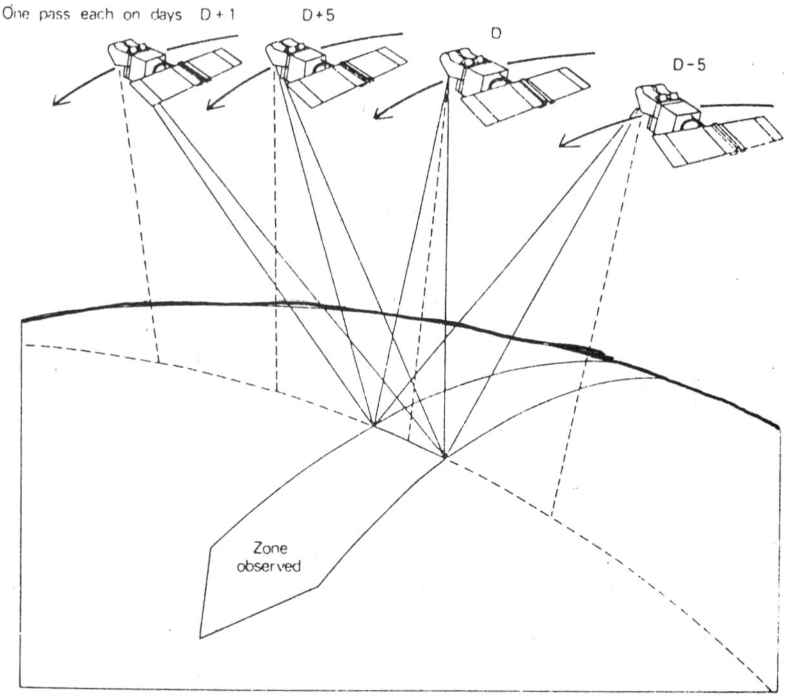

Fig. 3.6: An illustration of how the SPOT satellite can be used to obtain images of an area with greater frequency than its 26-day repeat cycle. In this case, the area that is viewed vertically on one day (D), was viewed obliquely five days earlier (D - 5) and will be viewed obliquely in five (D + 5) and ten (D + 10) days time. (Modified from CNES 1980)

three bands in the multispectral mode implies the accuracy with which corresponding points in the three MSS scenes can register.

The SPOT grid reference system (GRS) has a world-wide coverage and organized in columns (K) and lines (J). GRS is available in 14 sheets. The Indian coverage is shown in Fig. 3.7. In the panchromatic mode, the number of pixels per line is between 6,000 and 10,400. The number of lines per scene varies from 6,000 to 9,800 depending on the method of surveying (vertical to oblique). Hence the volume of data per scene or frame varies from 32 to 100 MB. Further, in MSS mode, the pixels per line varies from 3,000 to 5,200 while the number of

TABLE 3.6: Technical characteristics of SPOT images

Technical characteristics	Preprocessing levels	Initial specifications	Measured value
Location accuracy	1B	1,500m	832m
	2	50m	30m
Scale distortion	1B	1%	0.14%
Superimposition of	1A	0.3 pixels	0.2 pixels three bands

lines per scene ranges from 3,000 to 4,900 for each the three spectral bands. The volume of data required for a MSS scene thus varies from 27 to 76.5 MB. The spectral range of the SPOT panchromatic and MSS bands is as follows:

TABLE 3.7: Spectral range of SPOT data

Bands	Pixel size	Spectral range µm
Panchromatic (PA) B/W	10m	0.51-0.73
Multispectral (XS or MSS)	20m	
Green		0.50-0.59
Red		0.61-0.68
Near Infrared		0.79-0.89

Indian Remote Sensing Satellites

After initial success in launching communication and meteorological satellites, the Indian Space Research Organisation (ISRO) launched an ambitious programme to design and develop operational remote sensing satellites basically for the management of the natural resources of the country. The principal components of the Indian Remote Sensing Satellites or the IRS mission are to develop a three axis stabilized polar sun-synchronous satellite with multispectral sensors along with a ground based data reception, recording and processing system for the multispectral data. The ground systems for the in orbit satellite control includes the tracking network with the associated supporting systems. Further, the hardware and software elements for the generation of users oriented data products, data analysis and archival were also aimed.

The primary objectives of the IRS mission are:

- to design and deploy a three axis stabilized polar sun-synchronous satellite carrying a near state of art multiple solid state pushroom cameras in visible and near infrared bands for acquiring imagery for earth resources applications on an operational basis.
- to establish and routinely operate ground based systems for data reception, recording, processing, generation of data products, analysis and archival as well as mission control facilities.
- to use the data from IRS in conjunction with supplement/complimentary information from other sources for survey and management of resources in important areas such as agriculture, geology, and hydrology in association with the user agencies, that will additionally enable characteristics of a future operational system for the country at the optimum level.

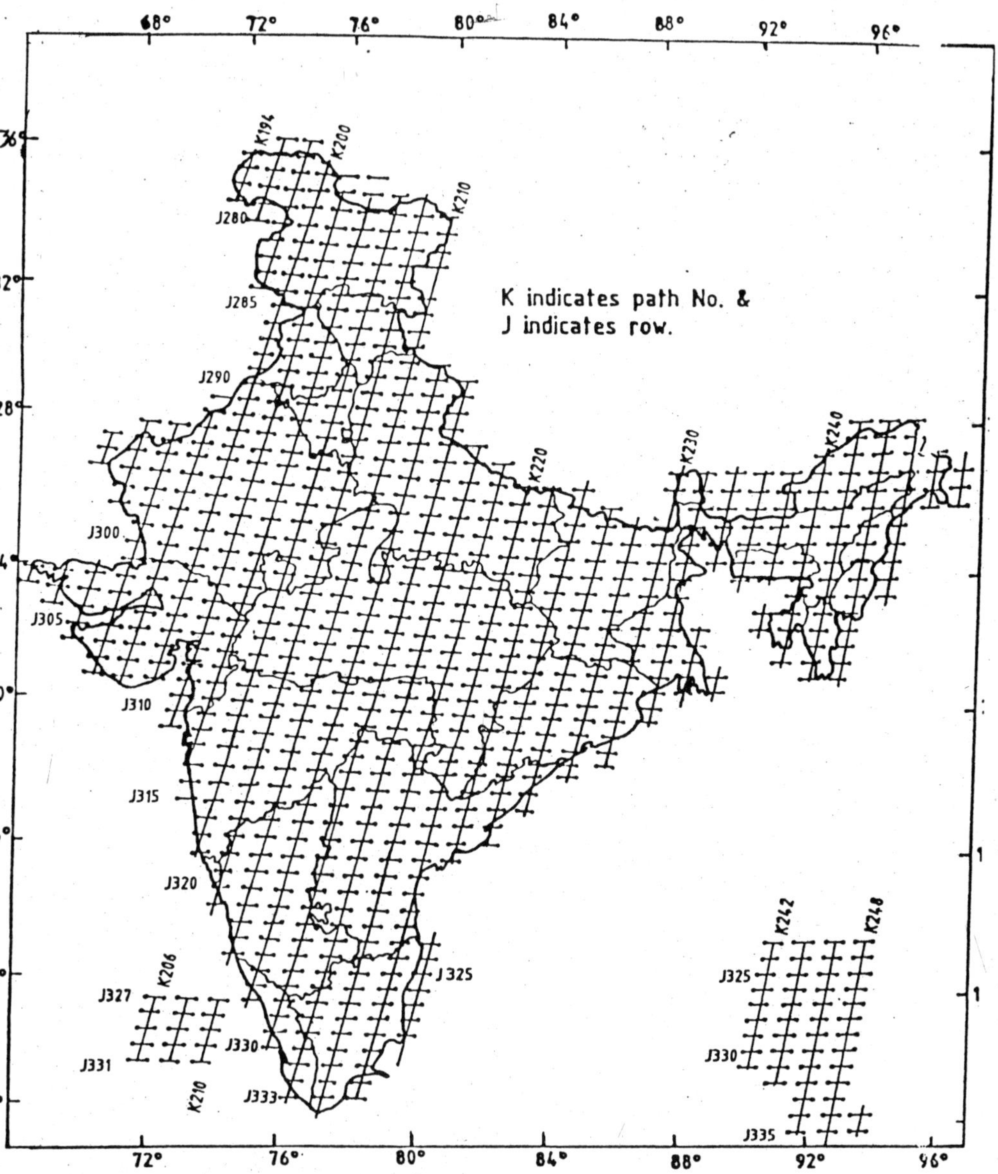

Fig. 3.7: SPOT Grid Reference System

There are also secondary objectives related to assess specific applications in different disciplines, and to build indigenous capability in a variety of technologies both for space and ground segments. IRS was considered to be an essential part of the National Natural Resources Management System (NNRMS) having an aim of conceptualizing an integrated resource management system for optimizing the utilization the natural resources of the country.

The first satellite, IRS-1A was launched into near circular orbit on 17th March 1988 which was followed by an identical satellite, IRS-1B launched on 29th August 1991. Both of them are operated in sun-synchronous near polar orbit at an inclination of 99° and at an altitude of 904 km (Table 3.1). It takes 103 minutes to complete an orbit and completes 1 orbit per day. The repeat cycle is of 11 days. Almost similar to SPOT, IRS 1A/1B use charge coupled devices (CCD) in the pushbroom scanning mode using satellite movement to generate successive lines. Both the satellites have two types of imaging sensors known as Linear Imaging Self-scanning Sensors or LISS. The first sensors, LISS-I has spatial resolution of 72.5m, and LISS-II with two separate imaging sensors with spatial resolution of 36.25m. Both the LISS-II sensors have swath of 148 km on ground, and the composite swath of 146.98 km. IRS 1A and 1B have four sensors operating in four spectral bands. The details of both the Indian satellites are given in Tables 3.8 and 3.9.

TABLE 3.8: IRS imaging sensors characteristics

Features	LISS-I	LISS-II
Focal length (mm)	162.2	324.4
Field of view (degree)	9.4	4.7 + 4.7
Instantaneous field of view, IFOV (microrad)	80	40
Detector (No. of CCD elements)	2048	2048
Ground resolution (meters)	73	36.5
Number of bands	4	4
Swath (kms)	148	2 × 74 (145 composite)
Radiometric resolution (grey levels)	128	128
Data rate (Mbps)	5.2	2 × 10.4
Weight (kgs)	38.5	2 × 80.8
Power (watts)	34	2 × 34

In addition to IRS-1A and 1B, IRS-P2 satellite was successfully launched by indigenously built PSLV (D2) on 15th October 1994. This satellite was placed in a polar sun-synchronous orbit at an altitude of about 817 km. The payload is similar to onboard IRS-1A/1B except that the combined swath is realized using a single

TABLE 3.9: IRS Spectral bands and applications

Band	Spectral range µm	Applications
1	0.45-0.52	Coastal environment studies Soil/vegetation differentiation Coniferous/deciduous vegetation discrimination
2	0.52-0.59	Vegetation vigor Rock/soil discrimination Turbidity and bathymetry in shallow waters
3	0.62-0.68	Strong-chlorophyll absorption leading to discrimination of plant species
4	0.77-0.86	Delineation of water features Landform/geomorphic studies

Source: NNRMS (1991).

electro-optical module. The CCDs are staggered in a way that there will be a gap of 6.3 km on ground along the track direction between the two individual image strips and the combined swath is 131 km. The revisit cycle is 24 days and crosses the equator at 10.30 a.m. The spatial resolution of this satellite is 32m across track and 37m along track. However, this data is resampled in order to provide continuity to the IRS-1A/1B data users. Further, IRS-P3 satellite was launched by PSLV (D3) on 21st March, 1996. It carries two imaging sensors, viz. Wide Field Sensor (WiFS) and Modular Opto-electronic Scanner (MOS) and for X-ray astronomy (NRSA, 1995b).

IRS-1C/1D

IRS-1C/1D projects have been planned as continuation projects of IRS-1A/1B with improved spatial resolution. These projects have been approved by the Government of India in May 1989. The objectives of this mission are to design, develop, launch and operate state-of-the-art three axis body stabilized satellite for providing continued space based remote sensing services to the user community with enhanced resolution and capability of IRS-1A/1B in order to be used for new application areas. The IRS-1C and IRS-1D satellites have been planned with similar payloads and with a mission life of three years. The principal components of the mission are:

- — a three axis stabilised polar sun-synchronous satellite with three sensors.
- — a ground system for in-orbit satellite control
- — hardware/software elements for the generation of user oriented data products, data analysis and archival.

The objectives of the mission are:

— To design, develop, launch and operate a state of art three-axis body stabilised satellite providing continuous space based remote sensing services to user community with enhanced resolution and capability compared to IRS-1A and IRS-1B.
— To establish and operate ground based systems for data reception, recording, processing, generation of data products, analysis, archieval and mission control facilities.
— To develop new areas of user applications to take full advantage of the enhanced resolution and capability of IRS-1C/1D sensors (NRSA, 1995b).

The IRS-1C has been successfully launched in December 1995 and preliminary results are now available. It will have following three sensors (Interface, 1995):

— PAN camera is a high resolution (5.8m) single band sensor (0.50-0.75 μm) with ground swath of 70 km. It can be steered upto ± 26° for obtaining stereoscopic data and 5 day revisit.
— LISS-III is a medium resolution sensor which will provide data in four spectral bands, viz. two in visible (0.52-0.59 and 0.62-0.68 μm), one in near infrared (0.77-0.86 μm) and one in short wave infrared (SWIR, 1.55-1.70 μm) regions. LISS-III will have a spatial resolution of 23m (69m in case of SWIR band) and a swath of 141 km. (148 kms in the case of SWIR).
— WiFS is a two band (visible: 0.62-0.68 μm and near infrared: 0.77-0.86 μm) sensor with a large swath (810 kms) having a resolution of 188m) (Fig. 3.10).

TABLE 3.10: Salient features of IRS-1C/1D imaging sensors

Features	LISS-III	PAN	WiFS
No. of spectral bands	4	1	2
Spectral bands (microns) μm	B/2 0.52-0.68 B/3 0.62-0.68 B/4 0.77-0.86 B/5 1.55-1.70	0.50-0.75	B/2 0.62-0.68 B/4 0.77-0.86
Spatial resolution (m)	23.5 for B/2, B/3, B/4 and 70.5 for B/5	Better than 10 (5.8)	188
Swath (km)	142 for B/2, B/3, B/4 and 148 for B/5	70, nadir steering range >> 4 ± 26 degree	
Radiometric levels	128	64	128

Source: NNRMS, Satellite Remote Sensing.

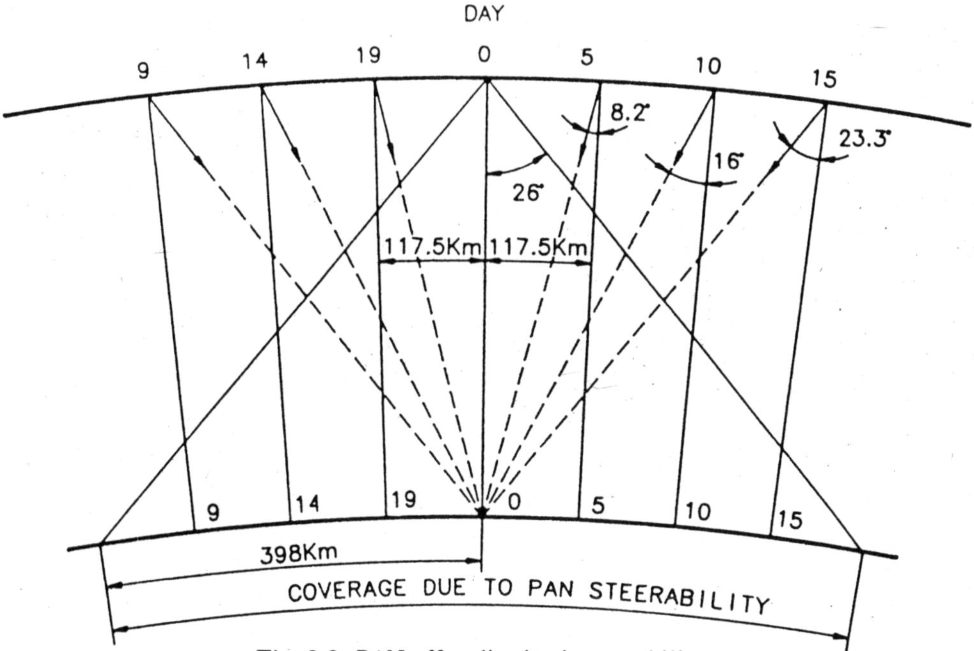

Fig. 3.8: PAN off-nadir viewing capability.

The IRS-1C referencing system is slightly different from IRS-1A/1B. IRS-1C has been designed with the following features:

(a) The number of orbits per cycle is made to correspond to the swath of LISS-III whose swath and resolution lie in between those of PAN and WiFS.
(b) The high resolution PAN camera, with a smaller swath, is made steerable in order to cover the gaps which occur if the camera were capable of only nadir operations (Fig. 3.8).
(c) The coverage pattern is such that the first orbit of successive days are shifted by five paths. This ensures that the WiFS covers newer areas on successive days instead of needlessly producing overlaps.

IRS-1C will take 101.35 minutes to complete one orbit thereby completing approximately 14 orbits in a day. It will have 341 orbits in a 24 days cycle in order to cover the entire globe. Further, two band carriers are introduced to transmit the data from the payloads. Onboard tape recorder is added in order to provide capability to record 24 minutes of either panchromatic or LISS-III data for payback in the subsequent orbit over Indian ground stations.

In addition to IRS-1C/1D having 5.8m resolution, India has planned IRS-P3 which will carry a modular opto-electronic scanner (MOS) provided by DLR, Germany specially designed to meet oceanography requirements as well as WiFS similar to IRS-1C payload with an additional SWIR band (Table 3.11). This

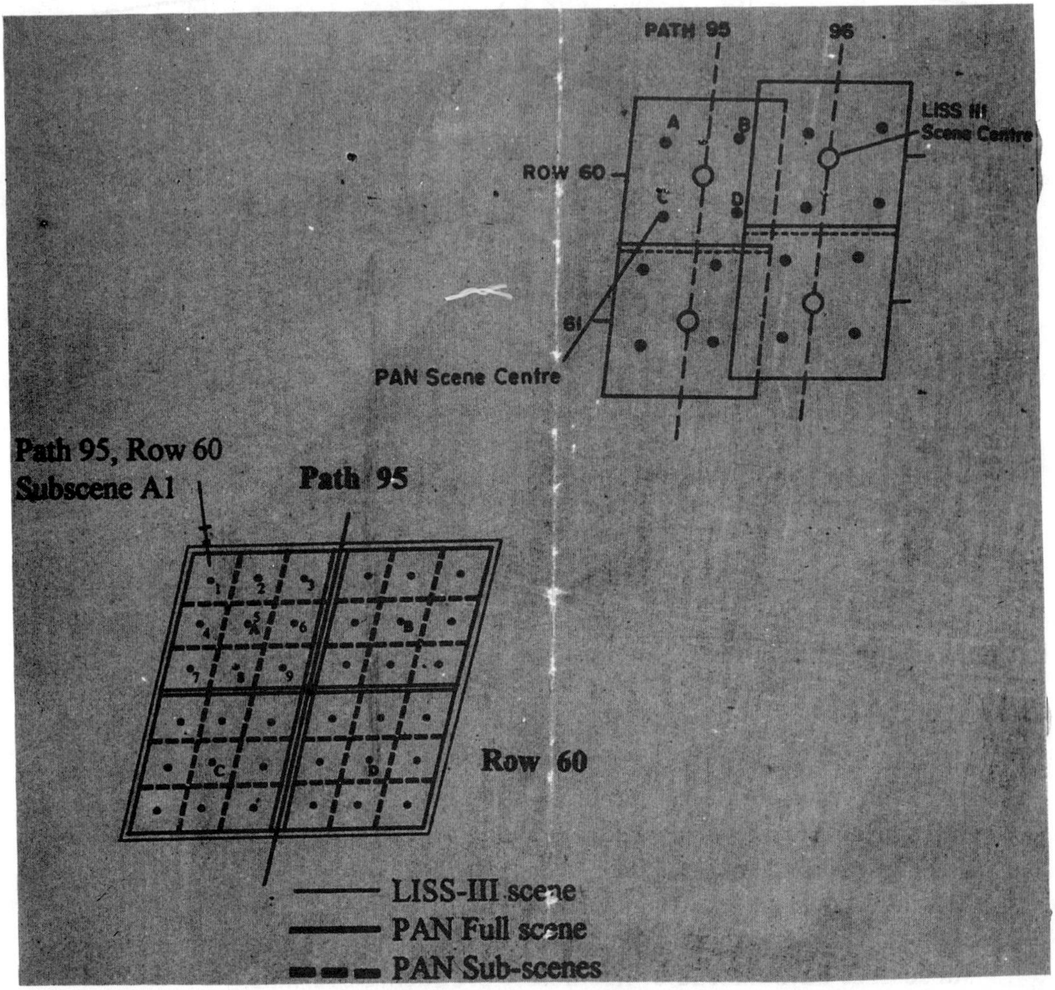

The IRS-1C is a three axis body stabilised satellite designed for an operational life of three years in a polar sun synchronous orbit at an altitude of 817 Km. Each path is repeated after a period of 24 days.

The referencing scheme of LISS-III consists of 341 paths numbered from west to east. Each path consists of 141 rows. Each LISS-III scene covers an area of 141×141 Km., in the case of LISS-III SWIR band, the area covered is 148×148 Km. The referencing scheme of PAN has been evolved around the LISS-III scene centre. A PAN full scene is of 70 Km × 70 Km and four PAN full scenes are fitted in a LISS-III scene. A LISS-III scene is divided into four parts which are designated as A, B, C and D and the PAN scenes will be referred to by the same path and row numbers as that of LISS-III along with the suffixes A, B, C and D as shown in the figure.

It is possible to process the data CCD array wise. This has led to the concept of Pan subscenes. A PAN scene consist of nine subscenes, each covering an area of 23.5 sq km at nadir and can be selected as shown in the diagram.

Fig. 3.9: Indian Remote Sensing Satellite-IC (IRC-1C) Reference Map

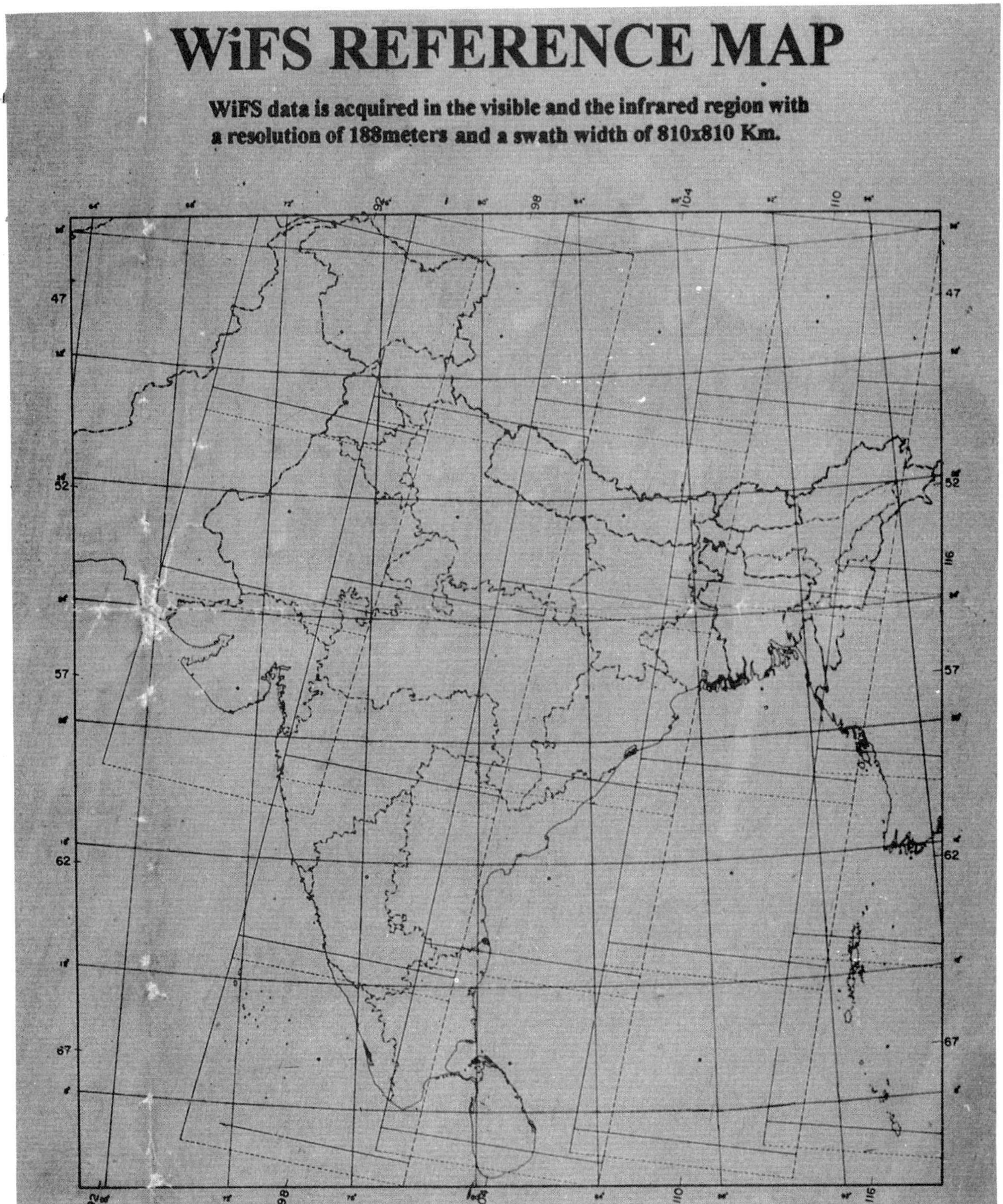

Fig. 3.9 (Contd.) : WiFS Reference Map

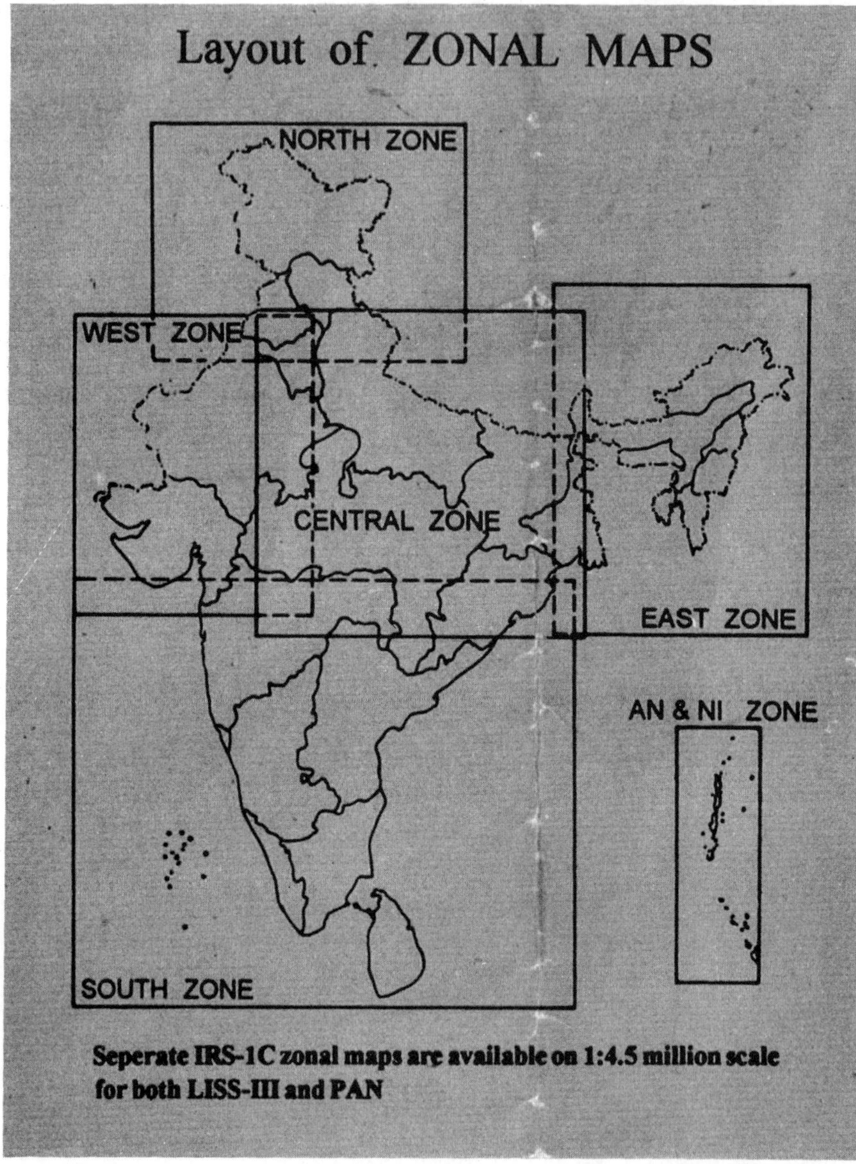

NRSA Data Centre, National Remote Sensing Agency,
Dept. of Space, Govt. of India, Balanagar,
Hyderabad - 500 037, Andhra Pradesh, India

Fig. 3.9: (Contd.) Layout of Zonal Map

TABLE 3.11: Features of IRS-P3

A. Modular Opto-electronic Scanner (MOS)

	MOS-A	MOS-B	MOS-C
Ground resolution (m)	1569 x 1395	523 x 523	523 x 644
No. of spectral bands	4	13	1
Spectral range (μm)	755-768	408-1010	1500-1700
Swath (km)	195	200	192

B. Wide Field Sensors: WiFS

Band 3	0.62-0.68 μm
Band 4	0.77-0.86 μm
Band 5	1.55-1.70 μm
Ground resolution	182 x 188m (red and near infrared)
Swath	246 x 188m (shortwave infrared)
	810 km

Source: ISRO.

additional band, being sensitive to moisture content in vegetation, provides additional data for crop condition assessment. Also, this band has applications like snow and cloud cover differentiation, snow-melt run-off studies, flood damage assessment and geology. This satellite will also have X-ray astronomy payload which is to be used to study time variability and spectral characteristics of cosmic X-ray sources and detection of transient X-ray sources. Further, there are plans to launch IRS-P4 at the end of 1996 which is an ocean satellite designed to meet the ocean and marine data needs (Gopalan, 1996). The payloads proposed for the Oceansat include 9-band ocean colour monitor (OCM), and a Multi-frequency scanning microwave radiometer (MSMR). In addition, initiatives have been made for IRS-P5 and IRS-P6 missions for cartographic and environmental applications to be launched between 1996 to 1998 (NNRMS report).

Conclusion

In the near future several sensors will be able to provide resolution close to 5m or even less. It has been claimed by several agencies that by the turn of the century, close to 1m resolution would be possible. These agencies are Early Bird (Pan: 3m; MSS: 15m), Earthwatch (Pan: 1m; MSS: 3m & 4m); Space Imaging (Pan: 1m; MSS: 4m); and Orbview (Pan: 1m & 2m; MSS: 8m). Such resolution will open new applications, such as urban mapping, environmental monitoring, topographical and large scale mapping. For example the Japanese High resolution satellite, Advanced Land Observing Satellite or ALOS, planned to be launched in 2002, will be having both optical and micro wave sensors. The former sensor will have panchromatic and multispectral parts. On the other hand the panchromatic part will have three telescopes in order to obtain three different views simultaneously. The horizontal resolution will be 2.5m with a swath width of 35 km. The MSS part will have four channels with 10m spatial resolution at nadir. It will also be equipped with a GPS receiver for precise position determination. Other satellites are also offering exciting features in terms of resolution, revisit capabilities etc. The finer resolution will certainly help cartographers to reduce the field work component in mapping. The Indian satellites do have plans to make their own contributions in this field. The Cartosat (IRS-P5) has been planned to be launched by PSLV by the turn of the century. It will have a cutting edge technology in terms of sensor systems and will provide state-of-art PAN camera with about 2-5m resolutions with fore-all stereo capability. This mission will cater to the needs of cartographers and terrain modelling applications. The satellite is likely to provide cadastral information upto 1:5,000 scale and should be useful for making 2-5m contour maps as well (NRSA, 1996).

REFERENCES

Gopalan, A.K.S. (1996), Payloads for P4 satellite, *SAC Courier*, vol. 21, No. 1, pp. 7-10.

IRSO (u.d.), *Indian Remote-Sensing Satellite-IRS-1A*, Indian Space Research Organisation, Bangalore (leaflet).

Konecny, Gottfried (1996), International mapping from space, in the *Proceedings of the International Workshop on International Mapping from Space*, Nov. 27-1 December 1995, Madras, pp. 1-7.

NNRMS (1991), *Remote Sensing in India: Facilities and Infrastructure*, National Natural Resources Management System, Department of Space, Technical Report No. DOS-NNRMS-TR-87-91, Bangalore.

NRSA (1995a), *Interface*, National Remote Sensing Agency, vol. 6, No. 3, July-September.

NRSA (1995b), *IRS-1C: Data Users Handbook*, National Remote Sensing Agency, Hyderabad.

NRSA (1996), *Interface*, National Remote Sensing Agency, vol. 7, No. 3, July-September, Hyderabad.

FOUR

VISUAL INTERPRETATION

REMOTE SENSING is the process of sensing and measuring objects from distance without directly coming physically into contact with them. Remote sensing is largely concerned with the measurement of electro-magnetic energy from the sun which is reflected, scattered or emitted by the objects on the surface of the earth. Different surface objects return different amounts of energy in different wavelengths of the electro-magnetic spectrum. Detection and measurements of these spectral signatures enables identification of surface objects both from the air borne and space borne platforms.

Remote sensing through air borne and satellite based sensors covers surveying as well as monitoring which are essential for the planning and management of national natural resources. This technology, integrated with traditional techniques is emerging as an efficient, speedy, cost effective and important tool for the national development efforts. The study of landuse, land cover or land form from the air has been a focus of interest since the early days of aerial photography and has been gaining momentum again with the availability of new remote sensing techniques using aircraft and spacecraft as platforms with a capacity for operating outside the visible part of the electro-magnetic spectrum using microwaves (radar) and thermal rediation. Thus photo-interpretation is now complemented by the interpretation of other kinds of imagery which are different from photography. The limitation of 'photo-interpretation' has now been changed to broad spectrum of 'image interpretation'. These remotely sensed data have relevance in major sections of the economy such as agriculture, forestry, irrigation, human settlements, geology, ecology and oceans through ensuring the optional use of land, water, mineral resources etc. and this calls for a multidisciplinary approach with active interactions by users at every level. Competence has been achieved in the use of this technique in areas like land use mapping and evaluation of ground water, water quality monitoring, geological survey, topographic mapping and numerous other areas. In order to derive the maximum benefits by the use of these new techniques, it is important to ensure that planners and managers and other potential users appreciate properly the value of this new tool through proper application of principles and methodology of image interpretation either by visual interpretation or by digital analysis. This data can be collected by the remote sensing devices including passive and active systems and employ different bands in the visible, near infrared, middle infrared and far infrared as well as microwaves regions. In the passive remote sensing, the reflected or emitted electromagnetic energy is measured by sensors operating in different selected spectral bands where the original source of energy is the sun. But in active remote sensing method the earth surface is illuminated by an artificial source of energy. The emitted and reflected energy detected by the sensors on board platforms are transmitted to the earth station. The data then processed after various corrections and are made ready for the users.

Types of Data Products

The remote sensing data products are available to the users in the form of (a) photographic products such as proper prints, film negatives, dia positives of black and white and false colour composite in a variety of scales and (b) digital form as computer compatible tape (CCTs) after necessary corrections (vide chapter 13 for details).

The visual interpretation of the remote sensing data is based on hard copies, such as the false colour composites (FCCs). Even after the digital techniques, the results are visually interpreted. Hence, this technique is very important and requires the specific training for the users. Scientists or analysts or other users will interpret the same scene for different purposes. In fact it is one of the rare source of information which can generate multiple themes such as water resources, soil, landuse, urban sprawl etc.

Image Interpretation Techniques

Image interpretation or analysis is defined as the "act of examining images for the purpose of identifying objects and judging their significance". Interpreters study remotely sensed data and attempt through logical process in detecting, identifying, classifying, measuring and evaluating the significances of physical and cultural objects, their patterns and special relationship. Image interpretation is a complex process of physical and psychological activities occurring in a sequence of time. The sequence begins with the detection and identification of images and later by their measurements. The various aspects of image interpretation are listed below in simpler form. Most of these have overlapping functions.

Detection

It is a process of 'picking out' an object or element from photo or image through interpretation techniques. It may be detection of point or line locations e.g. agricultural field or small settlement etc.

Recognition and Identification

It is a process of classification or trying to distinguish an object by its characteristics or patterns which are familiar on the image. Sometimes it is also termed as photo reading e.g. water features, streams, tanks, sands, etc.

Analysis

It is a process of resolving or separating a set of objects or features having similar set of characters. In analysis, lines of separation are drawn between group of objects and the degree of reliability of these lines can also be indicated, e.g. sands as that of river etc.

Classification

It is a process of identification and grouping of objects or features resolved by analysis.

Deduction

It is a process where references are drawn about the objects based on direct or indirect evidence of the information or phenomena under study. Deductions may be firmly confirmed by ground checks to avoid misclassification.

Idealization

It is a process of drawing ideal or standard representation from what is actually identified and interpreted from the image or map e.g. set of symbols or colours to be adopted in waste land maps etc.

Elements of Image Interpretation

There are certain fundamental photo-elements or image characteristics seen on image which aid in visual interpretation of aerial photo or satellite imagery. These elements are mainly (i) tone or colour, (ii) pattern, (iii) texture, (iv) size, (v) shape, (vi) shadow, (vii) location, (viii) association, and (ix) resolution.

Tone or Colour

Different surface objects reflect or emit different amounts of radiant energy. These differences are recorded as tonal/colour or density variation on the imagery. In black and white images, objects appear in different gray tones. These gray tones often fail to provide the interpreter a clear perception of objects. Whereas, the colour of a false colour imagery increases interpretability by providing a subtle tonal contrast between them. Tonal contrast can be enhanced or reduced optically or by enhancement techniques using computers.

Pattern

It refers to the spatial arrangements of surface features which are characteristics of both natural and man-made objects. Similar features under similar environmental conditions reflect similar patterns of recurrence. Pattern is of several types e.g. linear, non-linear, contiguous, clustered etc. More often patterns also reflect associations e.g. intensity of drainage patterns show its relation with rock types, soil texture etc.

Texture

It is defined as a repetition of a basic pattern. Texture in the image is done to tonal repetition in a group of objects which are often too small to be discernible. It creates a visual impression of surface roughness or smoothness of objects and is useful photo element in image interpretation e.g. salt affected land: fine to mottled.

Size

It refers to the spatial dimension of objects on ground. Size of an object is a function of scale of the image and photo and is also measurable.

Shape

It refers to physical form of an object and is also a function of scale of the image or photo size or shape are interrelated. In the image, shape refers to plan or top view of the object, as seen through the satellite.

Shadow

They are cast due to sun's illumination angle, size and shape of the object or sensor viewing angle. The shape and profile of shadows help in identifying different surface objects e.g. clouds, nature or hill slopes aspect etc. They also help in arriving at tree heights or chimney/building heights on aerial photos.

Location

The geographical site and location of the object often provide clue for identifying objects and understanding their genesis e.g. salt affected land-inland river/desert plains etc.

Association

It refers to situation of the object with respect to other surface and neighbouring features e.g. canals with agricultural fields etc.

Resolution

It is of two types, spatial and spectral. The former refers to 'picture element' or pixel discernible on the image or smallest area resolvable and identifiable on ground. Spatial resolution allows the interpreter to detect and distinguish the smallest object on the ground.

Image characteristics comprises of the various elements of image interpretation as described above. By understanding the image characteristics of each of the thematic classes an image interpretation key is involved. The visual interpretation key simplifies and facilitate interpretation of the image in a systematic way. However, it may be mentioned here that visual interpretation is subjective and differ from individual to individual and upon the season, scale and resolution of the imagery. A preliminary interpretation key is prepared and later finalised after ground varification. An example of visual interpretation technique is shown in Table 4.1 for identifying wastelands using interpretation key.

The instrumentation, necessary for image interpretation is equally important for the results obtained as that of the expertise and experience of the interpreter. A brief review to the equipments actually available for interpretation has been done below:

(i) *Magnifying Glass*

A very simple instrument but very useful in examining a single photo or imagery. One can perceive details in the image which cannot readily be seen with the naked eyes. In-built precise engravings may be used with advantage for the measurement of distances and angles which of course more crudely can also be made with a ruler and protractor.

(ii) *Light Table*

It is basic instrument required by all the photo/image interpreters for studying remote sensing photographic products of either in paper prints, or films/diapositive prints. The uniformity in illumination helps the interpreter in better diagnosis the various features in details. The stereoscopic study of

TABLE 4.1: Wastelands Image Interpretation Using TM Data

Sl. No.	Wastelands	Tone/Colour	Size	Shape	Texture	Pattern	Location	Association	Season
1.	Surface water logged land and marsh	Light in dark blue, subject to water spread and organic matter	Varying in size	Irregular	Fine in mottled	Scattered	Coastal low-lands and flood plains	Proximity to flood plain, tidal flats and lagoons	Dec. to March
2.	Degraded pastures/ grazing	Light brown to light red subject to	Varying in size	Irregular	Coarse to mottled	Non-contiguous	Inland plains/ uplands and valleys	Valley lands in hilly terrain; long fallow lands and other lands having thin soil cover	Oct. to December
3.	Strip lands	Light yellow to red	Small in width	Linear	Fine	Contiguous	Along roads/ railway lines/ canals	Strip Plantations	Jan. to March
4.	Sands	White to light yellow subject to moisture content	Varying in size	Regular to irregular	Fine	Contiguous to linear	Desert, river bed and coastal (on-shore plains)	Shifting desert sands, sand dunes, coastal beach sands	Oct. to March
5.	Mining industrial wastelands	Dark blue to light blue	Small	Irregular	Mottled	Scattered	Near mining area	Large indus-trial complex	Jan. to March

Source: Nag *et al*. (1990).

imagery adds essentially to its interpretability and is common practice with vertical aerial photographs. With some modified complexity it can also be applied to oblique photographs, radar images and orbital photographs. The simplest form of stereoscope is lense or pocket stereoscope.

(iii) *Pocket Stereoscope*

The most popular stereoscope with two plano-convex lenses, upper side flat with a focal length of 100 mm. The rays entering the eyes are now parallel and converge at infinity having been accommodated at 100 mm distance. The result is a magnification of 2.5 X. The instrument has got changable eye base range of 60 - 68 mm. It is cheap, transportable and has a large field of view with good clarity but has two disadvantages-- (i) limited magnification and (ii) distance between corresponding points is also 65 mm causing inconvenience in getting distant points in a pair of photograph. These drawbacks have led to develop the mirror stereoscope.

(iv) *Mirror Stereoscope*

The normal size photos (23 mm × 23 mm) can be separated and seen under this stereoscope without folding them. With the help of reflecting mirrors, the path of the bundle of rays has been diverted and brought to the eyes at 65 mm separation. The corresponding points on the photographs are kept at 240 mm distance so that the photographs are placed separately. In fact it effectively increases the eye base from 65 mm to 240 mm. The focal length of the instrument is 300 mm. Thus reduction of the size of the image occurs which is overcome by additional lenses.

(v) *Double Scanning Stereoscope*

This type has been developed for the purpose of simultaneous study of a photo pair by two individuals. The instrument has suitable provision to scan the stereomodel in both x and y direction without moving the photographs.

(vi) *Zoom Stereoscope*

It helps for stereoscopic analysis of a photo-pair on negative, dia positive or plates. This instrument has provision for continuous change in magnification accomplished by the zoom lens system and focussing and rotation of field of view. Independent zooming of the left hand and right hand binocular solves the problem in zoom stereoscope.

(vii) *Zoom Transferscope*

This instrument is useful for transforming the graphic information from a photograph or image to the base map when both are at different scales. In its anamorphic optical system different magnification ratios can be applied in the x and y directions. The zoom magnification provides continuous variations in magnification from 1 to 7 X. It has image rotation system by means of rotating prisms. Black and white or colour transparencies or hard copies can be utilized without any problem.

(viii) *Stereo Zoom Transferscope*

It is identical instrument with that of zoom transferscope having adjustable distances between eye piece and the left and right mirror respectively have been devised for stereoscopic studies of aerial photographs having different scales. The advantage of this instrument over zoom transferscope is stereovision, the rest operation systems are identical.

(ix) *Sketch Master*

It is a simple instrument for plotting planimetry. To get a rectified image of a photograph with tilt, this instrument is used. Sketch master consists of an optical device which allows one eye to receive two superimposed images, one from the photograph and the other from the base manuscript. It contains a plate carrier which can be tilted or rotated in any direction. A double prism with semisilvered surface which permits viewing of photograph and map sheet simultaneously through an oblique eye piece. Together with the prism the plate carrier is vertically adjustable above the map sheet on table top ie. photo-prism and prism-map distances can be veried to suit the scale solution. By lighting the photograph and changing the prism-photo and prism-map distance relationships the operator sees the photo image superimposed on the map. To compensate for the differences between the eye photo and eye map distanced the instrument is equipped with spectacle lenses in mounts. Appropriate ones can be inserted in the double prism in between the prism map and prism photo. The operator sees coincidence between the plotted control points on the map and the corresponding photo points.

(x) *Stereo Sketch Master*

The drawback of the sketch master is that the operator looks all the time with one eye only which is very tiring. This leads to the development of binocular instrument called stereo sketch master. A pair of photographs in seen stereoscopically. One of the mirrors is semi reflecting and hence the base map on the surface of drawing table is seen at the same time as the photographs. Photographic tilt up to ± 5° can be corrected in this instrument.

(xi) *Radial Line Plotter*

The main parts of the instrument are:

(a) A mirror stereoscope
(b) Two photo tables on which photos are centred with their bases collinear.
(c) Radial arms with thread engraved on transparent plastic pinned up at the centre of the phototable and can rotate about the pin. The length of these arms are longer than half the diagonal of photo.
(d) Parallel bar
(e) A linkage connecting the parallel bar to the pencil holder. The map sheet lies underneath the plotter and contains model control to fix the plotting scale. The operator sees the stereoscopic image and the radial lines appear to cross at a point. Its motion is conveyed to a pencil by means of parallel guidance. Scaling is done with the help of plotted ground control points on the map sheet and can be effected by changing the distance between sliding linkage. The plotting is suitable for mapping at scales less than the scale of photography.

(xii) *Parallel Bar*

For accurate measuring the parallel difference, the principle of floating marks is used in parallel bar or parallax wedge. A parallel bar consists of two glass plates engraved with identical measuring marks connected by a bar. The separation between the marks can be changed by a micrometer. Screw graduated so as to give reading up to 0.01 mm. The left hand side glass can be shifted along the rod and can be clamped by a screw. Graduations on the bar are arbitrary and do not refer to the actual separation of the measuring marks. The graduations on the micrometer and the bar are usually numbered increasingly as the distance between the corresponding points i.e. the separation is decreasing. The parallax bar is in common use for height measurement.

(xiii) *Optical Reflecting Projector*

The optical reflecting projector is essentially an optical equipment to transfer information in the form of a photograph or map to another of different scale. This instrument, basically a projection system, is built into a table for convenient working. The equipment consists of a lens system which enlarges or reduces the input (photo, map) kept on the input window and projects it on to the output window through two path folding mirrors. A pair of illumination units illuminate the input photo or map (Plate 4.1).

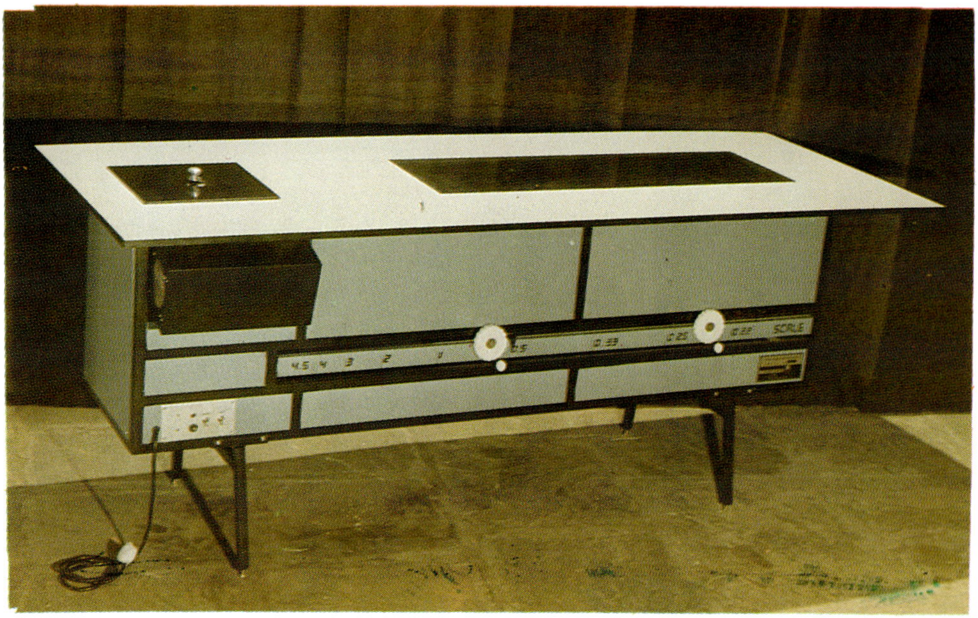

Plate 4.1: Optical Reflecting Projector

The projection system has got the following parts:

(a) Input window of 14" × 14" size provided with optically polished glass for keeping map or photographic print facing down. The maximum input object size is 10" × 10".
(b) Output window of 14" × 46" with polished glass top that serves as a screen to catch the image of the projected object on a translucent paper.
(c) Scale control handle that moves a projection lens carriage over a rack. This movement facilitates in changing the scale of enlargement or reduction. An index on this carriage reads the scale at which the input is projected. The scale can be changed from 0.22X to 4.5X.
(d) Focus control handle moves a mirror carriage over the rack helping the projected image to be focussed for the given scale.
(e) Locking knobs that lock both the carriages after final adjustments.
(f) Illumination units: The projector has got two illumination units one on

either side of the projector. Two tungsten lamps of 600W are provided in each unit with heat absorbing glass filters and cooling fans on each side. The photographic prints are kept over input window and covered with object cover. Then the scale adjustment is done with control handles. After focussing the projected image on the translucent paper one can get the required enlarged or reduced projection of the input image for his work.

Plate 4.2: Large Format Optical Enlarger

(xiv) *Optical Pantograph*

Optical pantograph is an instrument that helps changing scale rapidly. The instrument is designed specially to obtain geometrical rectification while transferring the data from aerial photographs/satellite imagery to maps by providing authorized tilt in x and y directions to copy platform. Large format inputs can be handled with this system. Continuous variable enlargement or reduction can be achieved from 4X to 0.25X range.

(xv) *Large Format Optical Enlarger*

Large format optical enlarger utilizes the principles of optical projection to form the real image of an illuminated object at the desired magnification (4X). Landset imagery are available mostly in 24 cm format on 1:100,000 scale with typical ground coverage of 185 × 179 km. The smallest scale topographical maps of our country are mostly on 1:250,000 scale. To facilitate easy comparison and reference it is desirable to convert the scale of Landsat imagery to that of topographical maps i.e. from 1:1 m scale to 1:250,000 scale. Large format optical enlarger developed for 4X enlargement of 24 cm format imagery has been found to be extremely useful for studying colour and B/W Landsat images. Basically, the system comprises of an optical assembly within a metallic housing mounted on an MS angle frame and a flat formica top table on which the image is projected (vide Plate 4.2).

(xvi) *Additive Colour Viewer*

It is a unique optical colour combining equipment for generating and viewing colour composite by introducing colour combination from black and white multi-spectral transparencies of aerial or satellite images. The multi-spectral imagery and photographs take simultaneously can be used for generating true colour compositing or false colour composites (F.C.C.) by using colour filters: blue, green and red or their complementaries like yellow, cyan and magenta. By varying colour, density and hue the interpreter can easily distinguish subtle details. The false colour composite or true colour composite thus prepared can be recorded through high power still camera (vide Plate 4.3).

(xvii) *Diazo Printer*

The diazo technique is based on subtractive colour process as the colours in a composite are produced by substraction of primary colours from white light in which the composite is being viewed—thus one has to use film of complementary colours to get the colour composites in primary colours. Diazo printer has been designed on the principle of contact printing where the imagery from which the print is to be made is kept in close contact with the film on which image is to form. Light passes through the original imagery and depending on the optical density of the imagery, diazonium compound coated on the film gets decomposed. Development of diazonium compound into colours requires alkaline atmosphere. Ammonia gas is a good media for this. This is achieved by putting liquid ammonia inside the chamber which is slightly hot and the ammonia from the solution is allowed to evaporate. Exposed diazo film is put into ammonia chamber where it develops. Unexposed diazonium compound develop into bright colour where as exposed diazonium i.e. decomposed diazonium remains colourness. The printer has got the following units:

(a) Printing unit: consists of two chambers.
(b) Light chamber: accommodates twelve equispaced U.V. tubes along with the chokes & starters helping in uniform irradiation of exposure plane.
(c) Vacuum exposure chamber: to achieve a perfect contact, so that no distortion is introduced, a vacuum printing frame, is used. It is evacuated by a vacuum pump.
(d) Developing unit: it consists of an ammonia chamber which is built within the table. Exposed transparencies are kept in the ammonia chamber for developing. An observation window facilitates the monitoring of prints. After proper exposure of the film (yellow, megenta and cyan) they are developed separately in the ammonia chamber. Once the films have developed they are put one over the other by matching the features of the scene and stuck with adhesive tape at one margin of the films. Thus the colour composite is ready. This is called also poor man's F.C.C. for its low generation costs.

(xviii) *Image Analyser*

It is an electronic colour combiner specially designed to provide colour composite an well as density sliced image display from B/W hard copy or transparency for interpretation and analysis. This is very helpful as colour tones of varied range and intensity can be assigned to each gray level of B/W imagery for evaluating better thematic information. Eight colours i.e. Blue, green, red, yellow, cyan, magenta, orange, violet can be assigned to the different density zones. For better analysis different colours as well as change of colours can be done by the operation and simultaneously it is possible to calculate the area of each colour density zone for classification of different features (vide Plate 4.4).

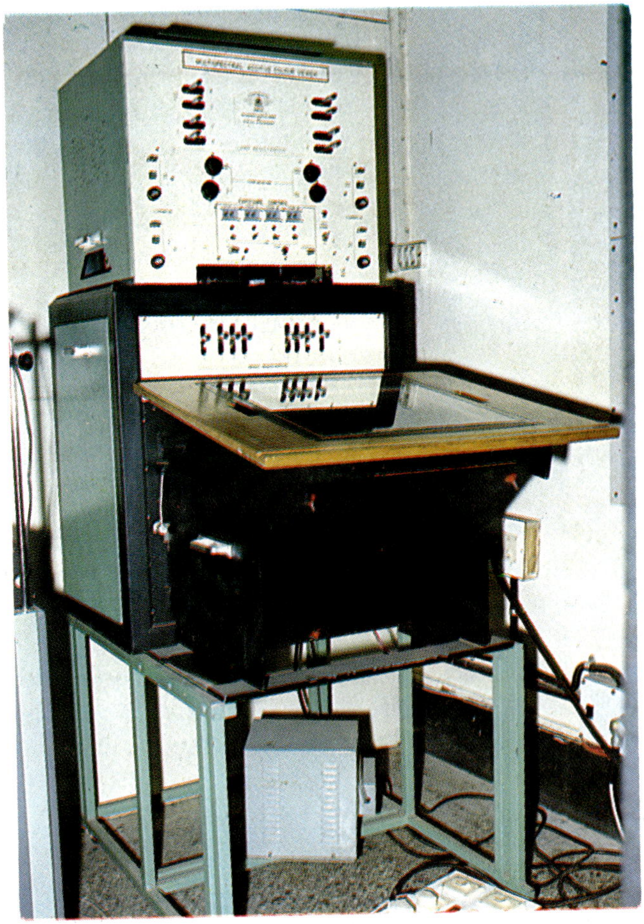

Plate 4.3: Additive Colour Viewer

(xix) *Duel Densitometer*

Both transmission and reflection densities of photographic films and prints in both B/W and colour can be measured through digital read out design provided in a basically photometer—is duel densitometer.

(xx) *Agro Photometer*

It is simple portable and battery operated instrument. Water content and chlorophyll concentration in the leaves can be measured directly with the

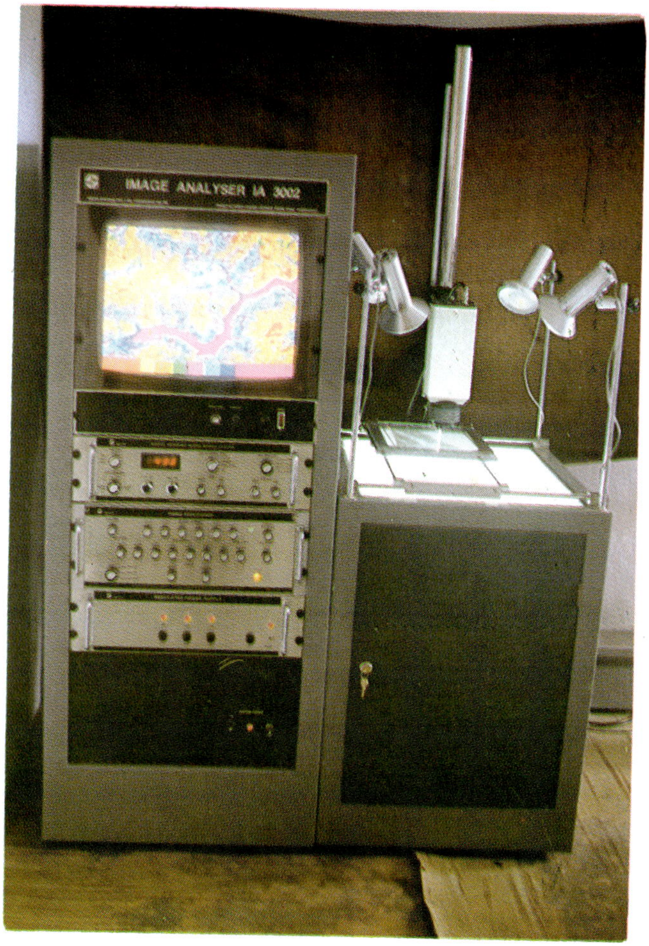

Plate 4.4: Image Analyzer

photometer under field conditions without destroying the specimen. It consists of a simple optical system to collect the energy and a silicon detector system for measuring the parameters--it is operated in any time of the day.

Conclusion

We have noticed that there is a wide range of remote sensing instruments now available for visual interpretation (Dubey *et al*., 1991). One has to understand his requirements before going for these options. Several such instruments have been developed by the National Remote Sensing Agency (NRSA), Hyderabad, and now manufactured by different companies with the approval or licence.

Most of such instruments are basically optical in nature. But at the recently, digital or optico-digital systems have also been developed. We would be discussing the capabilities of these system in details in the forthcoming chapters. Even with digital analysis, the results are finally interpretated visually. Hence, the remote sensing data is either visually interpretated directly or after digital processing.

REFERENCES

Dubey, R.P., N.D. Christan & K.D. Mallick (1991), Development of a high magnification enlarger, *Photonirvachak*, vol. 19, No. 4, pp. 276-86.

Krishnamurthy, J. (1987), *Role of Indian Space Research Organisation and National Remote Sensing Agency in the Development of Low Cost Remote Sensing Equipments in India*, ISO-NNRMS-SP-32-87.

Krishnamurthy, J., A.S. Padmavathy and M.G. Chandrasekhar (1992), Role of remote sensing instruments in extracting thematic information, in P. Nag (ed.) *Thematic Cartography and Remote Sensing*, Concept Publishing Company, New Delhi, pp. 107-26.

Nag, P., M. Kudrat, D.K. Pal. & B.K. Sinha (1990), Identification and mapping of wastelands from enlarged Landsat Thematic Mapper Imagery in Siwan district of Bihar, India, in S.C. Sharma, R.B. Chaturvedi and O.P. Mishra (eds.), *Utilisation of Wastelands for Sustainable Development in India*, Concept Publishing Company, New Delhi, pp. 123-8.

NNRMS (1991), *Remote Sensing in India: Facilities & Infrastructures*, National Natural Resources Management System, Department of Space, Bangalore, Technical Report DOS-NNRMS-TR- 87-91.

FIVE

HARDWARE AND SOFTWARE OPTIONS

FROM THE earlier chapters it is apparent that remote sensing can no longer be considered in isolation. It is becoming an integral part of the digital mapping system on one hand, and resource monitoring on the other hand, Geographical Information System (GIS), Land Information System (LIS), surveying, cartography, Geographical Positioning System (GPS) and the like are closely associated with it. Hence, for considering the hardware and software options, it is very important that we look into the possible linkages in associated application areas.

In order to provide a clear perspective on such issues, it is essential that some basic facts of computers should be provided so that the new entrants in this field can be familiar with the terms in use. Such a recourse is essential for the first generation users in India.

Computers have been introduced as a fast calculating device, replacing calculators. However, there are several other applications of this piece of hardware as well, such as recording, copying, comparing and analysis of even non-mathematical operations based on alpha-numeric details and symbols. It is also considered as a fast and accurate device for manipulation of data, which we will find, has a wider application in remote sensing. Such applications or results are produced with the help of programmes loaded on the computers, known as 'software'. It plays a vital role in integrating various functions and analysis.

All electrical, electronic and mechanical equipments of a computer which can be seen and touched by human being are known as 'hardware'. They are two types: (a) central processor, and (b) peripherals. On the other hand, software is a set of logical instructions using the hardware most efficiently for specific applications.

Computers can handle a large volume of data and can perform repetitive calculations with same accuracy and without getting tired. The data, fed in, are known as 'input'. After undergoing a process based on a software programme, become as 'output'. As mentioned earlier, a process can consist of calculation, comparison, decision-making or analysis or combination of above.

The data can reside in an internal storage for operation, or on external storage for future applications. Input relates to bringing data from external storage into the internal storage of the computer. The output or processed data can be transferred to devices like printers, or storage again which in turn may become input for other applications.

Generation of Computers

In the ancient Indian Vedic literature there are several Sutras which provide clues for large calculations. The Vedas are well known as four in number Ṛg, Yaju, Sama and Atharva but they have also the four Upavedas and the six Vedandas all of which form an indivisible corpus of divine knowledge as it once was and as it may be revealed. The four Upavedas are as follows:

Veda	Upaveda
Ṛgveda	Āyurveda
Sāmaveda	Gāndharvaveda
Yajurveda	Dhanurveda
Atharvaveda	Sthāpathyaveda

In this list the Upaveda of Sthāpathya or engineering comprises all kinds of architectural and structural human endeavour and all visual arts. The mathematics derived from such literature is known as 'Vedic Mathematics'. The sutras and sub-sutras in such books provide clues to solve mathematical problems. For example, 18, 28 or 42 steps required for solving fractions such as 1/19, 1/29 or 1/49 can be reduced to one single line even by young students (Jagadguru, 1965).

The earliest known instrument used for counting numbers is the Abacus which was used in China about 3,000 years ago. It has rods carrying round beads mounted on a wooden frame. These beads could slide along the rods for counting. It is still in use for teaching in nursery classes. The slide rule is a recent development based on the idea of logarithms. Till the advent of pocket calculators they were popularly used by the engineers.

The first mechanical method of calculation was introduced in 1642 by the great French mathematician Blaise Pascal. He was the son of a tax collector and developed this instrument to help his father. The limitation of this machine was that it could only be used for addition. Following this, a German mathematician called Gottfried Leibnitz produced a calculating machine in 1672. Since then there has been a drastic change in number crunching operations. This calculator could also subtract, multiply and divide. Leibnitz also invented Calculus and refined binary mathematics which became so important in the development of the digital system. The next important development was the invention of a Difference Engine, also known as Analytical Engine, by an Englishman, Charles Babbage who was professor of mathematics at Cambridge University. It could calculate the successive values by algebraic expression and arrange information in the form of tables.

Further, in 1853, he initiated still a more powerful machine known as Analytical Engine. It was a general purpose machine capable of doing mathematical operations faster with accuracy. Unfortunately Babbage was far ahead of his time and the work could not be completed, but there are remarkable similarities of its design with the modern computers.

Consider the fact that 1880 Census of the US was not yet tabulated, and the 1890 Census was approaching, the US Bureau of Census approached scientists and engineers to develop a device for faster tabulation and calculation of the population-related data. Dr Herman Hollerith took up this challenge and took three years to develop a machine which is known as Punch Card or Census machine. The holes were punched on a card for storage of information. Hollerith later formed the Tabulating Machine Company which ultimately merged with other companies in order to form the well known computer manufacturing company, the International Business Machine or IBM. The punch card machines used electricity for its operation. Till recently these machines were popular. Nevertheless, in 1930s the electric calculators were invented. Today calculators are similar to computers in terms of speed and accuracy. However, it has got its own limitations as it has got no logical abilities of its own. Almost at the same time (1934), Professor Howard Eiken from Harvard developed a automatic calculating machine which was known as Mark 1 digital computer. Its internal operations were automatically controlled and mathematical operations were mechanical. This electro-mechanical computer was in a fulfilment of Babbage's dream.

The first generation of computers were using vacuum tubes for the electronic components and electrostatic tube or mercury delay lines for storage. It was built in 1946 by a team from Moore School of the University of Pennsylvania and was known as Electronic Numeric Integrator and Calculator or ENIAC. It could do 300 multiplications per second. The second generation computers used the transistors. In the next generation, the large number of transistors were put together on a single silicon chip to save space and power. These chips were known as integrated circuits (IC) chips.

The current computers are known as fourth generation computers using large scale integration (LSI). The fifth generation will be radical departure from the earlier ones. It will be the era of artificial intelligence. There is also projection about the sixth generation computers using biological storage devices with natural language integration (Thiagrajan, 1992). The comparison of different generation of computers are shown in Table 5.1.

The memory or capacity of the computers, floppies or storage devices are defined in terms of bits, bytes, kilobytes (KB) or megabytes (MB). These are based on binary number system i.e. all the numbers are defined in terms of the 2 to the power of this or 2 to the power of that. The computer can only have following two situations : (a) ON/Yes/1, or (b) off/no/0. The only two numbers used are '0' and '1'. One byte contains 8 bits which is basically a series of ONs and OFFs. Further,

TABLE 5.1: Generation of Computers

Sl. No.	Facets	1st Generation	2nd Generation	3rd Generation	4th Generation	5th Generation	6th Generation
1	Year of introduction	1946	Early 50s	Early 60s	Early 70s	1982	2000 ?
2	Storage circuit components	Thermiosic valves	Transistors	Integrated circuit (ICs)	LSIs	VLSIs	Biological
3	Number of users	1	1	Multiple user community	Remote user	Same as IV generation	Not known
4	Language used	Machine language (low level)	Symbolic language FORTRAN (high level language)	COBOL, ALGOL, etc.	PASCAL, Query word processing	Kernal language, Query language, etc.	Natural language of the user ?
5	Nature of inputs	Torn tapes	Punched cards, MT, disk etc.	Cards, MT, disks	VDUs, floppies, optical disks, CDs	Optical disks, optical records, waves? Speech input	Speech, signs, thoughts
6	Nature of processing	Serial	Serial	Serial	Serial/ parallel array	Parallel/ array	Associative?
7	Cooling arrangements	Extensive	Moderate	Less	Room AC adequate	Room ACs	In-built liquid coolers ?
8	Space requirements	Large	Medium	Small	Normal office room. Table top Pc	Office room. Table top. Palm top	Hand-held?

Source: Modified from Thiagrajan, (1992).

1 KB = 1024 bytes, and 1 MB = 1024 KB (i.e. 1024 × 1024). Similarly, a capacity of 20 MB would be equal to 20 × 1024 × 1024 = 20,971,520. The maximum number of character generated by an 8 bit machine (generally used for remote sensing) is as follows:

$$= 2^7 + 2^6 + 2^5 + 2^4 + 2^3 + 2^2 + 2^1 + 2^0$$

$$= 128 + 64 + 32 + 16 + 8 + 4 + 2 + 1 = 255$$

It may be noted that in the digital data processing, every thing starts with '0'—not with '1'. The figure '110' will be stored as follows:

0 1 1 0 1 1 1 0

or $2^0 + 2^1 + 2^1 + 2^0 + 2^1 + 2^1 + 2^1 + 2^0$

or $64 + 32 + 8 + 4 + 2 = 110$

Computers are of different types depending on speed and capacity. The top of the line are super computers which are able to carry out 100 million instructions per second (MIPS). They are used for extremely complex analysis, such as weather prediction and war games. The largest computer is Cary which consists of 200,000 circuits with components per circuit, i.e. six million components. It will soon be possible to have higher capacity of components on a single chip. Next in the line are the mainframe computers because the parts were mounted on in large metal cabinet frames. But recent mainframes are more powerful and less cumbersome than they used to be about 25 years earlier. They are generally 32 bits machines with a CPU speed of 10 million instructions per second, and can be used for manipulating remote sensing data. This is followed by supermini range of computers which are used in a sizeable remote sensing laboratory. They are generally 32-bit word length as well. The speed of the CPU is more than one million instructions per second. They have excellent time sharing operating system. Further, down the line are mini, micro and personal computers (PC) with a speed of 0.5 to 1.5 MIPS and 16 bits word size. The first type of computers are smaller in size than mainframe but can do almost everything that mainframes can do. However, they cannot handle large volume of data and are not faster. Micro-computers are further smaller than minicomputer. They are now popularly known as Personal Computers (PCs). The PCs are available in different capacities. They are now available in a wide range based on the intel chip 286, 386 or 486. PC 386 or 486 are being used as host computer for digital remote sensing. They can do almost all the functions which can be done by a supermini computer however in a limited way. Further, there are solutions which can hook PC or work stations (WS) to mainframe computers for remote sensing data processing. The PC work standalone and have limited networking capabilities, but WSs can also work standalone and in network. WSs are like high-performance PC. While the hardware capabilities are similar, the difference is in the operating system. PC run on MS-DOS or OS/2, while WS work on UNIX or VMS as well.

Recent developments in work station technology have made it suitable for decision support and even remote sensing and geographic data management functions. Geoprocessing of information is computer intensive which requires fast disk input/output and high colour graphic displays. It also requires fast central processing unit or CPU functions for complex spatial analyses. However, it provides open multi-processing environment at a relatively low cost in comparison to mainframes or even superminis. A graphic work station which is required for the digital remote sensing should have a high performance CPU consisting of a processor, graphic screen, key board, mouse and local disk drive for data storage. Each work station can be used by a single user and generally has computing power more than minicomputer. However, this scenario is changing very fast. High-performance 32-bit work stations can be connected to a local area network (LAN). The LAN permits individual nodes to share data, and create single or integrated system. Some work stations have propriety operating system, such as VMS for Digital's VAX station. But, currently, they are mostly supporting UNIX operating system in order to make the whole system more open. Further, the architecture allows easy integration with other terminals. On the other hand, users at work stations can have access to extremely powerful CPUs and fast access to spatial databases and that too at a lower cost. Hence, it is best suited for multi-user environment. The popular work stations now available in the market are as follows:

— Apollo (UNIX)
— Data General MV 7000 series work station (AOS/VS)
— Data General A ViiON (88K) (UNIX)
— DEC VAX GPX and VAXstation series (VMS)
— DECstation (ULTRIX)
— Hewlett Packard 9000/300 series
— IBM RT and RISC System/6000 (AIX)
— Intergraph Interpro and CLIPPER (UNIX)
— Sun-3 and Sun-4 series (UNIX)
— Sun-SPARCstation (UNIX)
— Sub-386i (UNIX)
— Tektronix 4300 series (UNIX)
— Tektronix XD88 (UNIX)

Considering the power, functionality and flexibility of a work station, the GIS and remote sensing software programmes have been developed which can distribute the computing load over the network of high performance work stations. Such a system can take care of the data organisation in this computing environment, such as relational database interface (RDBI), and can expand as and when desired. The open architecture solution encourages better data management and freedom to integrate different data management technologies such as Oracle, Ingres, Informix, Sybase, Rdb or AS/400. In fact the development of the work station technology is somewhat in tune with the progress in the softwares related to GIS, image processing, digitization, digital mapping and the like.

Selection of Hardware

The options available in the market are enormous and fast changing, therefore confusing. Everyday new products are coming which are competitive, technically and financially. Their capacity and compatibility are important aspect to look into while selecting a hardware. However, the most important aspect is to have a clear objective of the proposed system.

The minimum configuration for a digital image processing system, which is suitable for a university department of any branch of earth science, like geography and geology, would be a personal computer, and a input/output (I/O) device. The selection of PC is based on the following factors:

— Central Processing Unit (CPU): Intel chip 286/386/486
— Clock speed in Mhz: 33, 50, 75
— Random Access Memory (RAM): Expandable or not
— Cache memory: Expandable or not
— Floppy Disk Drive (FDD): 5.25 (1.2 MB) or 3.50 (1.4 MB)
— CD Rom. Drive
— Hard Disk Drive (HDD): Expandable or not
— Monitor: Black/white, MGA, CGA, EGA, VGA, super VGA
— Monitor screen size: 14", 17", 19"
— Port(s): serial, parallel, mouse
— Key board: 84, 101, 102 keys
— Quality of screen: Flat, anti-glare, Liquid Crystal Display (LCD), Thin Film Transistors (TFT)
— Maximum resolution on screen
— Power supply: 150, 200, 250 watts
— Maximum colours or grey levels: 256 or 16
— Co-processor

The PCs are compared with that of the IBM (International Business Machines). It has been used for various purposes, varying from playing games at homes to a terminal hooked up with a mainframe. The earlier version of PC-extended technology or PC/XT was having 8088 intel chip, 640K of memory and 12 Mhz speed. Following this, PC-advanced technology or PC/AT came into the market with 80286 (or simply 286) intel chip, IBM memory which can be expanded to 8 MB and speed 20 Mhz. The PC/AT 386 (80386 intel chip) was further improvement. It has 4 MB memory which can be expanded to 32 MB, speed of 33 Mhz and cache memory of 64 KB. This PC can also have 80387 co-processor. The latest available PC in the market is PC/AT with intel 80486 chip. The memory or RAM varies from 8 to 64 MB. The clock speed remains the same, i.e. 33 Mhz. However, some PC 486 are also available with 50 Mhz speed. The cache memory of 64 KB can be expanded to 256 KB. These PCs can be used for image processing as well. The standalone system can be hooked to other such PC or mainframes through servers. Image processing WSs are also available which can do such functions in a better way.

Intel chips were available for 90 Mhz speed a year ago. The new Pentium version has 120 and 133 Mhz chips. The price of these chips are going down. Nevertheless, demand for the more powerful PC is being driven by home computer users' desire to run the latest educational and game software with sophisticated photo-like graphics. Intel Corporation is now coming out with even faster 150 and

166 Mhz chips. Hardware companies have plans to bring out PC based on such powerful intel chips.

Against the PC environment, there are solutions based on supermini, mainframe and super computers. These are expensive hardware and can only be purchased if the objective is justified. They require high level of maintenance as well as under strict working environmental conditions. VAX 11/780 based image processing system in RRSSCs is shown in Plate 5.1.

Plate 5.1: Image Processing System

Scanner

Scanner is not required for remote sensing as the data is generally available in raster format. However, for soft photography and associated work it may be useful. Scanners are becoming more and more powerful and hence playing important role in digital mapping (Deekshatulu *et al.*, 1977). The features for this sub-system are as follows:

- Resolution in dots per inch (DPI): 300, 400, 600, 800, 1000
- Media size: length and width in inches or cms
- Scan width in inches: 36 inches
- Scan length in inches: limited or unlimited
- Scan speed: time in seconds to scan specific size of paper at given DPI (e.g. seconds require to scan AO size at 400 DPI).

— Optical resolution: in DPI (e.g. 270)
— Scan accuracy in per cent: 0.10, 0.25, 0.50
— Environmental specification: temperature, humidity and voltage

Plotter

Plotter is an essential part of a digital mapping or image processing system. The results of the experiments are to be obtained in the form of hard copies for applications etc. There are several varieties of plotters which are now available in the market. They are inkjet, electrostatic or thermal. The basic specifications can be as follows:

— Resolution in dots per inch: 300, 400, 600, 800, 1000
— Plotting speed inch/cm per second at given DPI: 1, 2, 4 inches
— Accuracy in per cent: X and Y axis
— Media width in inches: 36, 48 inches
— Interface
— Manual cutter facilities
— Environment: Temperature, humidity and voltage

Storage Device

Remote sensing involves a lot of storage as the data is in raster format. The hard disk has limited capacity. Hence the data has to be stored in an external system. A variety of storage devices are now available varying from conventional floppies of 1.2 MB or 1.4 MB to optical juke box. Following specifications are of interest:

— Nature of device: external or internal
— Media size
— Capacity: in MB or GB (Gega Bytes)
— I/O (input/output) media: magenetic tapes, cassette tapes, drums, video-recording cartridges, floppies, CCTs, cartridges, optical disk, compact disk etc.
— Access time (average): 60, 90, 120 seconds
— Interface
— Environment: Temperature, humidity, voltage

Photowrite System

A process to obtain hard copy from the digital data is extremely vital as it can be carried from one place to another, and visual interpretation can also be performed. There are several techniques now available in the market which are based on different technologies : purely photographic (ordinary camera or dunn camera), thermal or photowrite system. Indian companies are now producing the latter system with arrangements with the Department of Space. This is a precision digital film recording system capable of producing continuous tone black and white and colour transpires from the digital data. Such system should be able to accept common remote sensing data including the IRS. It can be standalone or a part of the digital image processing system. Thee final output is in the form of hard photographic print such as false colour composites (FCCs).

In principle, the film to be exposed is mounted on a motor driven cylindrical drum. The exposing optics, compromising optical modular lens system, is mounted on an electronically controlled carriage, driven by a leadscrew. By this the digital image on the magnetic tape is transformed to analog form. After suitable signal conditioning and simplification, these signals intensity modulates a narrow beam of light, which in turn, exposes the film on the drum. Two dimensional scanning required to expose the entire film is obtained by combination of drum rotation and transverse motion of the carriage.

The top of the line is Fire 2240 which is 10 bit Color Digital Film Recorder. It can be attached either to a PC platform or a workstation. It provides resolution greater than 1,270 lines per inch (50 lines/mm) --pushing the maximum resolution obtainable with Kodak Ektachrome 100 professional film. The other features are as follows:

— Maximum Image size: 9.45 inches (240mm) along-scan, 8.66 inches (220mm) cross-scan
— Maximum Image area: 81.8 square inches (528 square cm)
— Maximum pixel format: 16,000 along scan X 22,000 cross scan; 32 megapixels at 100 lines/mm
— Intensity Resolution: 8 bits per colour input from host into look up table (LUT); 10-bit output from LUT to DAC for beam modulator
— Image view: normal or mirror

There are some Fire film recording installation in India as well.

SOFTWARE OPTIONS

Digital handling of remote sensing data requires appropriate software at every step, i.e. from the reception of the data to the desired output. But in the present context we are concerned with the different aspects of the remote sensing, image processing and patterns recognition softwares. It includes data handling, pre-processing, enhancements, classifications, post processing and mapping techniques as well. There are following solutions available in a standard software package :

A. Image rectification
B. Image enhancement:
 — Contrast manipulation: contrast stretching, density slicing
 — Spatial feature manipulation: edge enhancement, spatial filtering
 — Multi-image manipulation: spectral band ratioing, canonical analysis, spectral vegetation indices, intensity-hue-saturation (IHS), colour space transformation
 — Other enhancements

C. Image classification:
 — Supervised classification: minimum distance to means, parallelopied, maximum likelihood
 — Unsupervised classification

D. Digital change detection:
 — Image differencing
 — Image regression
 — Image ratioing
 — Principal component differencing
 — Post classification comparison
 — Direct multidate classification

The above classifications are available with a lot of options and default values. One has to understand the significance of such values in order to get the best results. It is likely that a series of operations may be required for obtaining the desired results. In the following chapters, each of the above techniques have been discussed in details. However, these techniques are hardware dependent as well. For example, a digitizer would be required for geometrical rectification of an image.

Integrated Solutions

Generally two types of solutions are being proposed by the vendors. The fist one is vendor-specific which is a hardware-cum-software integrated system. In many cases it is a closed solution implying that no software can be loaded or a piece of hardware be added to the system and translation of data file may be a problem. However, in a closed environment, they work excellent. Against this, the open system allows for integration of a large variety of hardwares and softwares. While prescribing such a solution, it is necessary to evaluate the compatibility of all the pieces of hardware and the integrating software. Vendors do provide a multiple company solutions for digital image processing system or GIS.

The minimum configuration for a image processing and GIS laboratory should be as follows:

— 80386 based PC/AT with coprocessor
— 640 KB memory
— 40 MB hard disk
— 1 × 1.2 MB floppy drive
— High resolution colour monitor with VGA card to display 256 colours in 480 × 640 resolution
— MS-DOS 3.0 or higher version operating system software
— A digitizer with 4 buttons cursor
— A dot matrix printer or inkjet printer
— A multipen plotter
— Application software for (a) image processing, (b) GIS, and (c) map editing.

The above prescription may be good for a small remote sensing set-up. But the larger institutions would like to work in a multi-user environment. One of the well known solutions is of the five Regional Remote Sensing Service Centres or RRSSCs which were established under the national programme of the National Natural Resources Management System (NNRMS) of the Indian Space Research Organisation. They were originally planned to have the following solution :

— Host computer: VAX 11/780, 4 MB, built-in floating point accelerator
— 3 Winchester disk drives, 456 MB with array processors of 12 M Flops
— 5 Video terminals
— Line printer
— Benson Model Digitizer
— 2 Tape drive, 1600/6250 BPI
— 3 PC-2000 terminals
— 3 NRSA Intelligent display terminals (IIGT) PDP 11/23 based
— Dunn camera with 35mm and 70mm options

In addition, Photowrite system on off-line mode was kept optional. The interactive system is based on VIPS-32 Concept which has PC-2001 functions, and local functions as well for image processing. Apart from the VIPS-32 software, the integrated solution consists of VAX/VMS operating system, Fortran 77, Pascal, Cobol, Scientific Subroutine packages, and Data Base Management System (DBMS). Further the FPS software consists of the communication software between host and array processor and also the programme development package.

Recently, work stations have been added as they have better capabilities and can do almost all the functions as that of the VAX 11/780 super minicomputer. On the software side, EASI/PACE raster based image processing software is also preferred which is having tools for remote sensing, geometric correction, multispectral analysis, photogrammetry, radar processing, raster modelling, terrain analysis, atmospheric correction and map projection. This software is open to all hardware, data types and open to exchange data with GIS system. It supports open architecture which provides enormous flexibility. Its Generic Data Base (GDB) format gives maximum access to the other important file types such as ERDAS, ARC/INFO, TIFF, GRASS cell and binary vector, Intergraph COT, SIF, and DGN, DLG, DXF, HDF (NASA EOSDIS formet), TYDAC, Siemens, Genasys, GDS and several others. Though VIPS-32 claims some GIS functions as well, RRSSCs have gone for ARC/INFO vector based GIS software. Further, some of the RRSSCs have diversified from core image processing activities to digital photo-interpretation and photogrammetry.

The two major institutions in the field of remote sensing are (a) National Remote Sensing Agency, Hyderabad; and (b) Space Applications Centre, Ahmedabad. In the former institution, apart from the aircraft based remote sensing and the facilities for the reception of data from different sensors, it has perhaps the largest infrastructure for the processing of satellite based remote sensing data. There are several data processing system based on VAX 11/780, VAX 11/750, Micro VAX-II, PDP-11/55 with array processors, Gould displays system Frame Sync, tape drive and several other peripherals. Combinations of the above peripherals around a host computer is dedicated to specific remote sensing data. Since NRSA is also engaged in data production including the false colour composites (FCCs), it has two Micro VAX-II based Colour FIRE 240, and one PDP 11/34 based colour filming system Colour FIRE 240. For black and white filming, there are three filming systems of which two are PDP 11.23 based and one is VAX 11/730 based B & W FIRE 240s. It has a well equipped photoprocessing facility as well. The data analysis facilities at NRSA consists of a Micro VAX II based (MIDAS), PDP 11/23 system (MIDAS) and APPOLO Domain workstation for developing expert system. In addition, there are several PC/AT, microprocessor and work station based data analysis systems for image processing and GIS applications.

Space Applications Centre (SAC) has VAX 11/780 system with digitizer and plotter; Micro VAX system with DIPEX-ARIES-III; and indigenous ISROVISION for image processing. It also has colour photowrite system for writing digital imagery in photographic form. The application softwares are ARC/INFO on Micro VAX and PC/AT and ISROVISION-GIS.

The other national institutions, such as Survey of India, Geological Survey of India, Oil and Natural Gas Commission (PDP 11/73) have some digital image processing facilities. National Atlas and Thematic Mapping Organisation (NATMO) has recently acquired a digital system which is meant for digital mapping, GIS and image processing. The configuration is as follows :

— 1 Server Sun Spark SS-20, 64 MB RAM, 3.0 GB HDD, CD ROM, CTD, DAT
— 4 nodes of Sun Spark SS-10, 32 MB RAM, 1.0 GB HDD, CD-ROM

— 1 TA-100 plotter
— 1 Context 8200 Map scanner
— 1 AO Inkjet plotter
— 1 Dot matrix printer
— 1 Tape drive 1600/6250
— Two 3-user node locked license for ARC BASR plus ARC SCAN
— One float license for TIN, GRID, NETWROK analysis
— 3 User license for ERDAS image processing software.

Among the state centres, the Uttar Pradesh Remote Sensing Application Centre, Lucknow and the Institute of Remote Sensing, Madras have image processing facilities around VAX 11/780 and PC based systems. Other such centres do have visual interpretation facilities. However, some have acquired digital facilities as well or in the process of doing so, such as West Bengal and Assam.

ASSOCIATED TECHNOLOGIES

As mentioned in the beginning of this chapter, image processing of the remote sensing data should not be considered in isolation. It has widened its scope and can be linked with applications of various nature, ranging from defence to environmental monitoring. Some of the emerging thrust areas have been mentioned here.

The digital solution should provide connectivity to the proposed image processing and related system for transfer of data. It is important that the proposed installation of a network and the associated hardware and software upgraded and updated to the system in the near future should be decided upon with a lead time. This is significant because of the varieties of options in digital remote sensing are likely to increase in the near future.

Geographical Information System

Geographical information system or GIS has become a multi-discipline. It has become the most acceptable method for the collection, collation and processing of spatial or spatially referred data from a variety of sources. Obviously, remote sensing data becomes an essential part of such an activity. Since, the data is easily available at a repeated interval, and the same data can be used for different applications, it has opened a wide scope for GIS application in resources management, environmental assessment, and for developmental initiatives. Other data can be superimposed on the same geographic base and the changes can be related to the remote sensing data. Almost all the GIS packages now have the option to include the remote sensing data, and vice versa. Both have come closer due to the availability of powerful software and almost similar hardware requirements. The main five components of GIS are as follows:

— data encoding and input processing
— data management
— data retrieval
— data analysis, modelling and cartographic manipulation
— data output

Further, information technology is a rapidly changing field and its innovative ideas and accomplishments will effect the design and use of future GIS. It has been estimated that the development rate of new computer hardware tends to be underestimated, however, expectation about improved software systems are usually higher than what the industry can deliver (Frank *et al.*, 1991).

Considering the growing importance of GIS, a separate chapter (11) has been included here. All the functions available in a GIS package may not be oriented for the raster data. However, the traditional GIS packages which were primarily vector based now have facilities for processing raster data. In fact the software vendors specifically mention about the live link between the vector based GIS package and raster based image processing package (Appendix II).

Land Information System

Based on GIS principles, Land Information System (LIS) can be developed which is basically for the following purposes:

— Land settlement
— Land survey
— Land registration
— Land valuation and assessment
— Land control and management
— Infrastructure utilities

Since it involves cadastral mapping at a large scale, the satellite based remote sensing has limited role. However, in near future, with better resolution, such a source would be very valuable for mapping land related information (Dale & McLaughlin, 1988).

Geographical Positioning System

In order to find out the geographical co-ordinates of a place on the earth surface, NAVASTAR Global Positioning System (GPS) was developed. Initially 18 satellites were planned but now there are 24 orbiting satellites to transmit the identification and positioning information. The waves are programmed to include coded messages and a precision code (P) and a course/acquisition code (C/A) from which the ranges to the satellites can be determined. At least four such satellites are to be observed in order to calculate three-dimensional co-ordinates of the point where the receivers have been kept. The precision code (P) will provide higher accuracy to the order of one metre or better. Nevertheless, application of the C/A code can provide accuracy within 20m. This can be effectively used for relative position. Hence it is able to densify the control points for even large scale cadastral mapping. After the initial cost of GPS, the operation cost for fixing control points becomes extremely cheap.

GPS are basically two types. The receiver is to be placed on the desired point and after receiving data from at least four satellites, the data can be down loaded or processed by a PC. There are programmes available which can process the GPS reading directly for GIS activity or mapping. For example, the Geolink software connects the GIS directly to GPS receivers to obtain digital position information in the field in real time. Such a programe can also accept attribute information in the field and position the attributes in a coverage with accurately georeferenced co-ordinates. For better precision and cross checking, two such GPSs can be kept at the same place. With P code and observation for over an hour, the precision up to one metre is expected. Further, the other type of GPS is the handsat one. They can provide co-ordinates instantly, however the inaccuracy is higher and with the present technology up to 100m is predicted. For the extension of control can be carried from known points. In such a case receiver is placed on one known point or primary station and then on the point of interest. Either the geographical (log-lat) co-ordinates or UTM grids are available for such points. These points are available on a variety of datums as well.

GPS has facilitated the processing of the remote sensing data. For the recertification functions, the ground control points (GCPs) can be directly fed in the transformation models. The forthcoming remote sensing satellites will also be having onboard GPS facilities. This will enormously help in relating the subscenes to the geographical co-ordinates on the ground.

Conclusion

The hardware and software options now available provide a wide range of choice for the working out an acceptable or required solution. This scenario is changing very fast. Every day better versions of hardware and newer versions of software are hitting the market. The option for the open system has made it possible to integrate several peripherals. In any case, remote sensing or GIS or even digital mapping cannot be considered in isolation. One has to be very clear about his or her requirements.

REFERENCES

Agarwal, N.K. (1995), Global positioning system: Realities and limitations, *GIS India*, vol. 4, No. 1, pp. 18-21.

Dale, Peter and John D. McLaughlin (1988), *Land Management System*, Clarendon Press, Oxford.

Deekshatulu, B.L. *et al.* (1977), Design and Fabrication of Colour Scanner'', *Indian Journal of Technology*, vol. 15, pp. 154-55.

Frank, Andrew U, Max J. Egenhofer and Werner Kuhn (1991), A perspective on GIS technology in the nineties, *Photogrammetric Engineering and Remote Sensing*, vol. 37, No. 11, pp. 1431-6.

Jagadguru Swami Sri Bharati Krsna Tirthaji Maharaja (1965), *Vedic Mathematics or Sixteen Simple Mathematics Formulae from the Vedas*, Hindu Vishvavidyalaya Nepal Rajya Sanskrit Series vol. 10, BHU Press, Varanasi.

Muller, Jean-Claude (1993), Latest development in GIS/LIS, *International Journal on Geographical Information System*, vol. 7, No. 4, pp. 293-03.

Rao, Mukund, V. Jayaraman and M.G. Chandrasekhar (1994), *Organising Spatial Information Systems around a GIS Core*, Indian Space Research Organisation, Special Publication, ISRO-NNRMS-SP-70-94, Bangalore.

RRSSC (u.d.), *Training Course in Digital Image Processing*, Regional Remote Sensing Service Centre, Dehra Dun (Mimeo).

SIX

PRE-PROCESSING AND RECTIFICATION

SATELLITE REMOTE sensing data has been used as basic inputs for the inventory and mapping of natural resources of the earth surface like agriculture, soils, forestry, geology etc. (Ajai *et al.*, 1985; Chavez *et al.*, 1977; Saha *et al.*, 1990; Mulder and Epema *et al.*, 1986). The final product in most of the applications (classified outputs) is likely to complement or supplement the map. Space borne remote sensing data suffer from a variety of radiometric and geometric errors caused by satellite motion, the sensor system (Stein & Van Wie, 1976). These distortions would diminish the accuracy of the information extracted and thereby reduce the utility of the data, if the data is not precisely corrected (Rifman, 1973).

In order to update and compile maps with high accuracy, it is essential to relate the satellite remote sensing data to ground coordinate system. For comparison of data from different sources and studying the temporal changes, geometric registration to each other must be guaranteed (Davidson, 1986). For correcting the image/data, the type and source of distortions must be determined. Geometric and radiometric distortions which are caused by sensor system (internal) or platform and external geometry of the earth surface. They are categorised as systematic or predictable, and non-systematic errors are constant and can be predicted in Fig. 6.1. Systematic errors are constant and can be predicted in advance. They include mirror scan velocity, detector sampling delay, scan skew, panoramic effect, earth rotation, detector response and so on, and are determined from prelaunch calibration measurements. But the non-systematic distortions are not constant because they result from variation in spacecraft velocity, spacecraft attitude, altitude and atmospheric effect. These are not constant and predictable. The brief description of all effects are discussed.

The preliminary corrections of the systematic geometric distortion and radiometric corrections are called pre-processing.

RADIOMETRIC CORRECTION

Landsat MSS has six detectors for each band. When it scans, it scans six lines at a time per band. It is also expected that all the six detectors should record similar

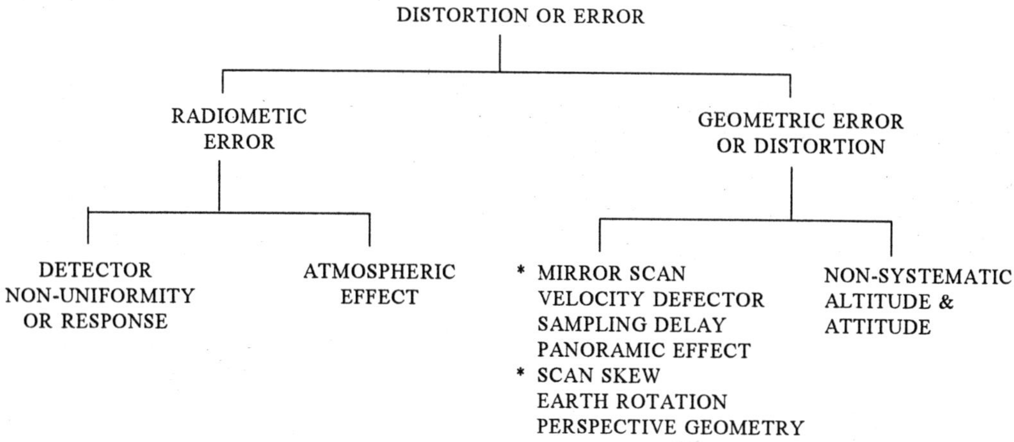

FIG. 6.1: Sources of Geometric & Radiometric Error of Remote Sensing Data
(Errors are not present in linear array defectors as in the case of SPOT, IRS and MOS satellite)

signature from same object on the earth surface. But in practice, it was noticed that the sensitivity of all the detectors are not uniform. This means sometimes detectors recorded irradiance (reflected) for same object may differ. This causes the stripping in the remote sensing imagery. Secondly, many times one of the detectors may fail to function. This causes loss of information of one complete line. The second phenomenon is called sixth line drop. Similar situation may also common in Landsat TM system. Here, it will be 16th line drop, since in TM there are 16 detectors per band.

The atmosphere is not vacuum. It comprises of particulates, gas molecules, water vapours etc. The gas molecules and particulates in the atmosphere scatter in the lower wave length while water vapour absorbs in the infrared regions. Both the methods modifies the actual remote sensing signals either by reducing or magnifying. So, the remote sensing signal may not be the true signal of a feature on the earth surface. The feature extraction (classification) algorithm is mainly dependent on the reflectance/emittance (gray values) and may lead to erroneous classification results until and unless the data is corrected (Rifman, 1973).

For modern linear array detectors like IRS LISS-I and II sensor stripping may be along the column instead of line, since the scanning mechanism is different from the Landsat MSS and TM sensor systems.

Line Drop Correction

The line dropout is usually overcome by replacing the zero value by the mean values of all the pixels of previous and the following line. Say for example, 10th line is the dropout line, then all the pixels in 10th line will be replaced by corresponding mean of 9th line and 11th line.

De-stripping

The stripping of the remote sensing data (MSS) are being corrected with the help of the detectors master calibration curve. This is discussed by Bernstein and Ferneyhough (1975), Bernstein (1983) and Jenson (1983). However, a simplistic method which is widely practiced in the digital technique is as follows (VIPs-32 Software documents):

$$Y_k\,(i, j) = \left(\frac{\sigma}{\sigma_k}\right) \times [X_k\,(i, j) - M_k] + M \quad \ldots \quad (1)$$

where $Y_k\,(i, j)$ = Output pixel gray value
$X_k\,(i, j)$ = Input pixel gray value
M = Mean of the full image
M_k = Mean of the k-th detector
σ_k = Standard deviation of the k-th detector
σ = Standard deviation of the full image

Atmospheric Scattering Correction

The atmospheric scattering or haze effect is very complicated phenomenon. Generally, scattering is more in the lower wavelength (visible) than higher wavelength (infrared band). Further scattering effect increases the signal value (Bias). According to USGS (1979), the haze correction one being practiced as following equation:

$$Y\,(i, j) = X\,(i, j) - \text{Bias} \quad \ldots \quad (2)$$

where $Y\,(i, j)$ = Output pixel value
$X\,(i, j)$ = Input pixel value

The bias is the amount of offset for each spectral bands (Bernstein, 1975). Bias can be determined by regressing the visible band vs. infrared bands. First of all, it is essential to identify some area like airport, deep homogeneous non-turbid water bodies (Jenson, 1983) and some shadow area in the scene. The brightness values of all the features from each band are then extracted. The gray values of the visible band (e.g. IRS-band 1) is plotted against corresponding values at the same pixel location in the infrared band (say IRS-band 4). The plot will result a scatter diagram (Fig. 6.2). Following the regression analysis, a line is to be fitted. It is assumed that if the data is free from atmospheric scattering, the best fit line should pass through origin. However, this is rarely happens. The line generally intercept the X-axis (Band 1). The X-intercept determines the bias, caused by the atmospheric effect. The procedure is to be repeated for all the visible bands.

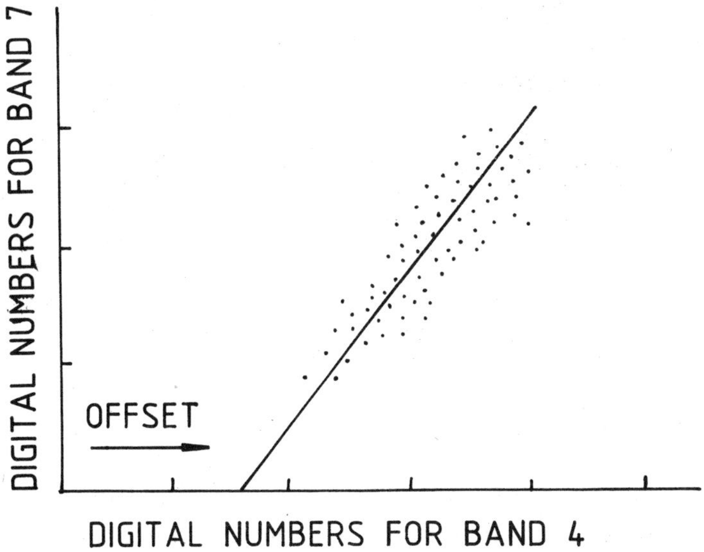

Fig. 6.2: Plot of Band 7 (IR) Versus Band 4 (visible)

GEOMETRIC DISTORTIONS

Scan Skew Distortion

The Mechanical Scanner (MSS) completes one full scan and it takes some time for one full sweep across the track. During this time, the spacecraft moves along the track. Because of this phenomena, the ground swath scanned is not normal to the ground track, but is slightly skewed. This produces cross scan geometric distortion (Bernstein and Ferneyhough, 1976).

This distortion is being corrected by the model as follows:

$$x_2 = x_1 + \sigma y_2 \qquad \ldots \qquad (3)$$

where x_1 = National scan line, if the spacecraft would have been stationary
x_2 = Actual scan line due to spacecraft movement
σ = Constant determined the amount of scan skewness
y_2 = Shift of the scan line from nominal

Spacecraft Velocity

If the spacecraft velocity deviates from the nominal, the ground track covered by a fixed number of successive mirror sweeps changes. This causes a cross scan scale distortion.

This kind of distortion may be corrected, according to the following model suggested by Bernstein (1983).

$$x_3 = \left(1 + \frac{dv}{v}\right) x_2 \quad \ldots \quad (4)$$

where x_3 = Actual ground track
x_2 = Nominal ground track
$\frac{dv}{v}$ = Normalised spacecraft velocity error
v = Space nominal velocity

Mirror Velocity

The scanning mirror of the MSS nominally moves at a constant angulaı rate (non-linear sweep). In practice, the mirror rate is not constant. Since the data samples are taken at regular interval of time, the changing mirror rate produces geometric distortion along scan line. Since the modern linear array scanner like SPOT, MOS and IRS system does not have this kind of error. The correction was not dealt here since IRS-1A satellite (LISS I, II & III) camera is free from this defect and this error is very minor.

CROSS TRACK DISTORTION OR PANORAMIC EFFECT

This defect is acquired due to finite detector field of view. The length of ground is actually proportional to the scan angle and therefore is greater at either margin of the scan line resulting in marginal compression. For applying this correction the width of the nadir pixel in the input subscene should be known. If the central pixel in an image does not lie on the satellite ground track i.e., either detector is offset with respect to the nadir or only part of the image line being used, the mean detector look angle must be specified. The formula to convert from an equi-angular representation to one in which all the pixels are of same size (Fig. 6.3).

$$R \sin(\theta + \phi) = (R + h) \sin\theta \quad \ldots \quad (5)$$

ϕ = Angle between pixel and nadir (measured from centre of earth)
h = Satellite height
R = Radius of the earth
FOV = Field of view of the detector
θ = Angle between nadir and the pixel (seen from the satellite)

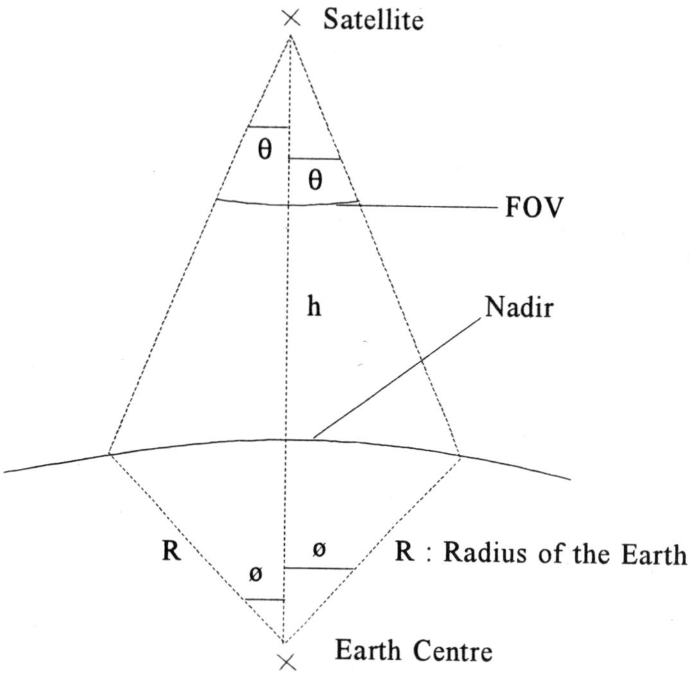

Fig. 6.3: Cross Track Distortions

$$\text{Sin}(\theta + \phi) = \left(\frac{R + h}{R}\right) \text{Sin}\,\theta \quad \ldots \quad (6)$$

$$\phi = \text{Arc Sin}\left[\left(\frac{R + h}{R}\right) \text{Sin}\,\frac{FOV}{2}\right] - \frac{FOV}{2} \quad \ldots \quad (7)$$

$$\theta = \frac{FOV}{2}$$

$$\text{Output line size} = 2R\phi \quad \ldots \quad (8)$$

$$\text{No. output pixels} = \frac{\text{output line in km}}{57/1000\ \text{km}} \quad \ldots \quad (9)$$

where 57 is pixel width for MSS.

However, the distortion is minimal in Landsat image because the scan angle is only 5.8° on either side of the vertical.

Earth Rotation Correction

The amount of earth rotation during the time required to scan one frame results in distortion in the scan direction that is a function of spacecraft latitude and orbit (Fig. 6.4).

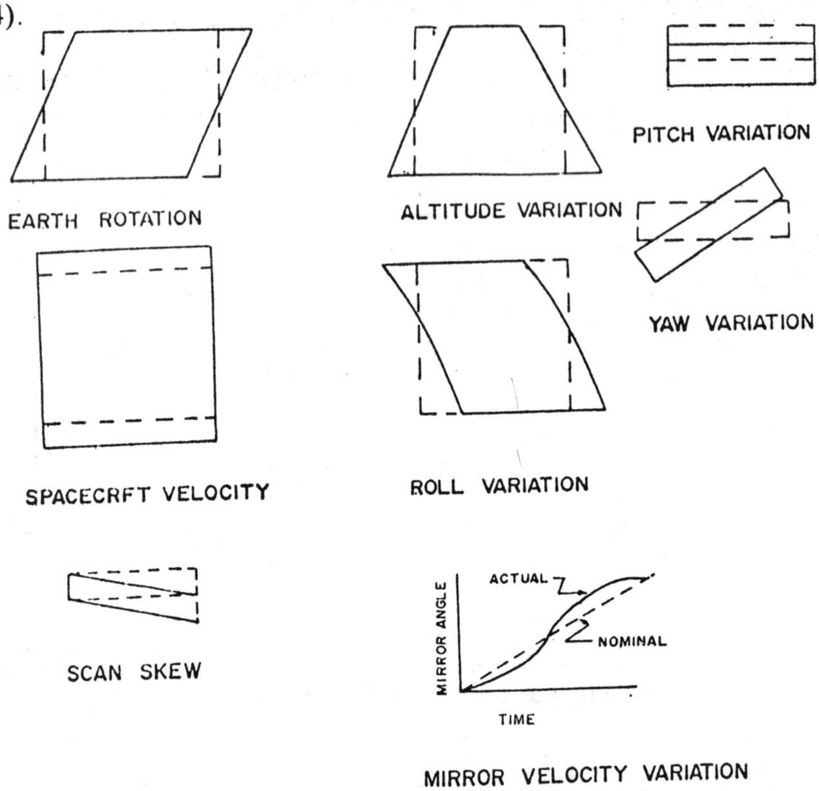

Fig. 6.4: Geometric Distortions

The angle through which the image has been shifted is termed as skew angle 'θ' and obtained from formula:

$$\text{Sin}(\theta) = C(\text{Sin}^2 i - \text{Sin}^2\phi)^{1/2} \quad \ldots \quad (10)$$

where C = inverse of number of orbits per day
i = orbit inclination
φ = mean latitude

The value of 'θ' which ranges from 0 to 4° (at equator) is used to calculate the number of zerofills to be inserted at the beginning of the first line i.e., N tan θ, when N is total number of lines.

NON-SYSTEMATIC ERROR AND CORRECTION

Altitude

Any deviation of the space craft from the nominal, altitude causes scale distortion in remote sensing data, because with the decrease in altitude pixel size will also decrease from the nominal size and vice-versa. For Landsat system (MSS) the distortion is primarily along scan line (Billingsley, 1983) and for IRS satellite, distortion will be both direction. Since the scanning mechanism is different from Landsat system.

Attitude

The satellite positioning is stable in the space and its axis system (roll, pitch and yaw) is normal to each other (Fig. 6.5). Rotation along the Y-axis (longitudinal axis) of the sensing platform is called 'roll'. Rotation along the X-axis (along the track direction) is called 'pitch', while the rotation on the axis orthogonal to the previous two axes i.e. along a line passing vertically through the sensing platform to centre of the earth is called 'Yaw'. Since the spacecraft depart from this normal position, geometric distortion inherits in the remote sensing data (Fig. 6.5). This type of distortion is uncertain and unpredictable.

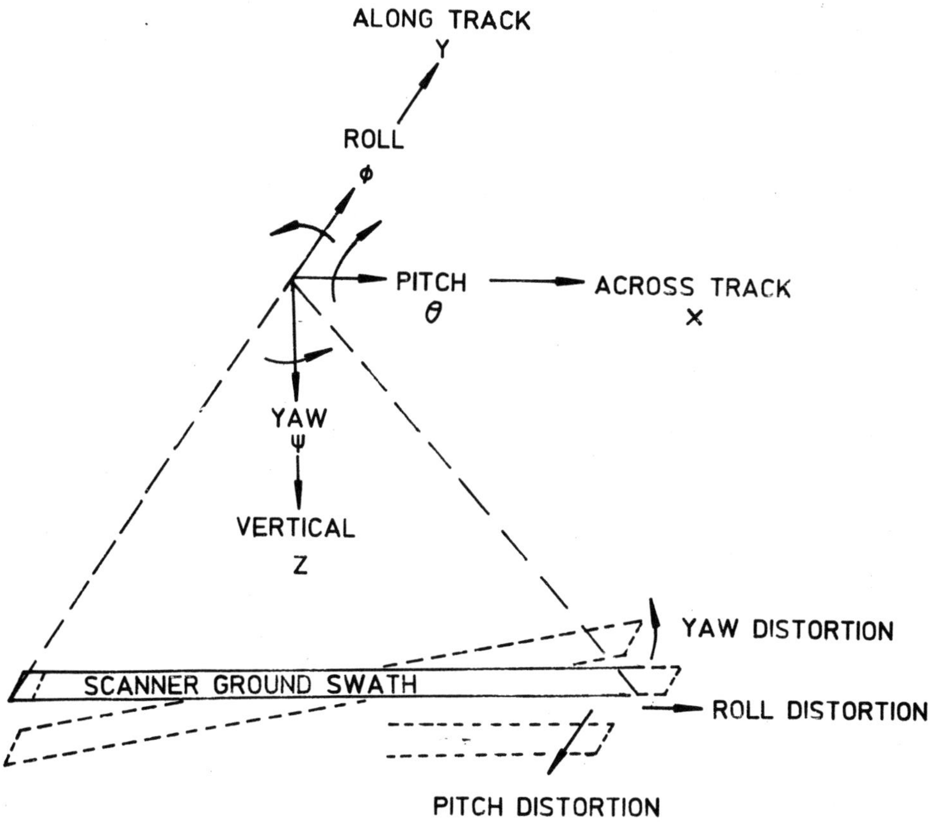

Fig. 6.5: Sources of geometric distortion due to platform attitude

The error caused by deviation of altitude and/or attitude can be corrected by modelling few known Ground Control Points (GCP) which are identifiable in the ground, map and in the imagery.

Ground Control Points

Some features natural or man-made are identifiable in the map as well as in the imagery. These include highway intersection, river bridge, intersection of two small streams etc. The minimum number of GCPs required for rectification depends on the order (P) of the transformation model and size of the scene.

$$\text{No. of GCP} = \frac{(P + 1)(P + 2)}{2} \quad \ldots \quad (11)$$

Rectification

It is a process of relocating the features in the input image to an output image which is in close agreement with a standard map, and the output image is made planimetric (Haralicks, 1973). The rectification are of two types: (i) image to map, and (ii) image to image. The latter type is called the registration. Here, we will first discuss the image to map rectification. Image to map rectification comprises two basic operations (Green, 1983):

— To establish the geometric relationship between the input image pixels in terms of Line and Column and the corresponding map coordinates (latitude and longitude). This operation determines the planimetry.

— Immediately after the geometric relationship established, interpolation of the brightness values of the output scene is to be performed by following some standard interpolation or resampling technique.

Generation of Image to Map Transformation Model

This process involves in the following operations:

— map-digitizer model
— acquisition of GCPs
— updating the transformation model
— calculation of rectified grid.

Map-Digitizer Model

For identification of GCPs in the map in terms of latitude and longitude, either in geographic or cartographic coordinates, a map-digitizer model must be generated. Normally, a second order polynomial are used, which is given below:

$$I = ao + a_1X + a_2Y + a_3X^2 + a_4XY + a_5Y^2 \quad \ldots \quad (12)$$
$$J = bo + b_1X + b_2Y + b_3X^2 + b_4XY + b_5Y^2 \quad \ldots \quad (13)$$

where (I, J) = map coordinates
(X, Y) = digitizer coordinates
a_0, a_1, a_2 .... a_5 = regression co-efficient
b_0, b_1, b_2, ... b_5 = regression co-efficient

With the knowledge of some known points, the map coordinates of which are known, co-efficients of the equations mentioned above will be determined by least square regression technique.

Acquisition of GCPs

With the help of map-digitizer model, maps on the digitizer board, and input image displayed on the display terminal, GCPs are being acquired as shown in Table 6.1.

TABLE 6.1: Rectification (GCP) List

GCP No.	Input Image (X, Y)		Map (X, Y)		Weight	Res. Error
1.	745	2234	3189696	320586	1	3.4688
2.	609	2368	3179990	320117	1	0.8188
3.	697	2240	3187452	328677	1	2.4149
4.	764	2101	3193707	337655	1	3.0337
5.	635	1869	3186625	355628	1	2.4368
6.	1270	1928	3231404	345273	1	1.3810
7.	1294	1957	3232764	342891	1	1.9810
8.	1364	1917	3237879	345250	1	3.8185
9.	1255	1667	3233021	364316	1	3.5478
10.	1403	2051	3239508	335072	1	2.7544
11.	1003	1783	3213568	358412	1	1.7981
12.	1531	2241	3247090	320282	1	1.7548
13.	1535	2351	3246307	312455	1	2.1901
14.	1587	2040	3253079	334210	1	2.8217
15.	1684	1944	3260560	340186	1	2.1009
16.	712	758	3202486	434914	1	3.5804
17.	738	1090	3201529	410777	1	1.6549
18.	526	1206	3185239	404520	1	0.7350
19.	692	1301	3197956	395708	0	0.0000
20.	934	1641	3210172	369647	1	3.0686
21.	446	1642	3175638	374126	1	2.7059

Calculation and Updating the Image-Map Transformation Model

The image distortion is empirically modelled as a mapping transformation from the desired map projection coordinates to the acquired image coordinates (Plate 6.1). The mapping function is generally chosen to be a pair of bivariate polynomials as suggested by Markavian *et al.* (1973), and Bernstein (1983). The general image to map transformation model is represented by the following equations:

$$I = \sum_{i=0}^{n} \sum_{j=0}^{n-1} aij\, X^i Y^j \quad \ldots \quad (14)$$

$$J = \sum_{i=0}^{n} \sum_{j=0}^{n-1} bij\, X^i Y^j \quad \ldots \quad (15)$$

where (I, J) = Input image coordinates of the acquired GCPs
(X, Y) = Map coordinates of the same points
n = Order of the polynomial transformation model
aij & bij = Unknown transformation co-efficients.

These co-efficients are determined in similar way as described for map-digiter co-efficients. Here 'n' is dependent on the degree of non-linearity of distortion. The order (n) must be large, if the distortion is highly localised (Ford and Zanelli, 1985).

The fitness of the model (map-digiter, image-map) is being checked by the residual error. If the computed image (or digiter) coordinates of any GCP is (X', Y') and the original image (or digiter) coordinates is (X, Y), then

$$\text{The residual error} = [(X' - X)^2 + (Y' - Y)^2]^{1/2} \quad \ldots \quad (16)$$

The residual errors for all the GCPs are displayed. The analyst is free, after examining the residual error, to remove or replace one or more points from analysis, if they appear to be anomalous (i.e. the residual error for the GCP exceeds some threshold value, generally within one pixel). The usual practice is to (a) delete the GCP from analysis (Table 6.1); (b) re-compute the model, and (c) re-calculate the residual error. This operation should continue until the desired results is attained.

Calculation of Rectification Grid

The image geometry is corrected by defining the rectangular grid in the map coordinates and applying the image-map transformation to each grid point to locate the points in the input image (Fig. 6.6). The user can determine the size of the grid (32 × 32, 64 × 64, 128 × 128 pixels). The rectification grid give the geometric correspondence between the pixels in the input image and the positions defined by a square grid in the rectified output image (SEP, 1986).

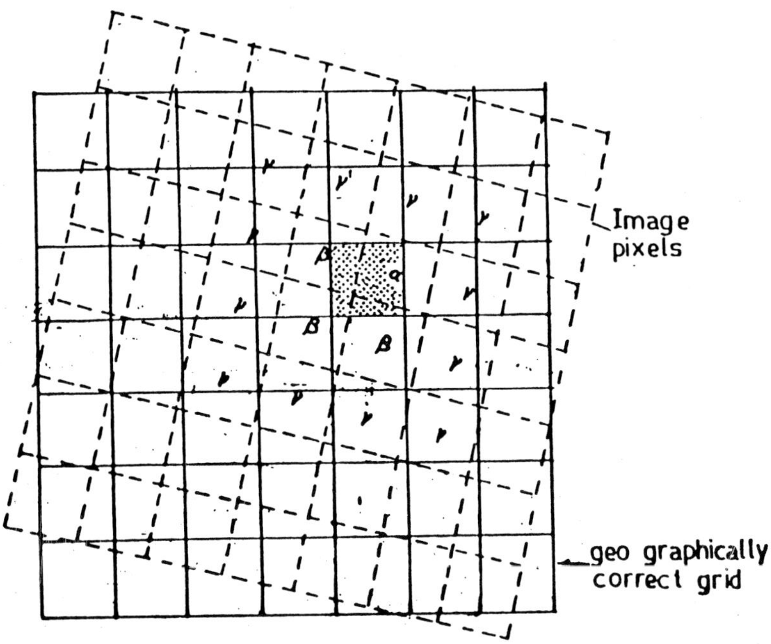

Fig. 6.6: Resampling procedures used to geometrically correct as Image (*Source*: Curran, 1985)

Resampling or Interpolation of Gray Values

Resampling is the means or technique of extraction of the gray values from a location in the original input image and its relocation to the appropriate coordinates in the rectified output image. The interpolation or pixel filling may be used to produce the output image line by line, column by column (Castlemen, 1978). Since there is no direct one to one relationship between the input pixel value to output pixel values, the intensity or brightness values can be assigned from the input image to rectified image by resampling technique (Green, 1983). Three resampling techniques, e.g. viz., (a) nearest neighbourhood; (b) bilinear, and (c) cubic convolution are most commonly used, depending on the kind, degree of precision and cost effectiveness (Billingsley, 1983). Each one is described below (Plate 6.1 and 6.2).

Nearest Neighbourhood (NNB)

The resampling technique assigns a gray value (DN) to the new output pixel

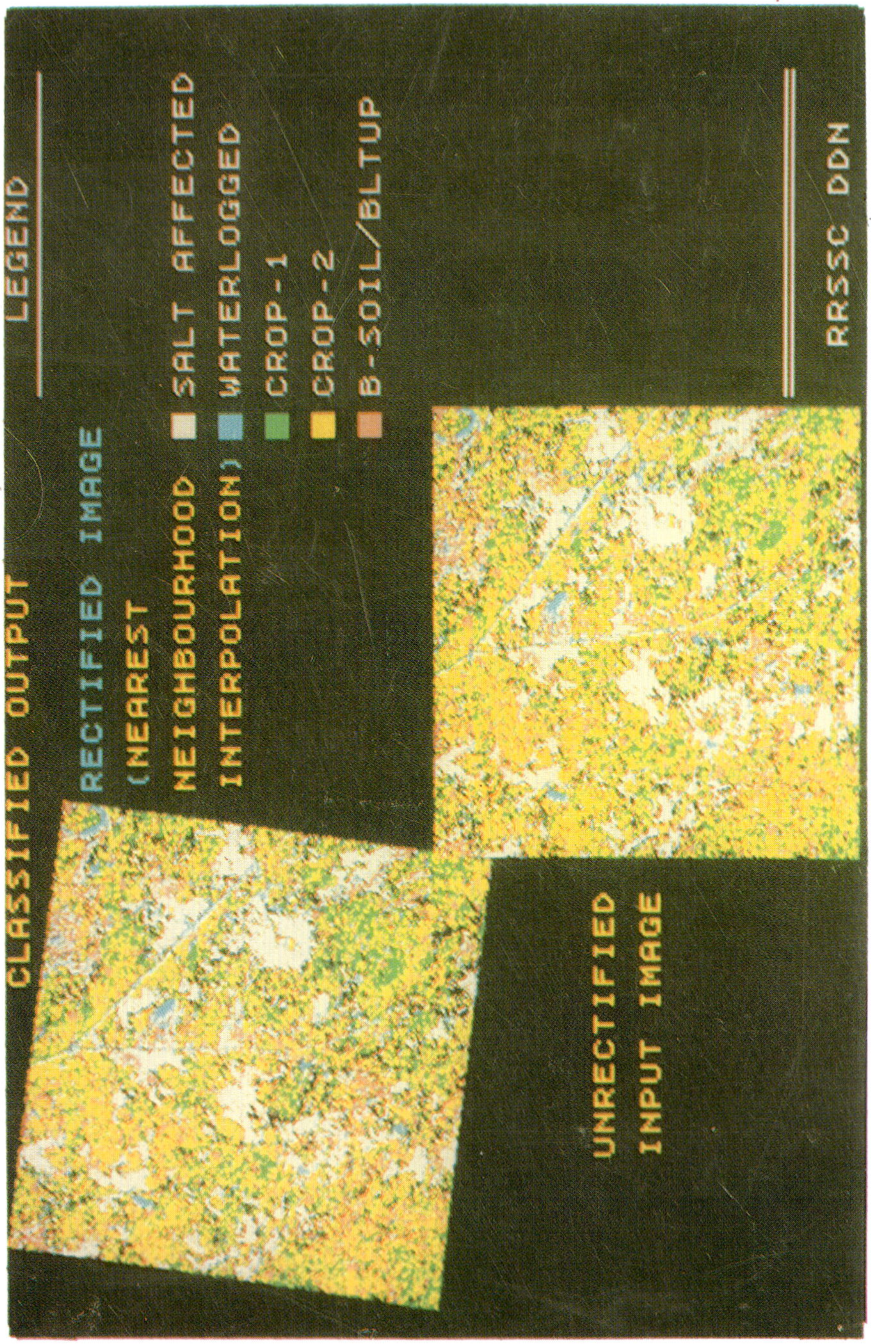

Plate 6.1: Rectification of Remote Sensing Data

Plate 6.2: Resampling of Remote Sensing Data

location (i, j) according to the DN value of the spatially nearest pixel (K, 1) in the input array to the precisely derived input location (X, Y) (see Fig. 6.7). This method preserves the original array values of the pixels and thus does not alter the thematic information. On the contrary, it introduces spatial shift error upto $\sqrt{2}$ times of the pixel size. The local geometry of the linear features (road, canals, railways, etc.) shows disjointed in the rectified output. This problem is undesirable during digital comparison because while the registration of details in two images may be perfect in one location, whereas there is a mis-registration in other locations (Billingsley 1983, Kudrat *et al.*, 1991) (Plate 6.1).

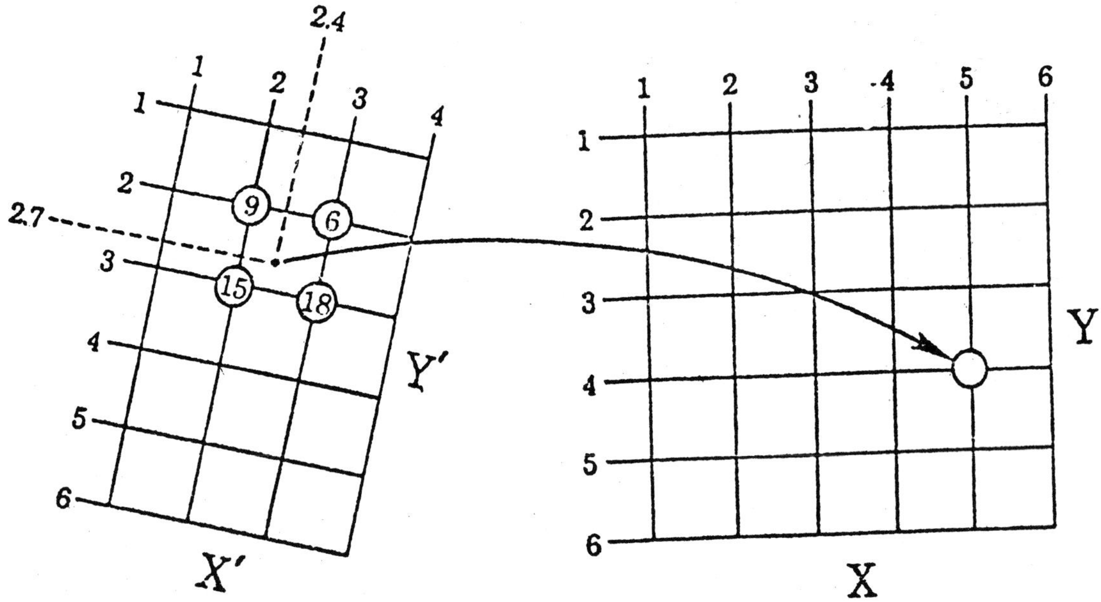

Original Input Image Rectified Output Image

Fig. 6.7: Nearest Neighbourhood

Bilinear Interpolation

In this resampling technique, the gray values (DN:ij) of the out pixels are assigned by using interpolation scheme in two orthogonal directions in the input image (say line and column). First of all, it determines the location of a grid points (X, Y) in the input image pixel location (X', Y'). Let us assume this pixel (X', Y') is exactly not matching with input image (Fig. 6.7), but lies between the pixels locations (1, K), (L, K+1), (L+1, K) and (L+1, K+1). The computation of gray values (X, Y) has been worked out based on the weighted distance to these four pixels. The weighted average of the new gray value (BVwt) is computed according to the equation, suggested by Jenson, 1983.

$$BV_{wt} = \frac{\sum_{k=1}^{4} (Z_K/D^2_k)}{\sum_{K=1}^{4} (1/D_K^2)} \quad \ldots \quad (17)$$

where Z_k = Gray values of the four pixels (1, K), (L, K+1), (L +1, K), (L+1, K+1), respectively

D^2 = Square of the distance between (X', Y') and the data point.

The resultant image is much smoother than that of NNB interpolation. The linear features like canal, road etc. is continuous. Bilinear interpolation output image has 1/4th of the mean square resampling error of NNB (Seidel *et al.*, 1983). This requires appreciably more computational time, because of four multiplication. In this method, gray values of the original image may alter, which may create confusion in classification to generate thematic maps (Kuglar & Tauch 1986, Kudrat *et al.*, 1991).

Cubic Convolution

This is comparatively the most sophisticated resampling technique to evaluate the gray values of several pixels, surrounding a given pixels in the input image to establish a new synthetic gray value (DNij) to be assigned to the corresponding pixels in the output image. Conceptually, this method is very much similar to bilinear resampling technique. The only difference is during computation of a gray value in the input image corresponding to output image, 16 surrounding pixels are involved rather 4 pixels.

The results of this resampling is very much comparable with bilinear resampling technique (Kudrat *et al.*, 1991), but the computation is time consuming. The comparison of three resampling technique is given in Table 6.2.

Registration or Image to Image Rectification

This operation is very much similar with map to image rectification. The basic difference is that the analyst has to develop a relationship between input image of an area and the reference image of the same area (rectified or unrectified). The basic steps are also similar to earlier rectification operations. Here, GCPs can be located on the reference image, as well as on the input image using the cursor by positioning in the display monitor. There is no need for any operation like map to digitizer model. This is also accomplished by auto correlation technique (SEP, 1986). In the latter operation, a window is to be selected for both input and reference image.

TABLE 6.2: A Comparison between Three Methods of Resampling of Remotely Sensed Images

Method	Technique	Example pixels to be transferred in Fig. 6.1	Relative computing	Advantages	Disadvantages
Nearest Neighbour	Transfer gray value of nearest pixel		1	Simple to compute. Gray values un-altered	Image disjointed pixel offset by upto half of pixel
Bilinear Interpolation	Transfer proximity weighted average of nearest 4 pixels		10	Smooth image geometrically accurate	Gray values ultered
Cubic Convolution	Transfer evaluation weight of nearest 16 pixels		20	Very smooth image	Complex to compute. Gray valued altered

(*Source*: Curran, 1985).

Within this window the pixel which shows maximum correlation will be considered as GCP. The image to image rectification is necessary for the analysis of temporal changes, updating the existing maps (Friedman *et al.*, 1983) and in a Geographic Information System (Williams, 1980).

Conclusion

The rectification and pre-processing methods described here are being widely used. The rectification and registration technique discussed in this chapter is conceptually simplistic in nature. However, care should be taken about the number of GCPs and its spread throughout the image/scene is very important. The root mean square GCP error is a function of the number of GCPs used (Bernstein, 1976). The root mean square GCP errors inversely proportional to the numbering GCP. But the optimum distribution is also important (Ford *et al.*, 1978, Foster 1980). Usually 25 to 40 GCP is the minimum requirement for the model to achieve the desired results. Many times, problem of inidentification of GCPs arises in snow covered mountainous terrain and coastal areas (Friedman *et al.*, 1983). The whole rectification process is tedious and time consuming. But with the advent of array processor, sophisticated hardware and software facilities, the resampling can be done within few minutes (Friedman, 1981). Friedman *et al.*, 1983 suggested a pseudo-physical model using physical spacecraft attitude orbit parameter alongwith few GCPs. In their method only 4 GCPs is sufficient to rectify the whole scene. Some workers (Siedel *et al.*, 1983, Naraghi *et al.*, 1983) advocated the utility of digital elevation model for image rectification.

REFERENCES

Ajoy, A.K., Gopalan, S. and Kamat, D.S. (1985), Discrimination of Winter Crop using Temporal Sectral Profiles. *International Journal of Remote Sensing*, vol. 6(09), pp. 1575-80.

Bernstein, R. (1983), Image geometry and rectification, Chapter 21 in R.N. Colwett (ed.), *Manual of Remote Sensing*, Falls Church VQ: American Society of Photogrammetry and Remote Sensing, vol. I, 875-81.

Bernstein, R. (1975), *All-Digital Precision Processing of ERTS Images*. Final Report, NASA Contract NASS-21716, April.

Bernstein, R. (1976), Digital image processing of earth observation sensor data. *IBM Journal of Research and Development*, vol. 20, pp. 40-57.

Bernstein, R. and Ferneyhough, D.G. (1975), Digital image processing. *Photogrammetric Engineering and Remote Sensing*, vol. 41, pp. 1465-76.

Billingsley, F.C. (1983), Data processing and reprocessing, Chapter 17, in R.N. Colwell (ed.), *Manual of Remote Sensing*, 2nd Edition, Falls Church VQ: American Society of Photogrammetry and Remote Sensing, vol. I, pp. 719-92.

Castleman, K.E. (1978), *Geometric Operation in Digital Image Processing*. Englewood Cliff, New York. Printice Hall, pp. 111-2.

Chavez, P., Berline, S. and Mitchel, J.L. (1977), Computer enhancement techniques of Landsat MSS digital image for land use/land cover assessments. *Proceedings of 6th Ann. Remote Sensing of Earth Resource Conference*, pp. 259-76.

Curran, Paul J. (1985), *Principle of Remote Sensing*, ELBS, Longman, Essex.

Davidson, G.J. (1986), Ground control pointing and geometric transformation of imagery. *International Journal of Remote Sensing*, vol. 7(1), pp. 65-74.

Ford, G.E., Algazi, V.R. and Agee, B.C. (1978), On the correction of geometric distribution in satellite acquired images. *Proceedings of Twelfth Asilomar Conference on Circuits, System and Computers*. pp. 341-4.

Ford. G.F. and C.I. Zanelli (1985), Analysis and quantification of errors in geometric correction of satellite images, *Photogrammetric Engineering and Remote Sensing*, vol. 51(11), pp. 1725-34.

Forster, B. (1980), Urban Control for Landsat Data. *Photogrammetric Engineering and Remote Sensing*, vol. 46, pp. 539-45.

Friedmann, D.E. (1981), Two dimensional resampling of line scan imagery by one dimensional processing. *Photogrammetric Engineering and Remote Sensing*, vol. 47(10), pp. 1459-67.

Friedmann, D.E., Friedel, J.P., Magnussen, K.L., Kwok, R. and Richardson, S. (1983), Multiple scene precision rectification of spaceborne imagery with very few ground control points, *Photogrammetric Engineering and Remote Sensing*, vol. 49(12), pp. 1657-67.

Green, W.B. (1983), Quantitative Image Processing Techniques, (Chapter 4), *Digital Image Processing*, New York: Van Nostrand Reinhold Company Inc., pp. 77-105.

Haralick, R.M. (1973), Glossary and index to Remote sensed image pattern recognition concepts, *Pattern Recognition*, vol. 5, pp. 391-403.

Jenson, J.R. (1983), Education image processing: An overview. *Photogrammetric Engineering and Remote Sensing*, vol. 49, pp. 1151-7.

Kudrat, M., Diwakar, P.G., Tiwari, A.K., Vohra, T.P.S., Saha, S.K. and Bhan, S.K. (1991), Effect of resampling on classification results. *Asian Pacific Journal of Remote Sensing*, vol. (2).

Kugler, B. and Tauch, R. (1986), Comparison of classification results of original and pre-processed satellite data in M.C.J. Damen, G. Sicco Smit and H. TH Verstappen (eds.), *Remote Sensing for Resource Development and Environmental Management*, vol. I, pp. 41-5, A.A. Balkema/ Netherlands.

Markarian, H., Bernstein, R., Ferneyhough, D.G., Gregg, L.E. and Sharp, F.S. (1973), Digital correction for high resolution images. *Photogrammetric Engineering and Remote Sensing*, vol. 39, pp. 1311-20.

Mulders, M.A. and Epema, G.P. (1986), Thematic Mapper: A new tool for soil mapping in arid areas, *ITC Journal*, No. (1), pp. 24-9.

Naraghi, M., Stromberg and Daily, M. (1983), Geometric rectification of radar imagery using digital elevation models, *Photogrammetric Engineering and Remote Sensing*, vol. 49, pp. 195-9.

Rifman, S.S. (1973), Digital rectification of ERTS multispectral imagery, *Proc. Symp. Significant Results obtained from Earth Resources Technology Satellite-I*, NASA Publ. SP-327: 1131-42.

Saha, S.K., Kudrat, M. and Bhan, S.K. (1990), *International Journal of Remote Sensing*.

Seidel, K., Ade, F. and Lichtenegger (1983), Augmenting Landsat MSS data with topographical information for enhanced registration and classification. *IEEE Transaction on Geoscience and Remote Sensing*, vol. GE-21, pp. 252-8.

Stein, M. and Van Wie, P. (1976), A Landsat Digital Image Rectification System, *IEEE Transaction on Geoscience Electronics*, vol. GE-15 (3), pp. 130-7.

SEP (1986), Pre-processing and rectification in *VIPs-32 Software Documents*, Vol. III, France.

Williams, O.W. (1980), Outlook on future mapping, charting and geodesy systems. Photogrammetric Engineering and Remote Sensing, vol. 46(4), pp. 487-90.

SEVEN

ENHANCEMENT TECHNIQUES

Introduction

LOW SENSITIVITY of the detectors, weak signal of the objects present on the earth surface, similar reflectance of different objects and environmental conditions at the time of recording are the major causes of low contrast. Other problem that complicates photographic display of digital image is that the human eye is poor at discriminating the slight radiometric or spectral differences that may characterise the features. The main aim of digital enhancement is to amplify these slight differences for better clarity of the scene. This means digital enhancement increases the separability (contrast) between the interested classes or features. The digital image enhancement, may be defined as some mathematical operations that are to be applied to digital remote sensing input data to improve the visual appearance of an image for better interpretability or subsequent digital analysis (Lillesand and Keifer, 1979). Since the image quality is subjective measure, varying from person to person, there is no simple rule as such that may produce a single best result. Normally, two or more operations on the input image may be suffice to fulfil the desire of the analyst, although the enhanced product may have a fraction of total information stored in the original image. This will be realised after seeing the different contrast enhancement techniques in this chapter. Thus, an enhanced image always highlights the useful information while suppress the other details usually (Siegel *et al.*, 1980).

In remote sensing literature, many digital enhancement algorithms are available. They are namely contrast stretching enhancement, ratioing, linear combinations, principal component analysis, spatial filtering etc. Broadly, the enhancement techniques are categorised as (a) point operations and (b) local operations. Point operations modify the values of each pixel in an image data set independently, whereas local operations modify the values of each pixel in the context of the pixel values surrounding it. Point operations include, contrast enhancement and band combinations, but spatial filtering is an example of local operations. Here in this chapter we will discuss the contrast enhancement and other techniques will be discussed in the subsequent chapters (Plate 7.1).

Contrast Stretch or Enhancement

The sensitivity of remote sensing detectors (TM, MSS etc.) was designed to record a wide range of terrain brightness from black asphalt and basaltic rocks to white sea ice under a wide range of lighting conditions. In general, few of scenes have the full brightness range (Sabin, 1986). To produce an image with optimum contrast ratio, it is inevitable to utilise the entire dynamic range (Jenson, 1986). Digital contrast enhancement is thus of prime importance. The objective of contrast stretching is to expand the narrow dynamic range of gray values (digital numbers) typically present in an input image over a wide range of gray values for the desired output image. A variety of contrast stretching algorithm are available. They are broadly categorised as (a) linear contrast stretch and (b) non-linear contrast stretching (Fig. 7.1). In this chapter most popularly contrast stretching algorithms have been dealt.

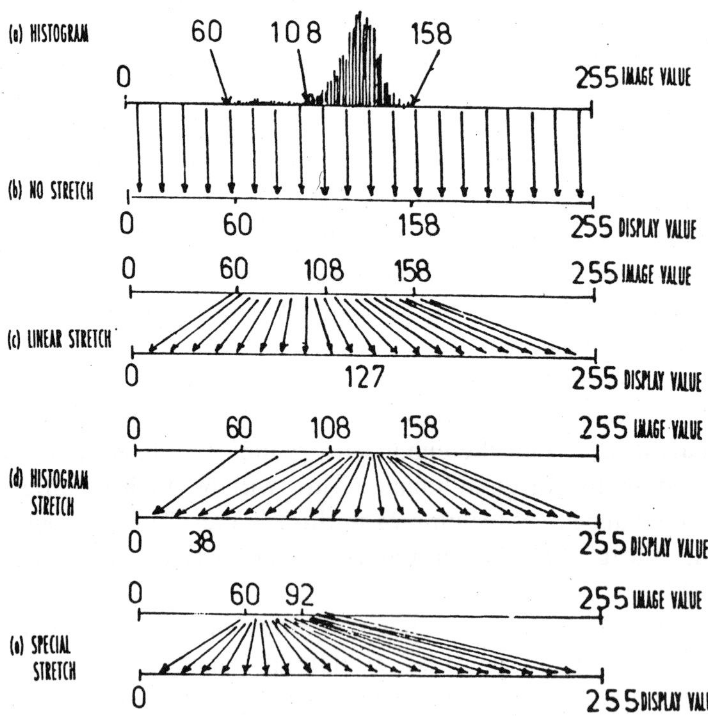

Fig. 7.1: Principle of some contrast stretch enhancements
(*Source*: Lillesand and Kiefer, 1987)

Linear Contrast Stretch

In this algorithm, the digital number (DN) value in the lower end of the original histogram is assigned to digital number zero (DN = 0) i.e. extremely black and a value at the higher end is assigned to extremely white (i.e. DN = 255). The intermediate values are interpolated between 0 and 255 by following a linear relationship, as given below:

$$Y = a + bx \qquad \ldots. \quad (1)$$

Where X and Y are the input gray value of any pixel and output gray value of the same pixel. a and b are intercept and slope respectively and they are determined as follows:

Let X_{min} and X_{max} be the minimum and maximum gray values of the input scene and Y_{min} and Y_{max} are the corresponding minimum and maximum gray values in the output scene.

So,

$$Y_{min} = a + b\,X_{min} \qquad \ldots. \quad (2)$$

$$Y_{max} = a + b\,X_{max} \qquad \ldots. \quad (3)$$

by solving (2) and (3), we have

$$b = \frac{Y_{max} - Y_{min},}{(X_{max} - X_{min})} \quad a = \frac{Y_{min} \bullet X_{max} - Y_{max} \bullet X_{min}}{(X_{max} - X_{min})}$$

Thus,

$$Y = \frac{Y_{min} \bullet X_{max} - Y_{max} \bullet X_{min}}{(X_{max} - X_{min})} + \frac{Y_{max}\ Y_{min}\ (X)}{(X_{max} - X_{min})} \qquad \ldots. \quad (4)$$

For

$Y_{min} = 0$ and $Y_{max} = 255$

equation (4) becomes:

$$Y = \frac{X - X_{min}}{X_{max} - X_{min}} \times 255 \qquad \ldots. \quad (5)$$

So, the intermediate values will be extrapolated as per equation (5).

The linear contrast stretch is widely used to improve the contrast of most of the original brightness values (Fig. 7.1c).

The linear contrast stretch or enhancement greatly improves the contrast of most of the original brightness values (DN), but there is a loss of contrast of the extreme high and low ends i.e. at the tail of the histogram (Sabin, 1986). Another advantage of digital enhancement is that the analyst can enhance any portion of the dynamic range of the histogram to achieve the desired results, since the digital image is flexible. Plate 7.1 is an example of linear contrast enhancement image of Dehradun area, India (IRS-1A LISS-II).

Histogram Equalisation

Histogram equalisation is widely used for contrast manipulation in digital image processing because it is very simple for implementation. It needs minimum information from the analyst. In this enhancement, the original histogram has been readjusted to produce a uniform population density of pixels along the horizontal gray value (DN) axis. Gonzalez and Wintz (1977) demonstrated the technique for the modification of the digital number (DN) or gray values from the original image. It involves in two steps: Firstly, it computes the histograms of the original image and the cumulative frequency density percentage (Fig. 7.1d and Table 7.1). Secondly, computation of transformation function based on which the contrast manipulation takes place in the output scene.

Computation of Transformation Function

This is accomplished by the cumulative frequency distribution (Table 7.1) of the original histogram (Fig 7.1a) using the technique suggested by Ganzalez and Wintz (1977).

$$Si = \sum_{i=0}^{i} Pi \qquad \ldots. \qquad (6)$$

where Si = Cumulative frequency per cent of the original scene = frequency per cent for the gray level (i) in the output image.

In the algorithm, immediately after the calculation of transformation function 'Si' (Eq.-6), it is to be compared with each of the original gray values (xi), and it will assign each of the transformation functions to the x1 values closest to it in value (Jenson, 1986). Table 7.1 showed original gray values corresponding relative frequencies (cumulative per cent) and corresponding gray values in the final output.

Let us take an example of a hypothetical image of size 64 × 64 pixels having eight gray level (xi). The first step in the procedure as mentioned earlier is to obtain the histogram equalization mapping as suggested by Gonzalez and Wintz (1977) (Table 7.1). In this example, we are interested to equatize the number of pixels in eight gray level. Therefore, we are to compare the 2nd column and the 5th column, and gray values are assigned for each transformation functions closely matching the gray level (xi) in the second column. If we take the first transformation function 'S0' which has a value of 0.19, closely matching with 'x1' = 1/7 = 0.14. So gray

TABLE 7.1: Transformation Function

Original gray values xi	Scaled value i/7	Frequency 'n'	Probability or relative frequency Pi = ni/N	Transformation function Si = Pi	Output gray l Si
x0	0	790	0.19	0.19	S0
xl	0.14	1023	0.25	0.44	Sl
x2	0.28	850	0.21	0.65	S2
x3	0.42	656	0.16	0.81	S3
x4	0.57	329	0.08	0.89	S3
x5	0.71	245	0.06	0.95	S4
x6	0.85	122	0.03	0.98	S4
x7	1.00	81	0.02	1.00	S4

level for first transformation function is x1. The same logic can be applied for all the transformation functions.

$$
\begin{array}{lcl}
(S0 = 0.19) & \simeq & (1/7 = 0.14) \\
(S1 = 0.44) & \simeq & (3/7 = 0.42) \\
(S2 = 0.65) & \simeq & (5/7 = 0.71) \\
(S3 = 0.81) & \simeq & (6/7 = 0.85) \\
(S4 = 0.89) & \simeq & (6/7 = 0.85) \\
(S5 = 0.95) & \simeq & (7/7 = 1) \\
(S6 = 0.98) & \simeq & 1 \\
(S7 = 1.00) & &
\end{array}
$$

Hence, only five district histogram equalised gray values are prominent and are presented in Table 7.2.

TABLE 7.2: Results of Histogram Equalization Mapping

xj - Sk	nk	Ps (Sk)
x0 - S = 1/7	790	0.19
xl - Sl = 3/7	1023	0.25
x2 - S2 = 5/7	850	0.21
x3, x4 - S3 = 6/7	985	0.24
x5, x6, x7 - S4 = 1	448	0.11

Source: Hall; 1978.

Histogram equalisation of IRS-1A LISS-2 data, band 4, for Dehradun area is presented in Table 7.3 and Plate 7.1. This Contrast Stretching applies the greatest

Plate 7.1: Enhancement of Remote Sensing Data

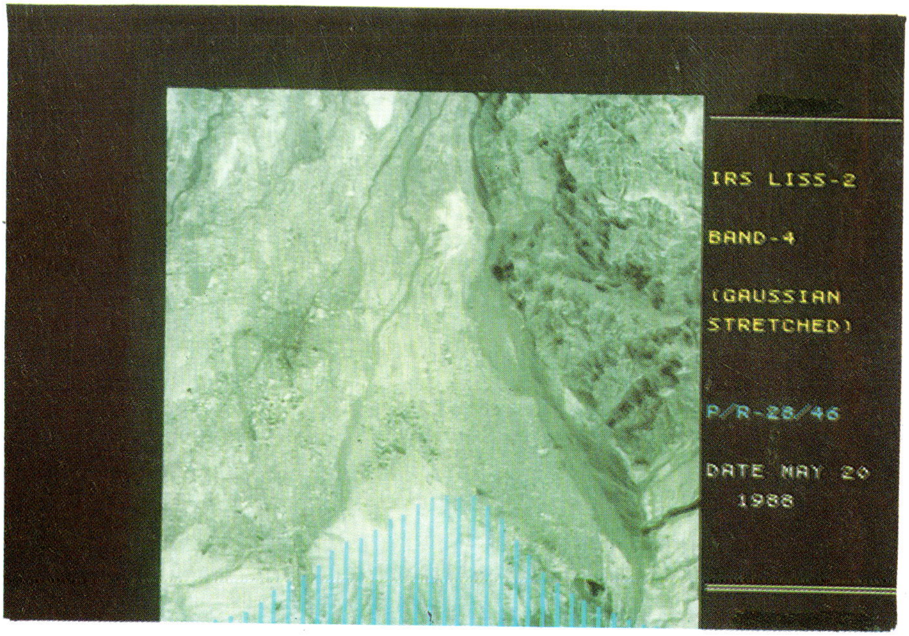

Plate 7.1 *(Contd.)*

contrast enhancement to the most populated range of DN values of the original image. Generally, the middle portion of the histogram is usually enhanced (Sabin, 1986).

Logarithmic Contrast Enhancement

This transformation is a very useful non-linear contrast enhancement. Here the output pixel gray values (Yij) will be generated from input pixel gray values (Xij) following some logarithmic expressions (eq.-7).

$$Yij = a \log (Xij) + b \qquad \ldots \qquad (7)$$

The co-efficients 'a' and 'b' are determined by taking the maximum and minimum gray values of the input image and the corresponding maximum and minimum values in the output image. The detail solution is very much similar to the linear contrast stretch, described earlier. The graphical relationship between input gray values and output gray values is shown in Fig 7.1. This transformation only highlights the features lying in the darker region of the histogram (Hall, 1979; Jenson, 1986). Plate 7.1 is the example of logarithmic enhancement.

The following are the advantages and important characteristics of logarithmic contrast enhancement (Hall, 1978).

— It makes the low contrast details more visible by enhancing low contrast edges.

— It provides a contrast signal to noise ratio.
— It somewhat matches the responses of the human visual system.
— It usually provides a more equal distribution of gray values.
— It transforms multiplicative noise into additive noise.

Exponential Contrast Enhancement

This transformation is included as a non-linear contrast enhancement. The gray values (Xij) in the input image transform to the gray values (Yij) in the output image is as follows:

$$Yij = a\, e^{bxij} + c \qquad \text{....} \qquad (8)$$

Where a, b and c are constant. And b is generally arbitrarily chosen between 0.01 to 0.1 to avoid higher value of e. Further, 'a' and 'b' are computed in the similar way to the previous method. The constants 'a' and 'b' scale the dynamic range of the gray values of the output image with 0 and 255. The effect of the exponential transformation on the edges in an image is to compress low contrast edges, while expanding high contrast edges. This generally highlights the feature having higher gray values. The effect is just reverse of logarithmic contrast transformation. This technique generally produces an image with less visible detail than the original and is thus of limited use for image enhancement (Hall, 1978). The Plate 7.1 is an example of this enhancement for Dehradun scene.

Gaussian Stretch

This is another type of non-linear contrast stretching and this enhances the contrast within the tails of the histogram. It is based on the hypothesis that the pixel gray values calculated by the function follows a Gaussian distribution in which distribution is centred around mean i.e. 128. The expansion or compression of the distribution uses the factor 'f' specified by the user, so that 2 half ranges 0-128 and 128-255. Here 'f' is nothing but the times of standard deviation.

The two sides of the mean (μ) μ – fE and μ + fE are represented on 256 gray values, where μ is mean, and E represents the standard deviation. If the factor 'f' = 2, 5 per cent or factor 'f' is 3, i.e. 270 of the pixel values are excluded from the histogram range and are truncated. Table 7.3 has an example of Gaussian stretch output.

TABLE 7.3: Digital Number of 10 × 10 pixels of IRS 1A LISS II Data of Dehradun: Raw Band 4, Histogram Equation, Exponential and Gaussian Contrast enhancement

A: Raw Band 4

57	59	59	58	58	55	55	62	66	65
56	58	59	58	58	56	55	61	66	66
59	59	59	59	59	56	54	57	63	65
60	61	59	59	59	57	57	57	60	66
61	60	59	59	59	59	59	60	59	64
60	60	59	59	59	59	58	59	60	63
60	60	58	57	57	58	57	56	58	63
59	58	58	58	59	58	56	55	58	63
59	58	59	59	59	59	57	55	58	61
59	57	57	58	58	60	58	57	57	60

B: Histogram Equalization Contrast Enhancement

Band No. 1

45	65	65	55	55	26	26	104	163	147
35	55	65	55	55	35	26	88	163	163
65	65	65	65	65	35	19	45	117	147
75	88	65	65	65	45	45	45	75	163
88	75	65	65	65	65	65	75	65	130
75	75	65	65	65	65	55	65	75	117
75	75	55	45	45	55	45	35	55	117
65	55	55	55	65	55	35	26	55	117
65	55	65	65	65	65	45	26	55	88
65	45	45	55	55	75	55	45	45	75

C: Exponential Contrast Enhancement

Band No. 1

9	12	12	10	10	7	7	16	25	22
8	10	12	10	10	8	7	15	25	25
12	12	12	12	12	8	7	9	18	22
13	15	12	12	12	9	9	9	13	25
15	13	12	12	12	12	12	13	12	20
13	13	12	12	12	12	10	12	13	18
13	13	10	9	9	10	9	8	10	18
12	10	10	10	12	10	8	7	10	18

(Contd.)

TABLE 7.3 (*Contd.*)

12	10	12	12	12	12	9	7	10	15
13	13	10	9	9	10	9	8	10	18
12	10	10	10	12	10	8	7	10	18
12	10	12	12	12	12	9	7	10	15
12	9	9	10	10	13	10	9	9	13

Band No. 2 : SBI Output

107	116	116	111	111	97	97	130	148	144
102	111	116	111	111	102	97	125	148	148
116	116	116	116	116	102	93	107	134	144
120	125	116	116	116	107	107	107	121	148
125	120	116	116	116	116	116	121	116	139
121	120	116	116	116	116	111	116	121	134
121	120	111	107	107	111	107	102	111	134
116	111	111	111	116	111	102	97	111	134
116	111	116	116	116	116	107	97	111	125
116	107	107	111	111	121	111	107	107	121

Band No. 3 : GVI Output

126	132	133	123	117	96	117	146	160	152
123	129	133	127	113	99	115	138	157	157
142	139	139	127	112	89	105	120	145	154
147	149	139	131	115	88	105	124	134	157
148	146	141	133	127	115	109	124	137	153
141	147	141	136	130	120	104	107	136	153
138	145	136	130	122	116	97	94	126	149
140	136	133	131	131	114	94	102	127	147
142	135	135	137	133	119	101	101	125	138
140	132	127	130	131	129	107	107	119	130

Band No. 4 : YI Output

133	140	133	143	133	131	162	140	128	136
136	143	133	135	130	128	143	148	129	140
128	129	129	133	125	130	142	147	138	143
132	129	129	136	127	139	152	149	147	140
135	129	138	133	133	137	126	148	146	141

(*Contd.*)

TABLE 7.3 (*Contd.*)

136	132	138	132	132	143	122	132	155	128
139	139	133	134	140	143	123	147	156	126
134	133	131	137	136	134	137	142	150	143
128	129	138	136	133	139	136	138	141	137
134	128	137	147	137	140	119	121	143	144

Band No. 5 : Soil Line Index Output

56	64	61	66	69	69	59	69	77	79
54	61	61	61	71	71	56	72	80	82
54	57	57	66	76	79	59	64	74	79
57	59	57	64	74	86	76	61	69	82
62	57	57	61	66	76	79	77	61	74
62	57	57	59	64	74	76	81	69	67
64	59	54	54	61	71	76	79	66	69
57	54	56	59	64	71	76	66	64	74
54	54	61	59	61	74	76	66	64	69
57	51	56	61	59	72	74	69	64	72

D: Gaussain Contrast Enhancement

94	108	108	101	101	80	80	128	155	148
87	101	108	101	101	87	80	121	155	155
108	108	108	108	108	87	74	94	135	148
114	121	108	108	108	94	94	94	114	155
121	114	108	108	108	108	108	114	108	141
114	114	108	108	108	108	101	108	114	135
114	114	101	94	94	101	94	87	101	135
108	101	101	101	108	101	87	80	101	135
108	101	108	108	108	108	94	80	101	121
108	94	94	101	101	114	101	94	94	114

REFERENCES

Gonzalez, R.C. and Wintz, P. (1977), *Digital Image Processing and Recognition*, Addison-Wesley Publishing Company. Reading, Massachusetts, pp. 123-6.

Gillespie, A.R. (1980), Digital techniques of image enhancement in Slegal and Gillespie (eds.) *Remote Sensing in Geology*. Chapter 6. John Wiley, New York, pp. 139-226.

Hall, E.L. (1978), *Computer Image Processing and Recognition*, Academic Press, New York. Chapter 4, pp. 159-84.

Jenson, J.R. (1986), *Introduction to Digital Image Processing*, Prentice-Hall, Englewood, Cliff, New Jersey, Chapter 7, pp. 117-76.

Lillesand, T.M. and R.W. Keifer (1979), *Remote Sensing and Image Interpretation*, John Wiley and Sons, New York.

Sabins, F.F. (1986), *Remote Sensing--Principle and Interpretation*, Freeman and Co., New York, Chapter 7, pp. 246-60.

Siegal, Barry S. & Alan R. Gillespie (eds) (1980), *Remote Sensing in Geology*, John Wiley and Sons, New York.

EIGHT

SPATIAL FILTERING

SPACE BORNE imagery have some inherent noise or interference due to sensor error or malfunctioning of the detector, channel transmission and sampling error. The noise caused by noisy sensor or channel transmission error appears as discrete isolated pixels variation that are not spatially correlated (Pratt, 1978) This means the pixels having error often appears markedly different from their surrounding pixels. At times, the edges or boundaries between the features are not clearly discernible and this creates problem for interpretation of images. Spatial filtering, a digital local enhancement technique, can be very much useful to overcome the problem.

Spatial filtering is a pixel by pixel transformation which depends on the digital number of surrounding pixels. So spatial filtering is a context dependent operation that alters the gray level of a pixel at any location according to its relationship with the digital counts of the pixels in its immediate vicinity (Sabins, 1986)

Spatial filtering refers to separate out some features. It is to emphasize or deemphasize image data of various frequencies. Essential spatial frequency unable the tonal variation per unit area of an image (Lillisand and Kiefer, 1986). So an image area of high frequency is totally rough. This signifies the digital number (DN) values in these area changes abruptly over a small number of pixel. Conversely the image area of low frequencies are tonally smooth and it signifies that digital number in the image changes gradually over a large number of pixel, e.g. water bodies. Spatial filtering is categorised as low pass filtering, high pass filtering (Fig. 8.1) and band pass filtering. Low pass filtering being used for smoothening or removing the noise. It is designed to preserve the low frequency details at the cost of high frequency details. High pass filtering, on the other hand, preserves the high frequency details at the cost of low frequency detail. The latter one is referred as edge enhancement or edge crispening operation. Band pass filtering is widely used in signal processing. It allows the spatial frequency in a limited region of the spectrum instead of all the frequencies above or below a cut off frequency. This is applicable only for isolating periodic noises from an image. It is of limited use for thematic applications.

In view of application, the spatial filtering operations are categorised as (i) noise removal, (ii) edge enhancement or edge crispening, and (iii) edge

-1	-1	-1	1	2	1	1	1	1
-1	16	-1	2	4	2	1	2	1
-1	-1	-1	1	2	1	1	1	1
High-Pass				Low-Pass				

Fig. 8.1 : Low pass filtering and high pass filtering

extractions. Each of them is separately dealt here (Plates 8.1 & 8.2).

How Filtering is Performed

Filtering is performed by using convolution windows. These windows are called mask, template filter or kernel. The window is moved over the input image from extreme top left hand corner of the scene pixel to pixel performing discrete mathematical function transforming the original input image digital number to a new digital value. First it will move along the line. As soon as the line is complete, it will restart for the next time for covering the entire image (Sabins, 1976; Lillisand and Kiefer, 1986).

The mask window may be rectangular (1×3, or 1×5 pixels) size or square (3×3, 5×7 or 7×7 pixels size). Each pixel of the window is given a weightage. For low pass filters all the weights in the window will be positive and for high pass filter all the values may be negative or zero, but the central pixel will be positive with higher weightage value. In the case of high pass filter the algebraic sum of all the weights in the window will be zero. Many types of mask window and different sizes can be designed by changing the size and varying weightage within the window. The simplest form of mathematical function performed in filtering operation is neighbourhood averaging. Another commonly used discrete function is to calculate the sum of the products given by the elements of the mask and the input image pixels digital number at the same point and use this sum as the output digital number of the central pixel digital number in the moving window. Figure 8.2 explains the filtering operations.

Noise Removal

Image noise caused by a noisy sensor or channel transmission error usually appears as discrete isolated pixels. These pixels are very much dissimilar with the surrounding pixels. To remove this spatially uncorrelated pixels, some suitable filtering technique is made use (Graham, 1962; Rosenfield, 1969). Different noise removal technique through spatial filtering are discussed here.

8	8	8	9	9	9	7	6	6	6	6	6
8	8	8	9	9	9	7	6	6	6	6	6
8	8	8	9	9	9	7	6	6	6	6	6
8	8	8	9	9	9	7	6	6	6	6	6
9	9	9	8	8	6	6	6	6	6	6	6
9	9	9	8	7	7	6	5	6	6	6	6
7	7	7	7	7	6	6	6	6	6	6	6
6	6	6	6	6	6	6	6	6	6	6	6
6	6	6	6	6	6	6	6	6	6	5	5
6	6	6	6	6	6	6	6	6	6	6	6
6	6	6	6	6	6	6	6	6	6	6	6
6	6	6	6	6	6	6	6	6	6	6	6

Image before filtering

8	8	8	8	7	7	6	6
8	8	8	7	7	6	6	6
8	8	8	7	7	6	6	6
8	7	7	7	6	6	6	6
7	7	7	6	6	6	6	6
7	7	6	6	6	6	6	6
6	6	6	6	6	6	6	6
6	6	6	6	6	6	6	6

Image after filtering

Fig. 8.2 : Filtering operation : Application of 5 × 5 filter on 8 × 8 image

Averaging

With this technique each pixel is sequentially examined and if the pixel digital number (DN value) is greater than the average brightness (DN) of its surrounding pixels by some threshold (t), it is replaced by the means of the 3×3 pixels window. The window may be of any size, such as 5×9, 7×7 and so on. The larger the window size, the more will be computational time. Fig. 8.3 demonstrates the comparison of computation time of varying window. The central pixel 'X' has no correlation with the neighbouring pixels and is replaced by the mean value of surrounding pixels. Plate 8.1 demonstrates output of noise removal.

a_1	a_2	a_3
a_4	X	a_5
a_6	a_7	a_8

$$x = \frac{1}{8} \sum_{a_i = 1}^{8} \quad \text{if} \quad [x - \frac{1}{8} \sum a_i] > t \ldots\ldots \quad (1)$$

Fig. 8.3: Averaging

Median Filtering

The main problem in above approach is that resultant image becomes blurs. So the edges between the features becomes fuzzy. The median filtering is an alternative approach to spatial averaging. In this filtering technique, the concept is very much similar to averaging. Here, the abnormal or spatially uncorrelated pixel digital count (x) will be replaced by the median within the 3×3 window. This method eliminates isolated values related to noise spikes, while giving rise to lower resolution loss than with averaging.

Edge Enhancement

Edge or boundary between two features is very important for separating the different features. Edge is characterized by high frequencies. In nature, edge between two features or objects may not always be distinct. Thus edge enhancement or edge crispening is required for better interpretation of an image. Edge enhancement filtering technique enhances the edges only. Without affecting the values of the rest of the pixel within the feature. High frequencies signify the large degree of tonal variation within small area or within a small spatial distance. This refers to those pixels digital number that occur at the transition between two categories. The edge crispening filtering technique is typically employed to exaggerate the edges between contrasting cover type or bring out linear trend of geologic significance (Pratt,1978; Sabins, 1986). It is also useful in mapping lineaments and drainage pattern. This can be accomplished by a number of ways.

Discrete Convolution Filtering

The input pixel array can be modified following the equation (2) to generate output pixel array.

$$Y(m_1, m_2) = \Sigma\Sigma\, F(i, j) \times H \qquad \ldots\ldots \quad (2)$$

where, F_{ij} = Input pixel array of an image.
$Y(m_1 m_2)$ = output image pixel array.
H = Convolution array (high pass filtering as shown below):

Some convolution mask (H) commonly used are given below:

$$H = \begin{matrix} 0 & -1 & 0 \\ -1 & 5 & -1 \\ 0 & -1 & 0 \end{matrix}$$

$$H = \begin{matrix} 1 & -2 & 1 \\ -2 & 5 & -2 \\ 1 & -2 & 1 \end{matrix}$$

$$H = \begin{matrix} -1 & -1 & -1 \end{matrix}$$

-1	9	-1
-1	-1	-1

Output image will be generated from the input image in steps. First all the digital number of an input image will be multiplied with the corresponding values in the mask (H) and then the algebraic sum of all the multiplied values will replace the central pixel of window or mask (3×3 pixel size)

For example Input image within 3×3 window

F_{ij}			H			Output image		
a_1	a_2	a_3	0	-1	0	x	x	x
a_4	a_5	a_6	-1	5	-1	x	y	x
a_7	a_8	a_9	0	-1	0	x	x	x

Where $Y = 5a_5 - (a_2+a_4+a_6+a_8)$ ..... (3)

Then in the output pixel a_5 will be replaced by Y. The important thing to note is that the sum of all elements in the mask (H) window is unity, whereas in other high pass filters like Laplacian operation the sum is zero. Here since the sum of all the coefficient in the mask matrix is equal to unity, the digital number would have been retained at the same value as the original, only change will take place at the edges (Galbaits, 1990)

Laplacian Edge Enhancement Filtering

Edge enhancement for one or more input bands in a scene can be performed using Laplacian operation on a 3×3 neighbourhood. This is being explained as follows:

a_1	a_2	a_3	0	-1	0	a_1	a_2	a_3
a_4	x	a_5	-1	4	-1	a_4	Y	a_5
a_6	a_7	a_8	0	-1	0	a_6	a_7	a_8
Input image			Filter or Mask			Output image		

The central pixel (Y) of the output image = $4x - (a_2 + a_4 + a_5 + a_7)$ ...(4)

This is conceptually very much similar to earlier convolution filtering operation. The only operating window is different. The sum of all the elements within the filtering window is zero. Hence it is a kind of high pass filtering.

Basically the Laplacian filtering operation computes the differences between the digital counts of the central pixel and the average of the DN values of four adjacent pixel in the horizontal and vertical location. The above equation can be explained as

$$Y = (X - a_4) + (X - a_5) + (X - a_2) + (X - a_7) \qquad \ldots\ldots \quad (5)$$

So the output image is nothing but the sum of the partial differences in the horizontal and vertical pixels within the operator (Hall, 1979; Gonzalez and Wintz, 1983).

Let us taken an hypothetical example of small 7×7 pixels image is shown below:

4	4	4	8	8	8	8
4	4	4	8	8	8	8
4	4	4	8	8	8	8
4	4	4	8	8	8	8
4	4	4	8	8	8	8
4	4	4	8	8	8	8
4	4	4	8	8	8	8

Input Scene

x	x	x	x	x	x	x
x	0	4	-4	0	0	x
x	0	4	-4	0	0	x
x	0	4	-4	0	0	x
x	0	4	-4	0	0	x
x	0	4	-4	0	0	x
x	x	x	x	x	x	x

Output Scene

Fig. 8.4: Input and Output Scenes

Laplacian filtering operation applied to an image containing a vertical edge (a) results in zero crossing identification of edge in output scene. In this operation, edges at any direction will be highlighted and it is sometimes called 'omnidirectional' filtering operation. This operation is very much sensitive to noise. Plate 8.1 is the example of Laplacian filtering for IRS-1A LISS-2 scene of Dehradun area.

Robert Edge Enhancement Filtering

This kind of edge enhancement technique is generally used to highlight the edges between the features. In this operation the central pixel of 3×3 subimage will be replaced by some out digital number. This is explained below:

$$\begin{matrix} a_1 & a_2 & a_3 \\ a_8 & X & a_4 \\ a_7 & a_6 & a_5 \end{matrix} \qquad \begin{matrix} 1 & 1 & 1 \\ 1 & 0 & -1 \\ -1 & -1 & -1 \end{matrix} \qquad \begin{matrix} a_1 & a_2 & a_3 \\ a_8 & Y & a_4 \\ a_7 & a_6 & a_5 \end{matrix}$$

Input image Filter or operator

The output central pixel $Y = (a_1 - a_5) + (a_2 - a_6) + (a_3 - a_7) + (a_8 - a_4)$. (6)

This operation uses the diagonal derivative to estimate gradient at the point. It highlights any feature extending along the diagonal of the scene. This is very useful for geologic lineament analysis and lithology (Sabins, 1986). All the edge enhancement filtering techniques are ascertaining the boundaries between agricultural fields, lakes, forests and other terrain regions, Plate 8.1 shows the result of Robert filtering operation in comparison with the raw image of Dehradun area.

Statistical Differences

This is a very simple operation for edge enhancement and was suggested by Ketcham (1976). This operation generates the output image (a c j k) by dividing each pixel digital number (F_{jk}) by its measured statistical standard deviation within the mask window of any size (3×3, or 5×5 etc) according to the following equation:-

$$G_{jk} = \frac{F_{jk}}{r_{jk}} \qquad \ldots\ldots \quad (7)$$

$$\text{Where standard deviation } r^2_{jk} = \frac{\Sigma\,(F_{jk} - F_{jk})^2}{N}$$

is computed over some neighbourhood of the pixel at coordinate (jk). F_{jk} is nothing but the digital count of the smoothed image (averaging) discussed earlier. The enhanced output G_{jk} will be increased in ample with respect to original image at the edge point (Pratt, 1978). The edge point deviate from other neighbourhood. The amplitude of the output image will be decreased elsewhere. As a result the edges will be enhanced. In this operation noise may increase (Wallis, 1976).

Local Optimisation

This technique performs filtering in the spatial domain by dividing the whole scene into a number of small regular grids. The size of the grid may be variable and depends on the choice of the analyst. Generally 39 × 39 pixels grid is being routinely followed. Each grid is enhanced in such a way that the output value of each grid will have a mean of 128 and standard deviation of 128/f, where f is the normalisation factor. In each grid mean and standard deviations are calculated. The detail computing algorithm is as follows (VIPS-32, 1986)

$$Y_{jk} = 128/f \times 6 \times x_{jk} + [128 - (128 \times m)/(f \times 6)] \qquad \ldots\ldots \quad (8)$$

Where x_{jk} = input gray values at the pixel location (jk)
y_{jk} = Output gray values at the pixel location (jk)
m = mean of the grid

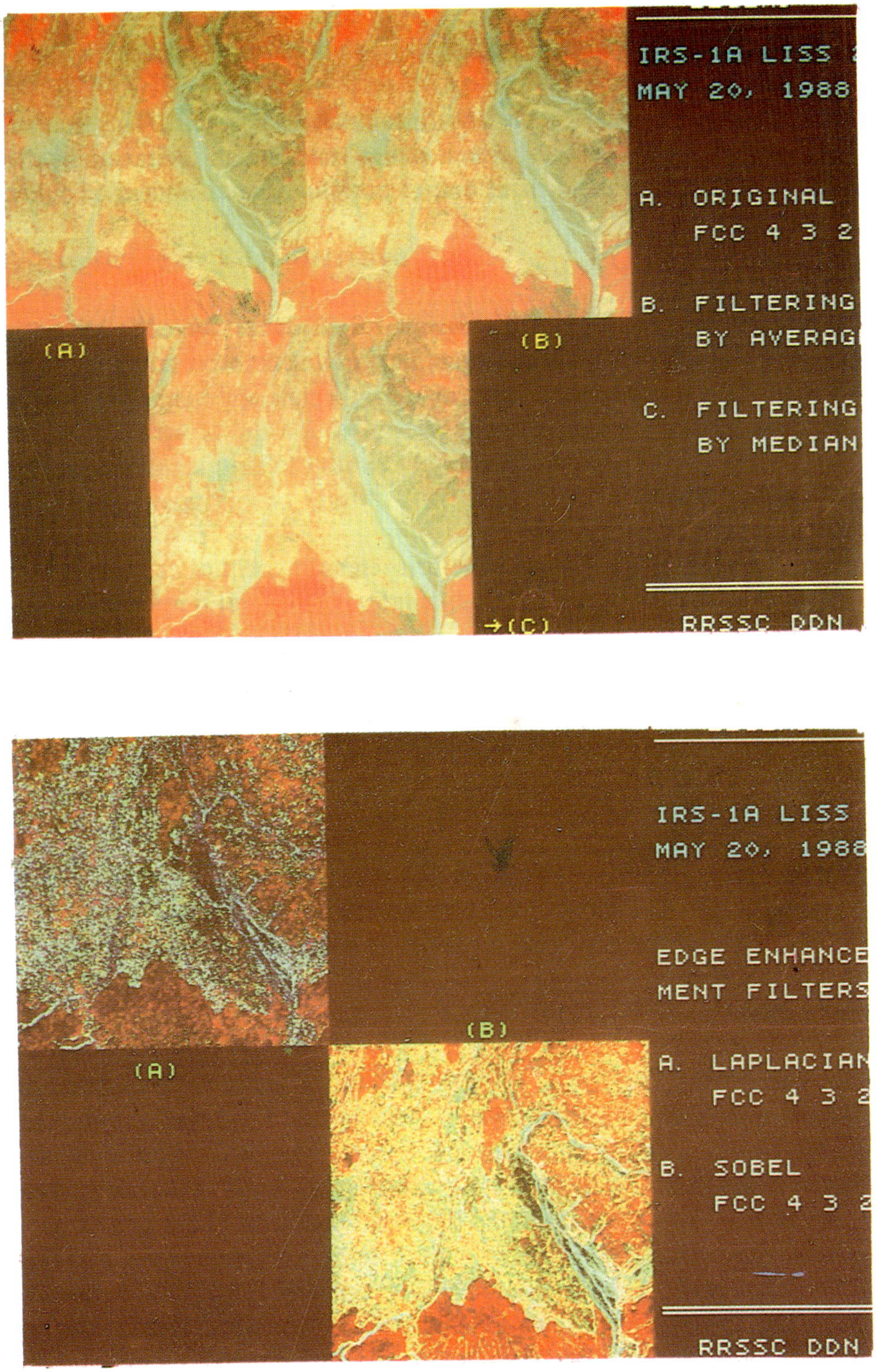

Plate 8.1: Some Examples of Filter Techniques

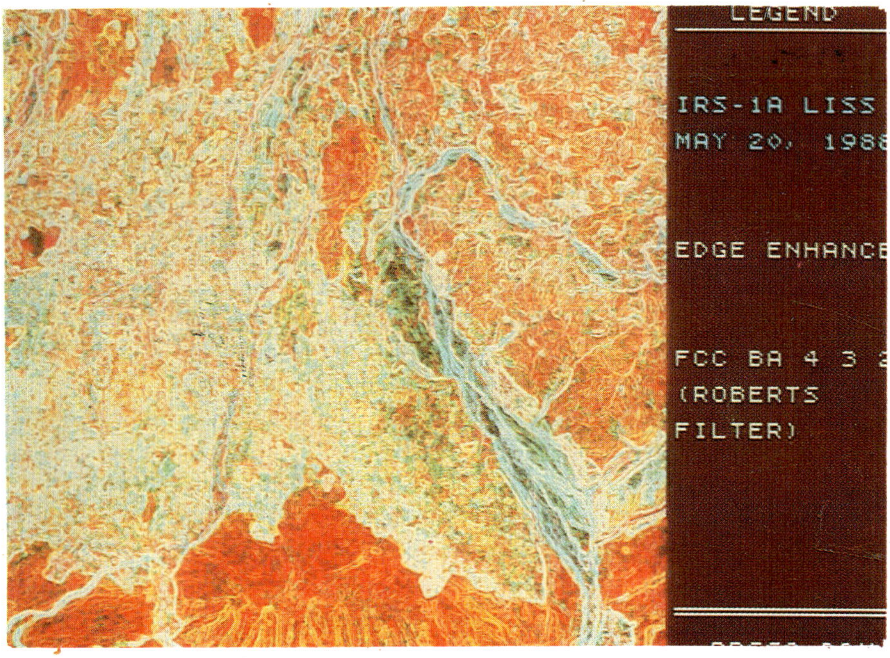

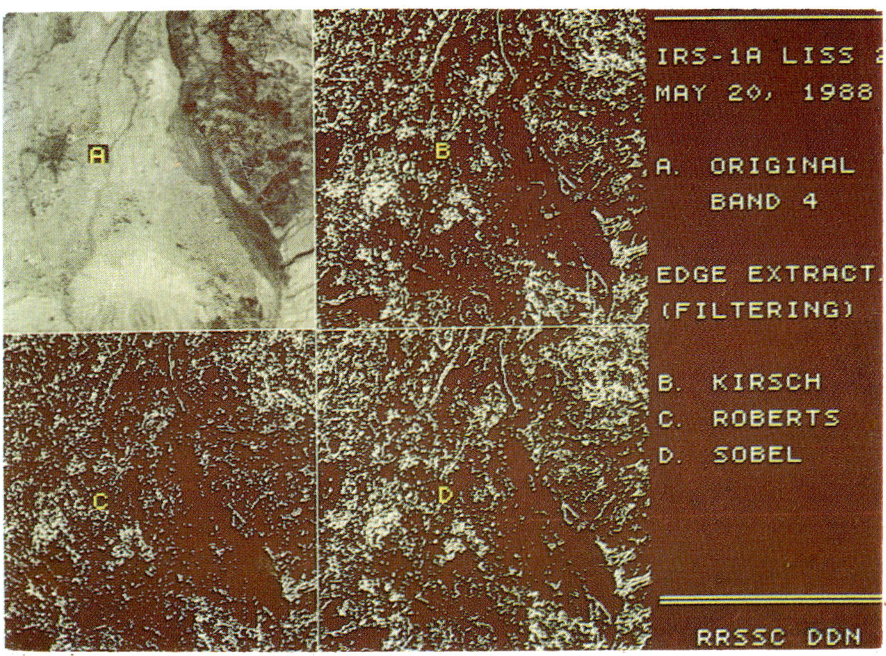

Plate 8.1: Some Examples of Filter Techniques

transformation, sine and cosine transformation (Jain, 1986; Pratt, 1978; Gonzalez Wintz, 1987; Hall, 1983). Detailed discussion of each transformation is beyond the scope of the present discussion. Only Fourier and Hadamard transformations in the frequency domain are considered in the context of spatial filtering (Plate 8.2).

Fourier Transformation

Fourier transformation is being used to convert an image from spatial domain to frequency domain. In digital image processing Fourier transformation is being used for enhancement, compression, textural classification, noise removal or image restoration, quality assessment, cross correlation etc. (Hall, 1983; Jain, 1989; Lillesand and Kiefer, 1987). In Fourier filtering operations are performed in frequency domain. There are three steps: (Plate 8.2).

Step 1

First forward Fourier transformation transfers the input image from spatial domain to frequency domain through the formula given below:

$$F(u,v) = \frac{1}{MN}\sum_{m=0}^{M-1}\sum_{n=0}^{N-1} f(m,n)\, e^{-2i\pi\left(\frac{mu}{M}+\frac{nv}{N}\right)} \quad \ldots\ldots \quad (13)$$

where (u,v) = New coordinate = spatial frequencies that represent gray value change with respect to spatial distance (x, y)

u = 0, 1, 2 ..... M-1

v = 0, 1, 2 ..... N-1

F(u, v) = gray values after transformation

m = Line number = 0, 1, 2... M

n = Column number = 0, 1, 2... N

f (m, n) = Input gray values.

i = -1

The transformed frequency spectrum is complex number. It has both real and imaginary parts. The real part represents the phase and the imaginary part is amplitude of the system at spatial frequency (u, v).

Step 2

The above complex transformed image can be manipulated by either low pass or high pass filtering operation following the convolution filtering operation. But both window (operator) and the input image must be in the frequency domain. The

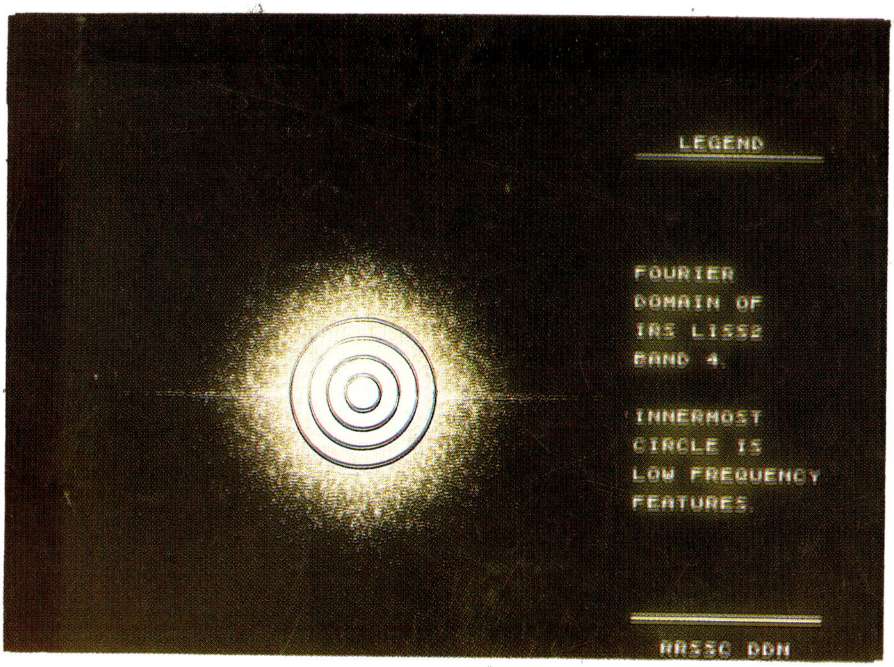

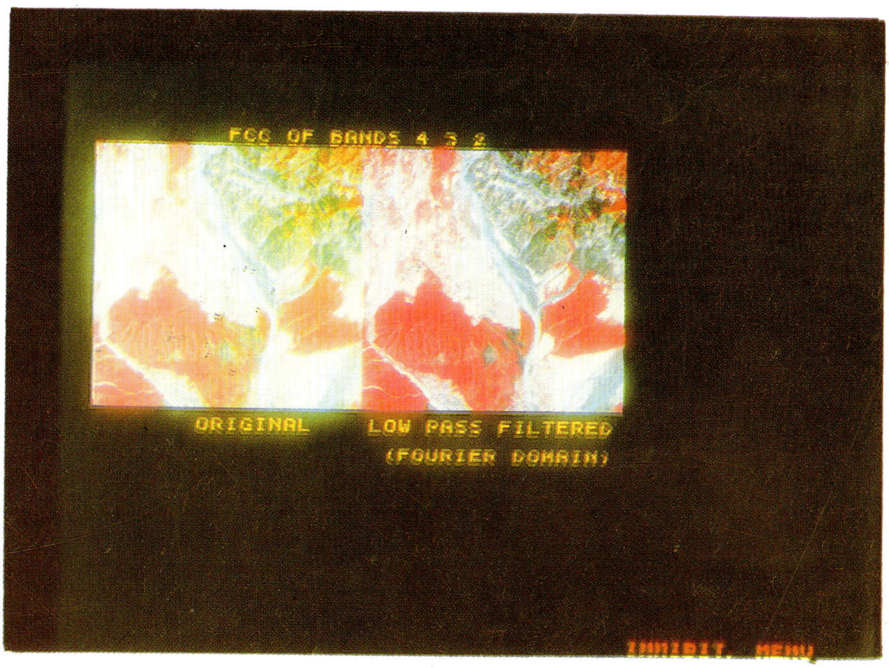

Plate 8.2: Some Examples of Filter Techniques

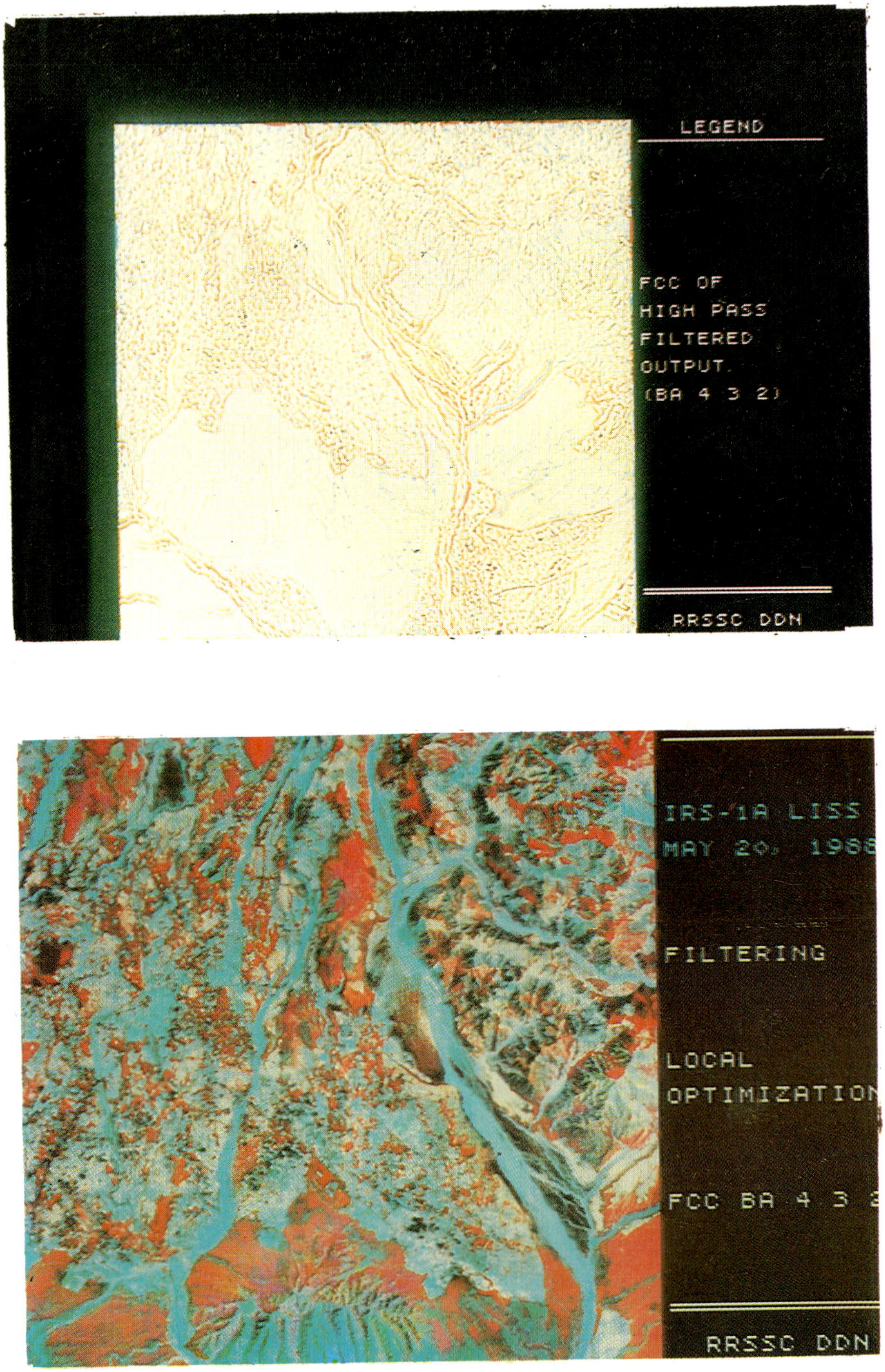

Plate 8.2: Some Examples of Filter Techniques

filtering must be chosen by the user.

The frequency spectrum of the frequency plane F(u,v) are the cluster of points around the centre of the image. The spectrum of the fourier frequency plane can be separated by a number of concentric circle. The inner circle have higher information and it represents low frequency details and the outer circle signifies for the high frequency detail. The superimposed circles which have the radii of 5, 11, 22, 36 and 48 encloses 90, 95, 98, 99.5 and 99.9 per cent information respectively. About 90 per cent of the information of the original scene can be explained by the smallest radius of 5 units (Gonzalez and Wintz, 1987).

A two dimensional low pass filter is one whose transfer function satisfies the following relations:--

$$H(u, v) = \begin{cases} 1 & \text{if } D(u, v) < D_0 \\ 0 & \text{if } D(u, v) > D_0 \end{cases} \qquad \ldots\ldots \quad (14)$$

Where D_0 is the cutoff frequency or radius of the circle of interest in frequency plane. This means beyond D_0, cutoff frequency or radius of all the frequencies will be attenuated.

Plate 8.2 describes the H(u, v) versus u and v. The ideal low pass filter signifies that all the frequencies inside the circle of radius D_0 are kept, while all the frequencies outside this circle are completely attenuated. The high pass filtering is just reserve of the low pass filtering. This means that all the frequencies outside the circle having radius D_0 will be preserved, but all the frequencies inside the circle will be completely attenuated (Hall, 1983; VIPS-32, 1986; Gonzalez and Wintz, 1987).

Plate 8.2 showed the outputs of low pass filtering and high pass filtering of band 4, IRS-1A LISS 2 sensor. The scene is part of Dehradun area.

Step 3

After the filtering operation is over, the complex filtered image in frequency domain should be converted back to spatial domain through the use of inverse Fourier transformation equation, given below:

$$Y(m,n) = \frac{1}{MN} \sum_{m=0}^{M-1} \sum_{n=0}^{N-1} F(u, v)\, e^{-2i\pi \left(\frac{mu}{M} + \frac{nv}{N}\right)} \qquad \ldots\ldots \quad (15)$$

This transformation is similar to forward transformation equation 13. Only difference in sign of the exponential part. This transformation results is real number. But the filter output is in 32 bit. So, it needs some normalization for display.

Normalisation or Range Compression

At times the dynamic range of the image data may be very large. The dynamic range of any transformation in the frequency domain is very large that only a few pixels are visible if displayed, because the output of forward and in Kirsch transformation will be 32 bit (Jain, 1987; VIPS-32, 1986). There is a need to rescale the output gray values for display. Gonazlez and Wintz, 1987 suggested useful processing technique that compensate the above difficulty. The display function is given below.

$$D\ (m, n) = \log\ [1 + Y\ (m, n)\ ] \qquad \ldots\ldots \quad (16)$$

Where F (m, n) = Inverse output of Fourier transformation.
D (m, n) = output gray values for display.

According to Andrew and Pratt (1970), the generalised logarithmic can be written as:

$$D\ (m, n) = k \ln \left\{ \frac{a + b\ [Y\ (m, n)]}{Y\ (m, n)} \right\} \qquad \ldots\ldots \quad (17)$$

Quantitatively, logarithmic transformation processing reduces the dynamic ranges within the transform domain and provides a stretching of the resultant dynamic range in the final image for display. Plate 8.2 showed the example of filtered output in Fourier domain.

Hadamard Transformation Filtering

Spatial filtering in Hadamard transform domain is very much similar to that in the Fourier domain. Only differences is transformation of two dimension discrete spatial image uses the Hadamard transformation for a subscene of M×N pixels is given below:

$$F(u,v) = \frac{1}{MN} \sum_{m=0}^{M-1} \sum_{n=0}^{N-1} F\ (m,n)\ (-1)^{P\ (m,\ n,\ u,\ v)} \qquad \ldots\ldots \quad (18)$$

$$\text{where } P^{(m,n,u,v)} = \sum_{i=0}^{M-1} m_i u_i + \sum_{j=0}^{N-1} n_j v_j$$

i.e. (bit by bit multiplication and summation)

where $M = 2^m$ and $N = 2^n$
$u = 0, 1, 2\ \ldots.\ M - 1$

v = 0, 1, 2 .... N - 1
mi = bits of binary conversion of m.
ni = bits of binary conversion of n
F(m, n) = Input gray value at the location (m, n)
F(u, v) = Transformed output in frequency domain.

Computationally it is simpler and advantageous over Fourier transformation. Conversion of frequency domain to spatial domain will be taken as follows:

$$F(m,n) = \sum_{m=0}^{M-1} \sum_{n=0}^{N-1} F(u,v)\,(-1)\,q^{(m,n,u,v)} \qquad \ldots\ldots \quad (19)$$

$$\text{where } q^{(m,n,u,v)} = \sum_{i=0}^{m-1} m_i u_i + \sum_{j=0}^{n-1} n_j v_j$$

All other steps are same as Fourier domain.

Conclusion

In this chapter, we attempted at emphasising various filtering techniques both in spatial domain and frequency domain. Here more emphasis was given to the concept with some examples, so that it can enable the reader better understanding and interpretation. This cover noise removal edge enhancement and edge extraction through spatial filtering. There is no thumb rule to mention that the part, particular spatial filtering, will serve all the purposes. The results of any kind of enhancement must generally be evaluated subjectively. It depends on the field of application and experience of the analyst.

REFERENCES

Galbiati, L.J. Jr. (1990), *Machine Vision and Digital Image Processing Fundamentals*. Prentice Hall, Englewood, New Jersey. Chapter 6, pp. 105-26.

Gonzalez. R.C. and Wintz, P. (1987), *Digital Image Processing*. Addison - Wesley Publishing Company. California, Chapter 4, pp. 140-205.

Graham (1962), Snow-Removal: A noise stripping process for Picture signals. *IRE Tran. Inform.* Theory IT-8 (I), pp. 129-44.

Hall, E.L. (1979), *Computer Image Processing and Recognition*. Academic Press. New York, Chap. 4, pp. 158-257.

Jain, A.K. (1989), *Fundamentals of Digital Image Processing*. Prentice Hall, Englewood Cliffs, New Jersey, Chapter 7, pp. 233-66.

Ketcham, D.J. (1976), "Real Time Image Enhancement Technique" Proceedings Spie/OSA Conference on Image Processing Pacific Grove, California, vol. 74, pp. 120-25.

Kirsch, R. (1971), Computer Determination of the consistent structure of Biological Images,

Computer and Biomedical Research, vol. 4(3), pp. 315-28.

Lillisand T.M. and Kiefer, R.W. (1987), *Remote Sensing and Image Interpretation*, John Wiley 4 Son. New York, Chap. 10, pp. 610-705.

Pratt, W.K. (1978), *Digital Image Processing*. John Wiley & Sons, New York, pp. 20I-344.

Robert, L.G. (1965), Machine Perception of three-Dimensional solids in J.T. Tippett *et al.* (eds), *Optical and Electro-optical Information Processing*, MIT Press, Cambridge, Mass, pp. 159-97.

Rosenfield A. (1969), *Picture Processing by Computer,* Academic Press, New York.

Sabins (1986), Remote Sensing Principle and Interpretation, W.H. Freeman and Company. New York, Chapter 7, pp. 235-78.

Sullivan, G. Trottier, L. Saitta B. and Iannuzzi, C. (1984), "Multibus image processor supports 512×512 displays, *Electronic Imaging*, October, pp. 42-47.

VIPS-32, (1986), Software Documents, SEP, France.

Wallis, R.H. (1976), An approach for the space variant restoration and enhancement of images, *Proceedings Symposium on Current Mathematical Problem in Image Science*. Monterey, California.

NINE

BAND COMBINATION

Introduction

SINGLE BAND remote sensing image may not be sufficient to extract desired information. Handling of multiple bands is also tedious and inconvenient. Multiple bands may be combined to generate one or more transformed combined bands following different mathematical operations. Multi band combination has the capability to enhance the feature of interest to the analyst. Band combination refers to some mathematical operations to be performed on two or more raw band to generate one or more outputs bands to achieve the desired results. Band combination includes addition, subtraction, ratioing, principal component analysis, intensity hue saturation and so on (Plates 9.1 & 9.2). In this chapter we will discuss different types of band combinations, their concept and advantages. The main aim of band combination is to increase information or highlight the feature of interest and to ease the interpretability of the image. For example, most of the systems can display only three bands in the form of false colour composite. Three bands may not always be sufficient to provide all the thematic classes. But if we take the principal component (1, 2, 3), they will provide more number of classes since the total variance of first three principal components is higher than that of any three sets of original bands.

Linear Combination

The multiband information can be added to form a single band image using addition because this has the potential to increase the thematic information than any of the original bands. Subtraction, on the other band may be used to eliminate the features with similar DN values in the bands and will highlight the differential information. The linear combination, one of the digital multiband transformations, generate output image by combining two or more bands following a linear operation (transformation) (eq. 1) to have the desired effect.

$$Y = a_0 + a_1 x_1 + a_2 x_2 + a_3 x_3 \ldots\ldots + a_n x_n \quad \ldots\ldots \quad (1)$$

where $x_0 + x_1 \ldots x_n$ are the gray values of a pixel in different bands, Y is the output gray value and $a_0, a_1, a_2 \ldots a_n$ are the coefficients. These coefficients are to be given by the users, based on their knowledge and experience. However, the method of determination of these coefficients was described by Jackson (1983). Huete *et al.* (1984) and Ezra *et al.* (1984) provided additional insight into the importance of calculating the coefficients on a site specific basis, if possible, rather than using global coefficients.

The most important linear combination, widely used by the remote sensing community is the "Tassel Cap Transformation" (Kauth and Thomas, 1976; Crist and Cicone, 1984). It is called "Tassel Cap Transformation" because of its cap shape (Figs. 9.1 to 9.3).

Based on the linear combination, they identified four new indices namely the Soil Brightness Index (SBI), the Green Vegetation Index (GVI), the Yellow Vegetation Index (YVI), and the Non-such Index (NSI). The coefficients of each combination index given below are taken from Kauth *et al.* (1979) and Thomson and Wehmann (1980).

$$SBI = 0.332\ MSS4 + 0.603\ MSS5 + 0.675\ MSS6 + 0.262\ MSS7 \quad \ldots.. \quad (2)$$

$$GVI = -0.283\ MSS4 - 0.660\ MSS5 + 0.577\ MSS6 + 0.388\ MSS7 \quad \ldots. \quad (3)$$

$$YVI = -0.899\ MSS4 + 0.428\ MSS5 + 0.076\ MSS6 - 0.041\ MSS7 \quad \ldots.. \quad (4)$$

$$NSI = -0.016\ MSS4 + 0.131\ MSS5 - 0.452\ MSS6 + 0.882\ MSS7 \quad \ldots.. \quad (5)$$

Each Index image is created by multiplying each of the four Landsat MSS brightness value (DN) for each pixel by the corresponding coefficients as shown in equations 2-5.

Crist and Cicone (1984) suggested the TM Tasseled Cap transformation coefficients as given below:

$$SBI = 0.33183\ TM1 + 0.33121\ TM2 + 0.55177\ TM3 + 0.42514\ TM4 + 0.48087\ TM5 + 0.25252\ TM7 \quad \ldots.. \quad (6)$$

$$GVI = 0.24717\ TM1 - 0.16263\ TM2 - 0.040639\ TM3 + 0.85468\ TM4 + 0.05493\ TM5 - 0.11749\ TM7 \quad \ldots.. \quad (7)$$

$$\text{Wetness Index or third index} = 0.13929\ TM1 + 0.22490\ TM2 + 0.40359\ TM3 + 0.25178\ TM4 - 0.70133\ TM5 - 0.45732\ TM7 \quad \ldots.. \quad (8)$$

Sharma *et al.* (1990) generated the coefficients for Green Vegetation Index and Soil Brightness Index for Indian Remote Sensing Satellite. They are presented in equations 9-10.

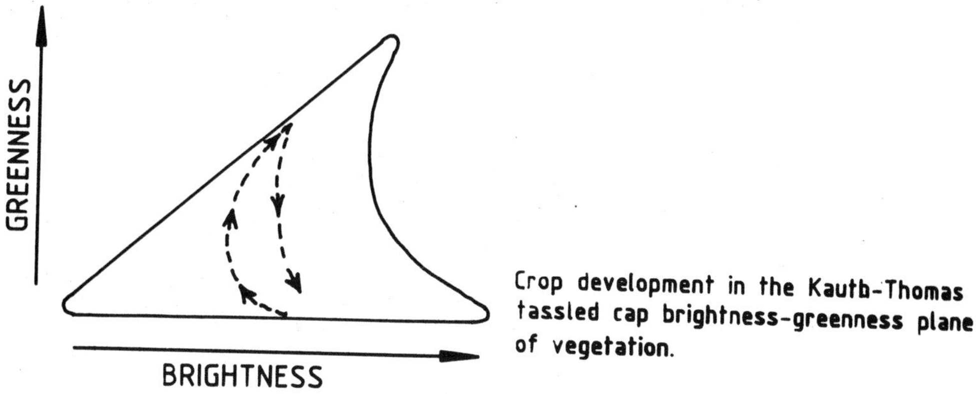

Fig. 9.1(a): Crop development in the Kauth-Thomas tasseled cap brightness-greenness plane of vegetation

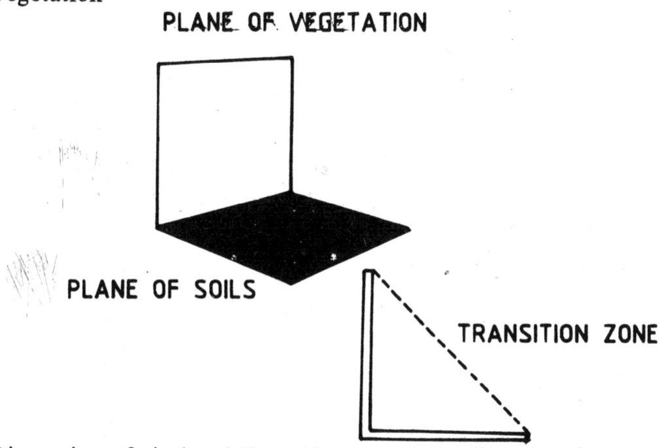

Fig. 9.1(b): Dispersion of six-band thematic mapper data based on the use of the Kauth-Thomas transformation

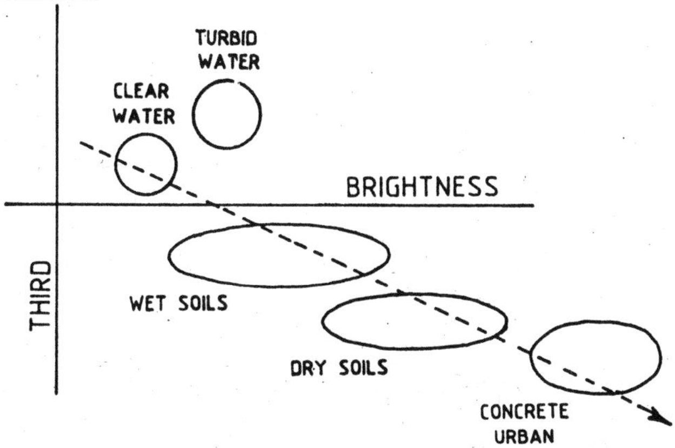

Fig. 9.1(c): Approximate direction of moisture variation in the plane of soils. The arrow points in the direction of less moisture

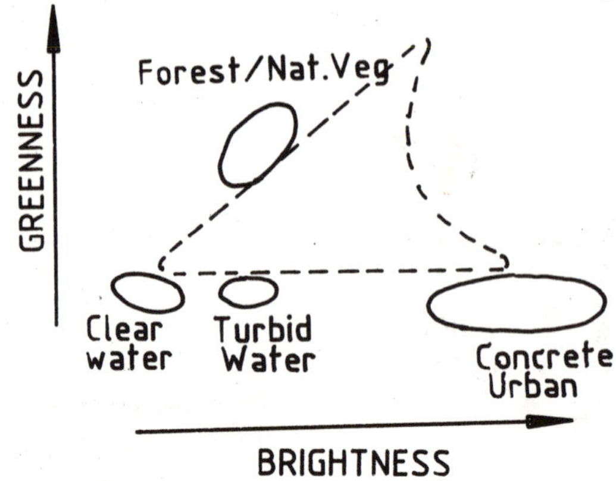

Fig. 9.2: Location of various types of land cover when plotted in the brightness-greenness, spectral space. Brightness is highly correlated with bare soil while greenness is highly correlated with leaf area index, per cent canopy closure and/or biomass

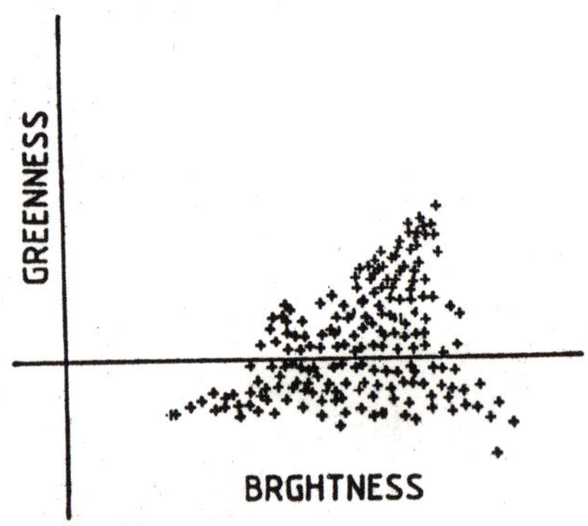

Fig. 9.3: Actual plot of brightness and greenness values for an agricultural area

$$SBI = 0.4233\ B1 - 0.4328\ B2 + 0.6490\ B3 + 0.4607\ B4 \quad ..... \quad (9)$$

$$GVI = 0.1072\ B1 - 0.1219\ B2 - 0.4341\ B3 + 0.8861\ B4 \quad ..... \quad (10)$$

where B1, B2, B3 and B4 are digital number in four bands of IRS-1A LISS-II data, respectively (Plate 9.1).

The Tasseled Cap Transformation captures the greatest amount of data variability in the fewest number of linearly combined output bands (features) and enhances the interpretability of data set with respect to actual scene characteristics. Kauth and Thomas (1976) showed that for MSS data most of the scene variability in agricultural scene can be explained by two dimensions defined by Soil Brightness and Green Vegetation Index. In TM data the scene characteristics can be captured in 3-dimensions namely Brightness, Greenness and Wetness (Crist and Cicone, 1984; Crist and Kauth, 1986). Generally, SBI and GVI together contains most of the scene information. Kauth and Thomas (1976) found that nearly 98 per cent of

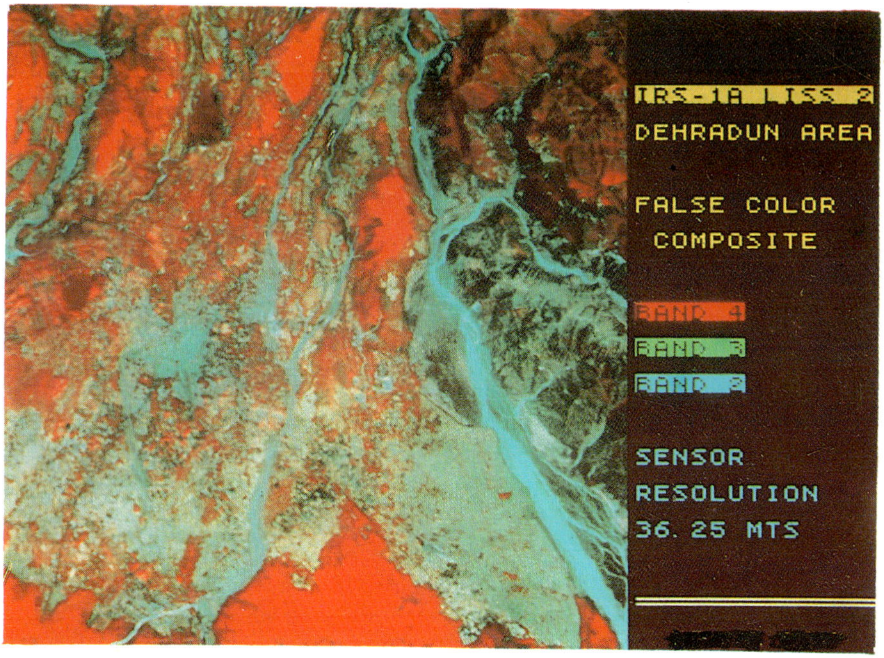

Plate 9.1: Some Examples of Band Combination

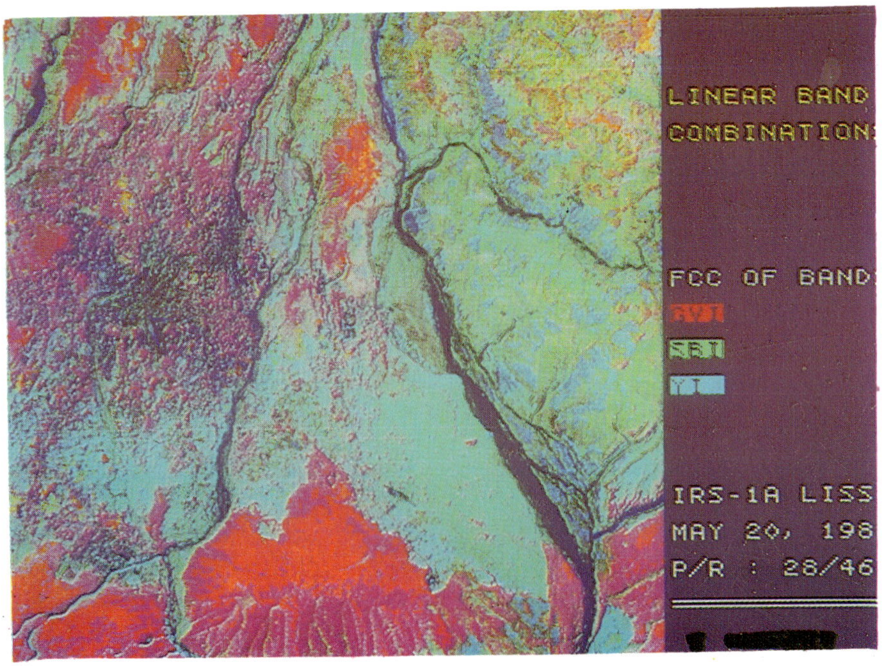

Plate 9.1: Some Examples of Band Combination

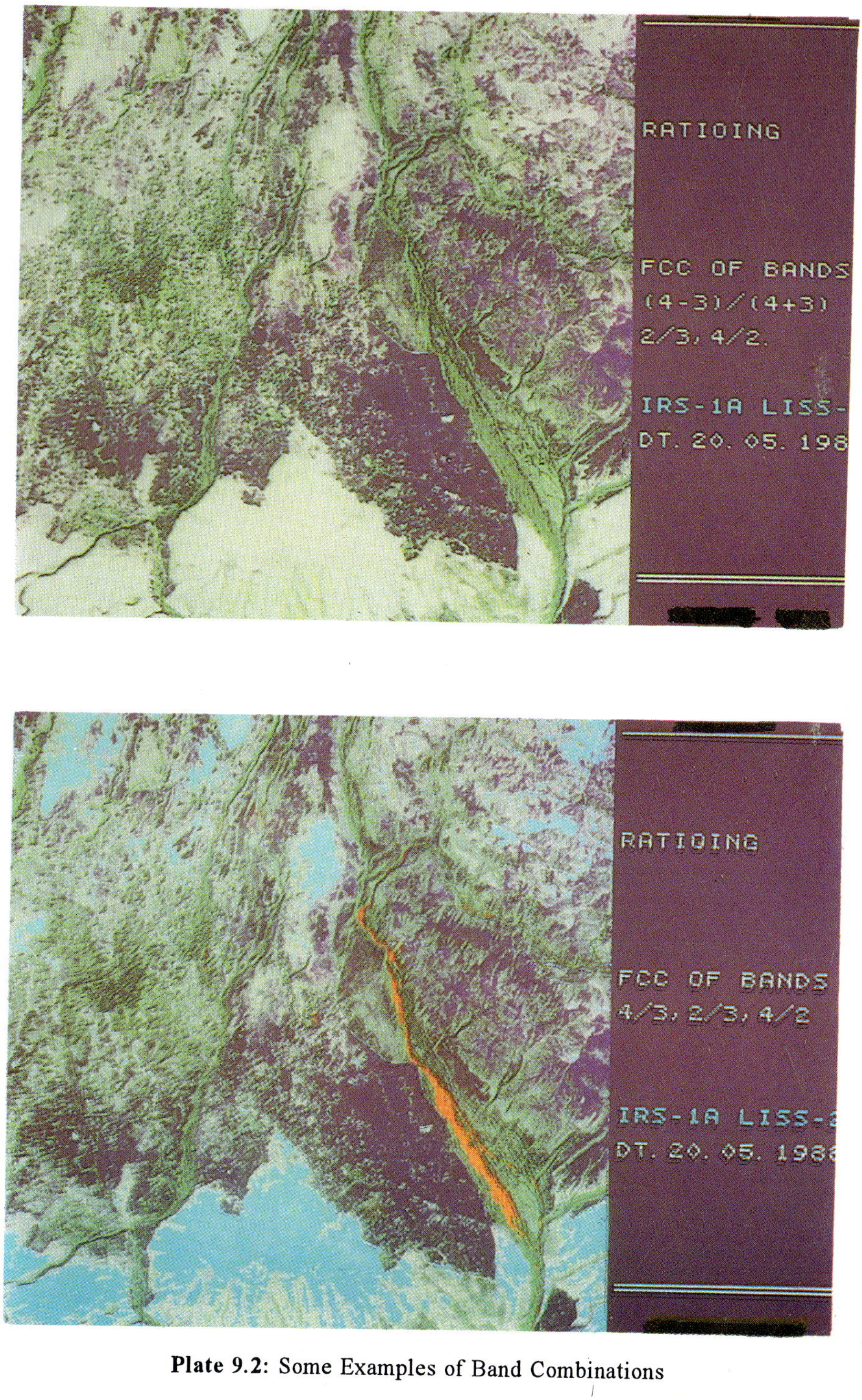

Plate 9.2: Some Examples of Band Combinations

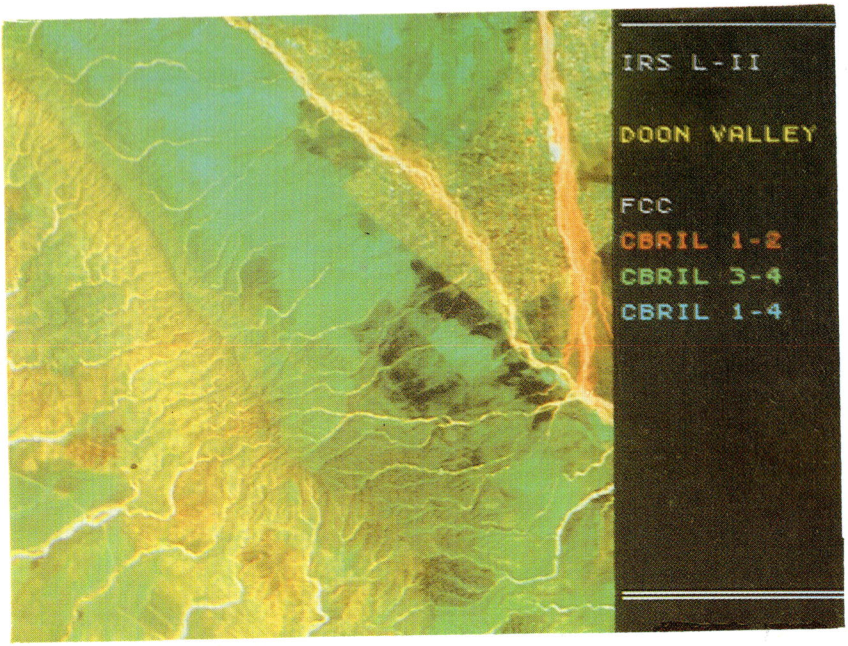

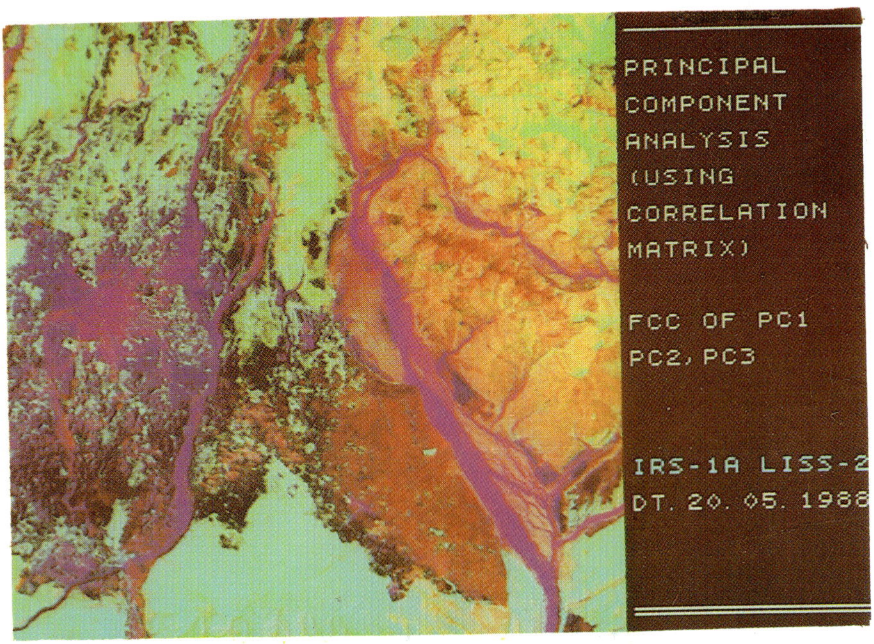

Plate 9.2: Some Examples of Band Combination

the variance in bare soil spectra from different soils could be explained by SBI. Crist and Cicone (1984) claimed that fully vegetated areas defined the "Plane of Vegetation" while bare soils data fall in a place of soils. But between the two planes, there are data with partially vegetative cover plots, where both soil and vegetation are visible. The third component (wetness index) provides subtle information concerning the moisture status of the soil and waterlogged and marshy land areas. The Tasseled Cap Trans-formation has been popularly used to a variety of applications, estimating moisture-stress detection (Thompson and Wehmanen, 1980; Bhat *et al.,* 1991), crop condition assessment (Ajai *et al.*, 1988), yield prediction (Barnett and Thompson 1982), identification of wastelands (Kudrat *et al.*, 1989; Saha and Kudrat, 1988). The linear combination of bands is very useful in feature selection during pattern recognition (Swains and Davis, 1978).

Ratioing of Bands

Ratioing, a combination of bands, generates a ratio image in which the DN value at any pixel represents the quotient of DN values of two raw bands or two combined bands (Plate 9.2).

$$Y_{ij} = \frac{X_{ijk}}{X_{ijl}} \qquad \ldots\ldots \quad (11)$$

where Y_{ij} = Output DN value of a pixel of the location (i, j)
X_{ijk} = DN of the input pixel at the same location for band 'k'
X_{ijl} = DN value for band '1' at the pixel location (i, j)

$$Y_{ij} = \frac{X_{ij}IR - X_{ij}R}{X_{ij}IR + X_{ij}R} \qquad \ldots\ldots \quad (12)$$

Here, numerator is the difference between DN values in the infrared and red channels and the denominator is the sum of the infrared and red bands. Generally, this called differencing and ratioing. Unfortunately, during computation of ratio, difficulties may arise, when the denominator in equation (11) is zero. However, it can be overcomed by adding some constant (say 1 or 0.1) in the denominator. The above problem can be tackled in an alternative way as well. Hord (1982) suggested that ratio of remote sensing data can also be computed by the following algorithm:

$$Y_{ij} = \tan^{-1} \frac{X_{ijk}}{X_{ijl}} \qquad \ldots\ldots \quad (13)$$

The ratio image will suppress detail common to the two images and will enhance details which are different. Ratio images are free from the effect of illumination across a scene (Fig. 9.4). It also removes the atmospheric effect to some extent. The possible number of ratios that can be developed from 'n' bands of data is n (n-1). Colour composite can be prepared using any three ratio images or using a combination of two ratio images and any individual band. Normally, for ratioing two uncorrelated bands are preferable (Jenson 1986). A number of

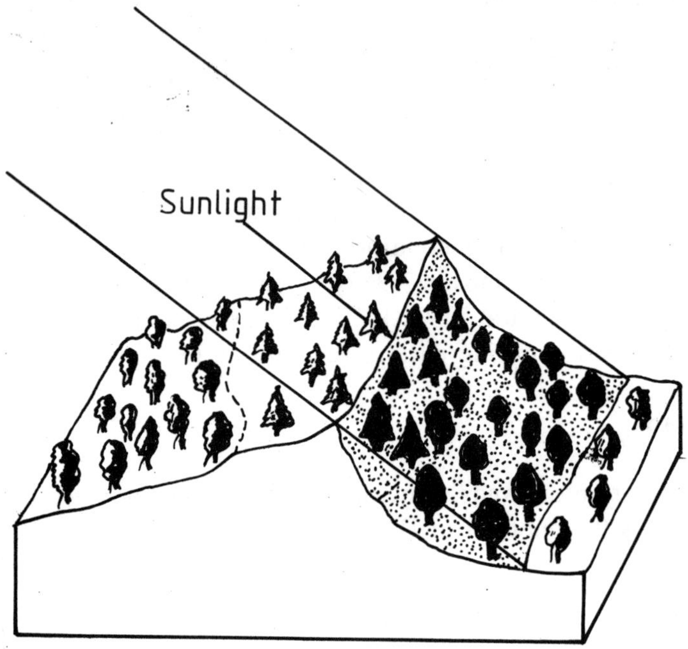

Land cover/illumination	Band A	Band B	Ratio (Band A/Band B)
	Digital Number		
Deciduous			
Sunlit	48	50	0.90
Shadow	18	19	0.95
Coniferous			
Sunlit	31	45	0.69
Shadow	11	16	0.69

Fig. 9.4: Reduction of Scene Illumination Effects through Spectral Rotation

Vegetation Index (VI) have been generated, which are of immense use in agricultural yield modelling (Burnett and Thompson 1982), Crop Growth Stage Monitoring, Temporal Change Detection (Colwell and Weber, 1981; Tiwari 1992), Crop Stress Identification (Thompson and Wehmanen, 1980; Sharma and Navalkund, 1989). Some of the vegetation index common in remote sensing are given below:

(i) Vegetation Index (VI) = $\frac{IR}{R}$ ..... (14)

(ii) Normalised Vegetation Index (NDVI) = $\frac{IR - R}{IR + R}$ ..... (15)

(iii) Transformed Vegetation Index (TVI) = $\sqrt{NDVI + 0.5}$ ..... (16)

The ratio images of IRS-1A LISS -II are given in Plates (3/4) below. Jenson advocated that the ratio of TM bands 2 and 5 is useful for identifying all water bodies and this will provide some subtle wetland information.

Brightness or Square Root of Sum of Squares

In this operation, the brightness index is being calculated for two raw or combined bands. The brightness index is nothing but the square root of sum of the square of two bands, as shown by the following equation:

$$Y_{ij} = \sqrt{(x_{ijl})^2 + (x_{ijk})^2} \quad \ldots\ldots \quad (17)$$

where Y_{ij} = Output pixel value

X_{ijl} = Digital number of pixel at the location (i, j) for band 1, 1 = 1, 2, ... n

X_{ijk} = Digital number of the same pixel at the same location for band k, k = 1, 2... n

In this operation, the user or analyst has to specify two bands. Generally, the bands which are highly correlated should be chosen. In this option, information of the raw bands can be brought to a single band and sometimes it can reduce the number of bands for feature extractions (Plate 9.2).

Post Normalisation

In most of the above operations, the output digital number may exceed 255, or it can be less than zero. But mostly the image processing system can display the pixel which has the DN value ranging from 0-255. Thus, it needs some kind of normalisation for bringing the dynamic range of digital number within the range of 0 to 255. The most commonly used post normalisation techniques are Minmax and Gaussian. The post normalisation is conceptually very much similar to contrast enhancement. Thus, to avoid repeatability the reader is advised to refer contrast enhancement in chapter 7.

Principal Component Analysis

Multispectral remote sensing data are used usually multidimentional in the feature space. This means MSS data have four bands; TM have seven bands, and IRS data also have four bands. These bands have high interband correlations. The available data processing system have the display unit, which can display three bands at a time in the form of false colour composite (FCC). FCC is the most effective means of visual presentation of three bands by following the additive colour principle. But the problem arises, when the analyst is interested to analyse more than three bands (seven bands of Landsat TM data). The problem becomes more severe with the multiplicity of channels (e.g. HIRIS or narrow band imaging sensors data). Principal component analysis may be used to compress the information content (variance) of a number of bands of imagery into just two or three transformed principal component images. This ability to reduce the dimensionality (i.e. the number of bands in the data set that must be analysed to produce desired results) from seven TM bands to two or three principal component analysis image is an important economic consideration especially if the potential information recoverable from the transformed data is just as good as that from the original remote sensor data (Anuta, 1977; Haralick and Fu, 1983). The principal component transformed images are often more interpretable than the original images (Kaneko 1978; Byrne *et al.*, 1980; Canas and Barnett 1985). The principal component analysis, a statistical dimension reduction transformation, converts the highly inter-band correlated data to decorrelated data.

The principal component analysis, known as Karhunen-Loeve (K-L) transformation in remote sensing and pattern recognition literature, is often used for determining the underlying statistical dimensionality of the image data set (Ready and Wintz 1973) for image enhancement (Gillespie 1980), for digital change detection (Byrne *et al.*, 1980) and for characterising seasonal changes in cover types (Townshend *et al.*, 1984).

The principal component analysis comprises the following stages.

(1) First of all, the variance-covariance matrix for all the bands of the images data is to be computed by the following formula:

$$\text{Cov}(x\,y) = \frac{(x_i - \bar{x})(y_i - \bar{y})}{N} \quad \ldots\ldots \quad (18)$$

where Cov (x, y) = covariance between (x, y)
x_i = gray values of any one band
$\bar{x}$ = mean gray value (x) of the same band
y_i = gray value of other band
$\bar{y}$ = mean gray value (y) of the same band
N = total number of pixels.

(2) The analyst is to compute the *eigen* values and *eigen* vectors of the variance-covariance matrix. The eigen vectors define the direction of each principal component (PC) and the eigen values measure the variances of the each principal component images (or length of each principal axis).

(3) As soon as the eigen vectors is known, final principal component images will be generated following the above equation (18).

Mathematics of Principal Components

If $X^T = [X_1, X_2, \ldots X_N)$ is an N-dimensional random variable (here digital numbers of different bands) with mean vector M and variance-covariance matrix 'C', then a new set of variables, say $Y_1, Y_2, \ldots Y_N$ known as Principal Components, can be expressed by:

$$Y_j = a_{1j} x_1 + a_{2j} X_2 + \ldots + a_{Nj} X_N \quad \ldots\ldots \quad (19)$$

$$= a_j^T X \quad \ldots\ldots \quad (20)$$

Where 'T' denotes transpose of a matrix and $a_j^T = [a_{1j}, a_{2j} \ldots a_{Nj})$ are the normalised eigen vectors [i.e. $a_j^T a_j = 1$] of the variance-covariance matrix. By denoting the (N × N) matrix of eigen vectors by A and the (N × 1) vector of principal components Y, we can write:

$$Y = A^T X \quad \ldots\ldots \quad (21)$$

Let us assume the (N × N) covariance matrix 'C' as follows:

$$C = \begin{bmatrix} P_{11} & P_{12} & -- & -- & -- & -- & P_{1N} \\ P_{21} & P_{22} & -- & -- & -- & -- & P_{2N} \\ -- & -- & -- & -- & -- & -- & -- \\ -- & -- & -- & -- & -- & -- & -- \\ -- & -- & -- & -- & -- & -- & -- \\ -- & -- & -- & -- & -- & -- & -- \\ P_{N1} & P_{N2} & -- & -- & -- & -- & P_{NN} \end{bmatrix} \quad \ldots\ldots \quad (22)$$

where P_{ij} = Covariance between the bands i and j (variables).

If λ_1 are the eigen values of matrix C, the matrix must be diagonal as the component have been chosen to be uncorrelated and $\lambda_1 > \lambda_2 > \ldots\ldots\ldots\ldots > \lambda_N$. But is typical characteristics.

$$\sum_{i=1}^{N} \lambda_i = \sum \sigma_1^2$$

The matrix 'C' can be diagonalised as equation of matrix

$$(C-\lambda_1)\ A^T = 0 \quad \ldots\ldots \quad (23)$$

Whereas, λ_i is the eigen value, I is the identity matrix. A^T are the eigen vectors. As per algebra, λ is the root of the characteristic equation (23) the determinant of $|C - \lambda_1| = 0$

Therefore

$$\begin{vmatrix} P_{11}-\lambda & P_{12} & -- & -- & -- & -- & P_{1N} \\ P_{21} & P_{22}-\lambda & -- & -- & -- & -- & P_{2N} \\ -- & -- & -- & -- & -- & -- & -- \\ -- & -- & -- & -- & -- & -- & -- \\ -- & -- & -- & -- & -- & -- & -- \\ -- & -- & -- & -- & -- & -- & -- \\ P_{N1} & P_{N2} & -- & -- & -- & -- & P_{NN}-\lambda \end{vmatrix} = 0 \quad \ldots\ldots (24)$$

The equation (24) is the polynomial of λ. This polynomial is the determinant of the "characteristic matrix" of 'C' which is zero because 'C' is symmetric i.e. $(P_{ij} = P_{ji})$.

If we consider a two dimensional case where N = 2, equation (24) becomes,

$$\begin{vmatrix} P_{11}-\lambda & P_{12} \\ P_{21} & P_{22}-\lambda \end{vmatrix} = 0 \quad \ldots\ldots (25)$$

or

$$(P_{11} - \lambda)(P_{22} - \lambda) - P_{21} \cdot P_{12} = 0$$

or

$$\lambda^2 - \lambda (P_{12} + P_{22}) + P_{11} \cdot P_{22} - P_{12} \cdot P_{21} = 0 \quad \ldots\ldots (26)$$

The equation (26) is a quadratic equation of λ and eigen values (λ_i) can be determined by solving the equation (26). With the knowledge of eigen values (λ_i) and putting the values in equation (24) and by solving, we will get eigen vectors (A^T) of the form:

$$A^T C A = I\lambda \quad \ldots\ldots (27)$$

$$A = \begin{vmatrix} a_1 \\ a_2 \end{vmatrix}$$

Therefore

$$\begin{bmatrix} P_{11} - \lambda_i & P_{12} \\ P_{21} & P_{22} - \lambda_i \end{bmatrix} \begin{bmatrix} a_1 \\ a_2 \end{bmatrix} = 0$$

By algebra,

$$(P_{11} - \lambda_1)\ a_1 + P_{12}\ \bullet\ a_2 = 0 \qquad \ldots\ldots \quad (28)$$

$$p_{21}\ \bullet\ a_1 + (P_{22} - \lambda_1)\ a_2 = 0 \qquad \ldots\ldots \quad (29)$$

By solving simultaneous equations (28) and (29), we can find out eigen vectors, i.e. a_1 and a_2. Similarly, for each eigen value, different sets of eigen values or coefficients (A^T) can be determined. These eigen vectors are nothing but the coefficients of equation (19).

Alternative Method for Determination of Eigen Vectors

We should note that there will be one vector solution 'v1' corresponding to each eigen value 'λi' of matrix 'C'. The vector is called the characteristic vector or eigen vector of 'C' associated with the characteristic as follows.

Step 1: For each eigen value 'λ_i', it is to write out the matrix 'C' by subtracting λ_i from each diagonal elements of variance-covariance matrix.

Step 2: It is to compute the adjoint adj(C) of the matrix 'C' in accordance with the step indicated by Tatsuoka (1971). It should be noted that the columns of adj(C) are all proportional to one another.

Step 3: The elements of any column of adj(C) is to be divided by square root of the sum of the squares of these elements. The resulting numbers are the elements of eigen vector (vi), with vi'v = I.

The principal component transformation has important characteristic very much relevant to remote sensing, as mentioned by Singh and Harrison 1985:

1. The eigen value is the measure of percentage of total variance in each principal component image. But the total variance is preserved in the transformation i.e.

$$N_{\substack{Pii \\ i=1}} = N_{\substack{\lambda i \\ i=1}} \qquad \ldots\ldots \quad (30)$$

   Where p are the variance of each scene or original variables.
2. It minimizes the mean square approximation errors.
3. This is only transformation that generates uncorrelated coefficients. In a geometrical sense it rotates the highly correlated features (bands) in N-dimensions to a more favourable orientation in the feature space, orthogonal to each other, such that the maximum amount of variance is accounted for in decreasing magnitude along the ordered components. This transformation process comprises the information content (variance) into a smaller number

of components from the large number of features (bands) by discarding redundant information into higher-order components. Table 9.1(a,b,c) shows eigen values and eigen vectors of correlation matrix and F are different PC products.

Geometrical concept of principal component analysis may be explained by considering a two bând data set. Two band data are plotted on a scatter diagram (Fig. 9.5) according to the digital number as originally recorded in band 1 and band 2. The principal component analysis defines a new axix (PC1) oriented new axis (PC2) perpendicular to the former axis. Band 1 and band 2 are the pixel co-ordinates system. From the figure it is interesting to note that the dynamic range of digital number for PC1 is greater than that for the either of the bands and the range of digital number for the second principal component (PC2) is relatively smaller (Lillesand and Keifer 1987; Sabin 1986). Geometrically it signifies that the principal component analysis [Table 9.2 (a,b,c)] transformation is nothing but the change or origin (mean of the distributions) of original co-ordinate system and rotation of axis to achieve maximum uncorrelation between the principal component (PC1 and PC2) in the new co-ordinate system.

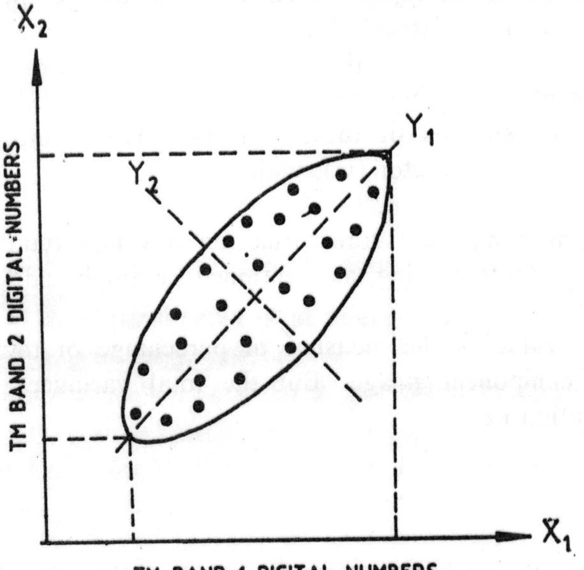

Fig. 9.5: Scattered plot of Landsat TM band 1 and band 2

TABLE 9.1(a): Variance-Co-variance Matrix

47.62674	34.23045	54.51388	-4.992673
34.23045	26.37058	40.22445	1.510928
54.51388	40.22445	72.91129	-8.655549
-4.992673	01.510928	-8.65555	39.637760

TABLE 9.1(b): Eigen Values Computed from Co-variance Matrix

	Components			
	1	2	3	4
Eigen Values \i	141.3080	39.84040	4.748404	0.6495916
Difference (\i+1 - \i)	101.4676	35.09199	4.098812	--
Total variance (%)	75.75	21.36	2.55	0.35

TABLE 9.1(c): Eigen Vectors (aj) Computed for the Co-variance Matrix

	1	2	3	4
1	-0.5672077	-0.4151276	-0.7065744	0.0837691
2	-0.0463428	-0.1661848	0.0207783	-0.9847859
3	0.6324961	0.3119414	-0.7022740	-0.0972228
4	0.5254295	-0.8382981	0.0844533	0.1185205

TABLE 9.2(a): Eigen Values Computed from Correlation Matrix

1.000000	0.9658892	0.9250907	-0:11499088
0.9658892	1.0000000	0.9173452	0.04633600
0.9250907	0.9173452	1.0000000	-0.16100620
-0.1149088	0.0467336	-0.1610062	1.0000000

TABLE 9.2(b): Eigen Values Computed from Correlation Matrix

	Components			
	1	2	3	4
Eigen Values \i	2.881632	1.015923	0.08267791	0.01976594
Difference (\i+1-\i)	1.865709	0.933245	0.06291197	--
Total Variance (%)	72.04081	25.39807	2.066948	0.4941486

TABLE 9.2(c): Eigen Vectors (ai) Computed for the Correlation Matrix

Components	Bands			
	1	2	3	4
1	-0.5806952	-0.5751593	-0.5718988	0.0701130
2	-0.0467487	-0.1671778	5.2177731	-0.9845338
3	0.4910593	0.3000124	-0.8121429	-0.0963169
4	0.6493319	-0.7424522	0.1031137	0.1284523

REFERENCES

Ajai, Bhatt, H.P. and Sharma, S.A. (1988), Monitoring drought impact on crop condition by remote sensing. Presented at Workshop on Agricultural Drought, NRSA, Hyderabad.

Anuta, P.E. (1977), Computer assisted analysis techniques for remote sensing data interpretation. *Geophysics*, vol. 42, pp. 468-81.

Badhwar, C.D. and Henderson, K.E. (1981), Estimating development stages from spectral data—An initial model. *Agron. Journal*, vol. 13, pp. 748-55.

Barnett, T.L. and Thomson, D.R. (1982), The use of large area spectral data in wheat yield estimation, *Remote Sensing of Environment*, vol. 12, pp. 509-18.

Bhatt, H.P., Sharma, S.A. and Ajai (1987), An attempt to detect moisture stress in Groundnut crop using Landset MSS data, Scientific Report No. IRS-UP/SAC/SN/07/87, Space Applications Centre, Ahmedabad, pp. 1-24.

Byrne, G.E., Crapper, P.F. and Mayo (1980), Monitoring land cover change by principal component analysis of multi-temporal Landsat data, *Remote Sensing of Environment*, vol. 10, pp. 174-84.

Canas, A.A.D. and Barnett, M.E. (1985), The generation and interpretation of false colour composite principal component images. *International Journal of Remote Sensing*, vol. 6, pp. 867-81.

Colwell, J.E. and Weber, F.P. (1981), Forest change detection. *Proc. 15th International Symposium on Remote Sensing of Environment*, An Arbor, MI May 11-15, pp. 839-852.

Crist, E.P. and Cicone, R.C. (1984), Application of the Tasseled cap concept to simulated Thematic Mapper Data, *Photogrammetric Engineering and Remote Sensing*, vol. 50, pp. 343-52.

Crist, E.P. and Kauth, R.J. (1986), The tasseled cap de-mystified, *Photogrammetric Engineering and Remote Sensing*, vol. 52, pp. 81-6.

Ezra, C.E., Tinney, L.R. and Jackson, R.D. (1984), Considerations for implementing vegetation indices for Agricultural Applications, Technical papers, *50th Annual Meeting of the American Society of Photogrammetry*, pp. 526-36.

Gillispie, A.R. (1980), Digital Techniques of image enhancement, in Siegal, B.S. and Gillispie, A.R. (eds.) *Remote Sensing in Geology*, John Wiley, New York, pp. 139-226.

Haralick, R.K. and Fu, K.S. (1983), Pattern recognition and classification. Chap. 18 in Colwell, R.N. (ed.) *Manual of Remote Sensing. The American Society of Photogrammetry and Remote Sensing*, Falls Church, vol. 1, pp. 793-805.

Hord, R.M. (1982), *Digital Image Processing of Remotely Sensed Data*, New York Academic, Press Inc.

Huete, A.R.; Post, D.F. and Jackson (1984), Soil spectral effects on 4 space vegetation discrimination, *Remote Sensing of Environment*, vol. 14, pp. 155-65.

Jackson, R.D. (1983), Spectral indices in N-space. *Remote Sensing of Environment*, vol. 13, pp. 409-21.

Jenson, J.R. (1986), *Introductory Digital Image Processing: A Remote Sensing Perspective*, Chapter 7. Prentice Hall, Englewood Cliffs, New Jersy, pp. 117-76.

Lillesands, T.M. and Kiefer, R.W. (eds.)(1987), *Remote Sensing and Image Interpretation*, John Wiley and Sons, New York.

Kaneko, T. (1978), Colour composite pictures from principal axis components of multispectral scanner data. *IBM Journal of Research and Development*, vol. 22, pp. 386-92.

Kauth, R.J. and Thomas, G.S. (1976), The tasseled cap--A graphic description of the spectral--Temporal development of agricultural crops as seen by Landsat. *Proceeding Symposium on Machine Processing of Remotely Sensed Data*, West Lafayette, Ind. Laboratory for the application of Remote Sensing, pp. 41-51.

Kauth, R.J.; Lambeen, P.F.; Richardson, W.; Thomas, G.S. and Pentland (1979), Feature extraction applied to agricultural crops as seen by Landsat. *Proceedings Technical Session, Lacie Symposium*, Houston, NASA, pp. 705-721.

Kudrat, M.; Nag, P. and Bhan, S.K. (1989), Identification of Wastelands in Sultanpur district, Uttar Pradesh, *Geographical Review of India*, vol. 50, pp. 71-8.

Ready, P.J. and Wintz, P. (1973), Information extraction, SNR improvement and data compression in multispectral imagery. *I.E.E.E. Trans. Commu.*, vol. 21, p. 1123.

Sabin, Floyd, F. Jr. (1986), *Remote Sensing Principle and Interpretation*, W.H. Freeman, San Francisco.

Saha, S.K.; Kudrat, M. and Bhan, S.K. (1988), Evaluation of digitally enhanced Landsat TM data for wasteland mapping in part of Aligarh district, U.P., *Proc. National Symposium on Remote Sensing in Rural Development*, pp. 333-42.

Sharma, S.A. and Navalkund, R.R. (1989), Estimation of growth stages of wheat from spectral data. *Jour. Indian Society of Remote Sensing*, vol. 17, pp. 1-6.

Sharma, S.A.; Bhatt, H.P. and Ajai (1990), Generation of brightness and green transformation for IRS-LISS-II Data, *Jour. Indian Society of Remote Sensing*, vol. 18, No. 3, pp. 25-31.

Singh, A. and Harrison, A. (1985), Standard principal components, *International Journal of Remote Sensing*, vol. 6, pp. 883-96.

Swain, P.H. and Davis, S.M. (1978), *Remote Sensing: Quantitative Approach*. Chap. 3, Mc Graw-Hill Book Company, New York, pp. 136-87.

Tiwari, A.K. (1992), Role of aerial photography and satellite remote sensing in thematic cartography: A case of forest mapping, in P. Nag (ed), *Thematic Cartography and Remote Sensing*, Concept Publishing Company, New Delhi, pp. 137-47.

Thompson, D.R. and Wehmanen, O.A. (1980), Using Landsat digital data to detect moisture stress in corn--soybean growing regions, *Photogrammetric Engineering and Remote Sensing*, vol. 46, pp. 1082-9.

Totsuoka, M.M. (1971), *Multivariate Analysis: Techniques for Educational and Psychological Research*, John Wiley & Sons, Inc., New York.

Townshend, J.R.G.; Goff, T.E. and Tucker, C.J. (1984), Multi-temporal dimensionality of images of the normalised difference vegetation index at the continental scale. *I.E.E.E. Trans. Geoscience and Remote Sensing*, vol. 22.

TEN

CLASSIFICATION TECHNIQUES

ONE OF THE most common image processing techniques for the application of remote sensing data has been the classification technique. The different bands of the multi-spectral data can be combined in different ways for generating a new set of derived data which is able to highlight additional features. Such techniques vary from a simple division of the whole range of DN values of a scene to most sophisticated statistical calibrations.

Graphical Presentation of Pattern Recognition

Different band classification techniques are meant to have better approaches for pattern recognition. In order to understand the graphical presentation of patterns on the earth surface, a very simple example may be taken. Let us assume that there are three types of cover classes, such as vegetation, soil and water. Each of them has spectral response (Fig. 10.1) which do not vary with season, atmospheric haze and sun angle etc. (Haralick & Fu, 1983). Further, it may also be assumed that two bands measurements have been taken for plotting them in measurement space (Fig. 10.2). Since, as mentioned earlier, they obviously form clusters which are separated by each other. It is possible to associate the above localized and distinct clusters or regions with specific ground cover classes. These ground cover classes should be able to discriminate otherwise it cannot be separable.

The main objective of designing a classifier is to (a) enable to take decision regarding regions each of which corresponding to a specific or thematic class, and (b) then construction of the classifier so that it will identify any pixel belonging to the class corresponding to the decision region to which it falls (Swain & Davis, 1978). Determination of a decision region is an intricate task. However, the decision region corresponding to each class can be determined by constructing a line passing through the points for equal observed discriminant functions because in practice two such functions can have equal values only at points on the boundaries between decision regions (Anderson, *et al*., 1973; Haralick & Fu, 1983; & Swain & Davis, 1978). Thus once identified, the discriminant function can successfully be used to construct a classifier as well as decision regions. Let $g_1(x)$, $g_2(x)$ and $g_3(x)$ be probability density function as in maximum likelihood classifier for a pixel such

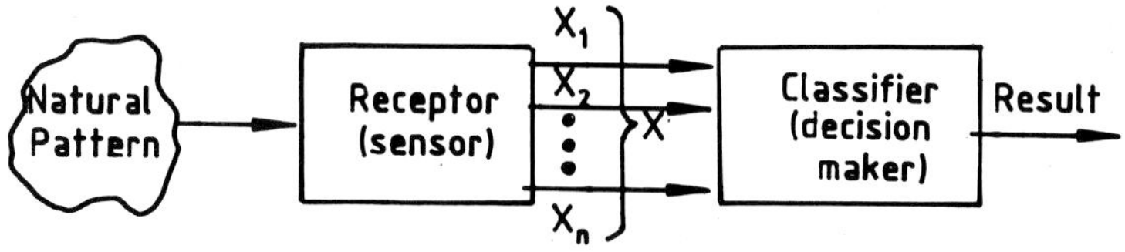

A model for pattern-recognition system

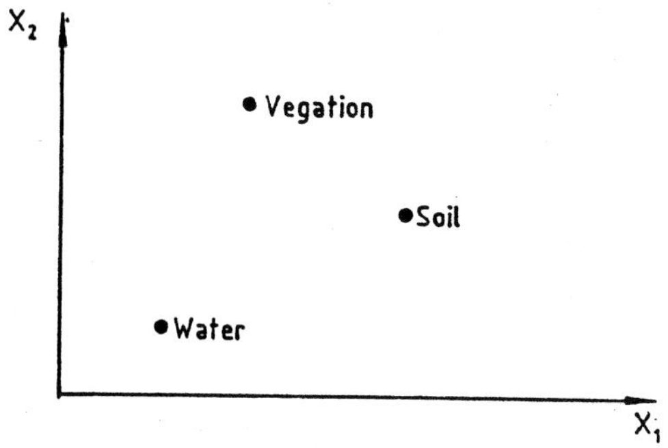

Fig. 10.1: Data in two dimensional space

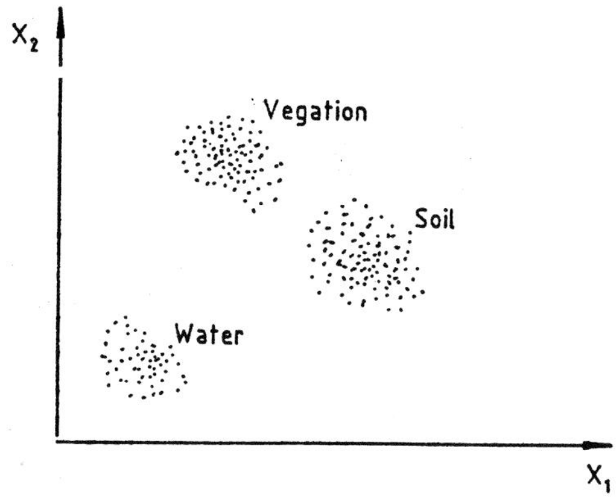

Fig. 10.2: Cluster of data points

that $g_1(x) > g_2(x) > g_3(X)$. Then the pixel will be classified as Class 1. But the discriminant function can be used in constructing the decision region.

Selection of Bands

Multi-spectral digital classification of remote sensing data with several spectral bands (e.g. Landsat-TM, airborne scanner data), requires a judgement prior to final classification about spectral bands which are most effective, accurate and economical in separating each class from all others. The problem of selecting the best N bands from P original bands is not new one, several researchers have brought out various approaches. A spectral band selection approach using maximum determinant and maximum entropy selection criteria with advantages and limitations was suggested by James (1985) and Sheffield (1985). Earlier, Swain and Davis (1978) proposed a different approach, which was by the selection of a set of features for which the maximum separability between any pair of classes is largest. The measurements of separability recommended by them were 'divergence' and 'J-M distance'. Both the above methods require complex computations resulting in higher efforts in analysis. A simplified approach using 'maximum information content' criteria based on optimum index factor (OIF) was developed by Chavezz *et al.* (1982 & 1984) for selecting suitable three band combination for the generation of FCC having optimum information. The combination of spectral bands having large values of OIF will have higher information and is most suitable. Recently, Oza

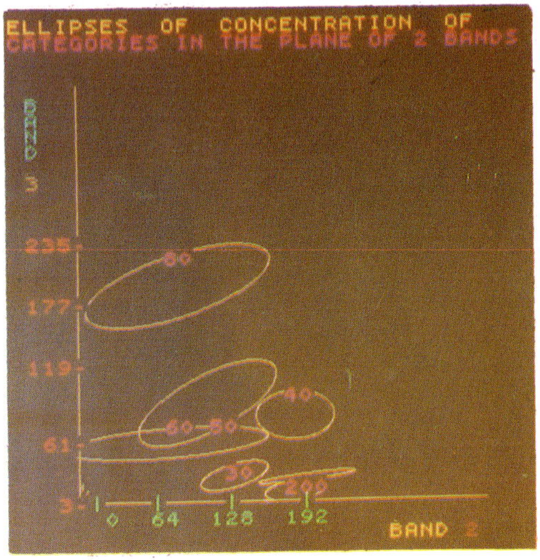

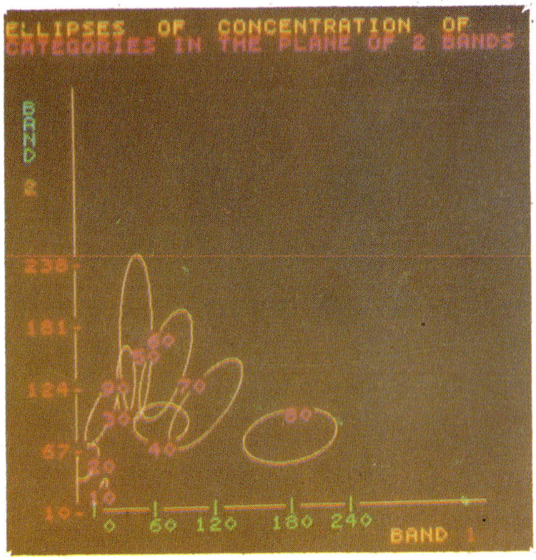

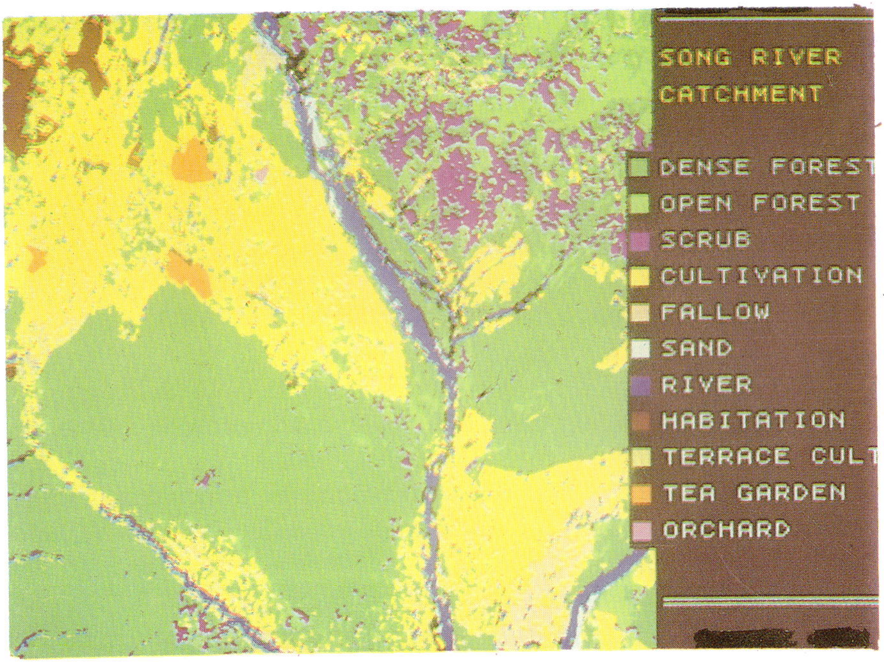

Plate 10.1: Classified Images

(1989) has demonstrated the utility of OIF approach for selecting band combination for generating colour composite of the IRS data. Ma and Olson (1989) outlined a simply defined and easy to compute brightness value overlapping index (BVOI) for an accurate and efficient way to identify spectral bands which will improve accuracy of land cover classification using remotely sensed data.

As mentioned earlier, remote sensing technology is concerned with the measurements of reflected or emitted electromagnetic radiation indicating the physical characteristics of an object which is different from the surrounding objects on the earth surface. The reflected or emitted electromagnetic energy is recorded by the sensor at attitudinal distance in the form of discrete digital number in various spectral channels. These digital numbers in different channels, known as 'bands', represent the number of measurement vectors. If we plot the digital numbers in multi-dimensional space. we will get scattered plots. All the similar types of land cover features but at different geographical locations will form a single cluster of points rather than a single point (Fig. 10.3). Actually each cluster represent separate spectral class. This phenomenon is the typical feature of the multi-spectral remote sensing data. It is caused by the inherent randomness, viz. random leaf orientation, or atmospheric condition and even noise of the sensor of the sensor system (Swain & Davis, 1978). However, each cluster or spectral class signifies the specific aspects of ground cover which leads to the discrimination of different features on the earth surface (Plate 10.1).

Classification leading to identification of the ground cover features can also be accomplished by the principle of statistical pattern recognition. A pattern is defined as the sets of measurements having spatial or geometrical character in two or more dimensions. A pattern recognition system classifies the different patterns based on a series of measurements and compares the same on the "typical" measurements in the interpretation key (decision rule). A match or closest match to a pixel in the interpretation key produce the desired classification. The most important aspect of pattern recognition is the interpretation key or classification rule and the method by which the measurements (discriminant functions) from the pixels to be classified are compared as per the classification rule. Fig. 10.2 shows the simplified model of classification of the remote sensing data in different bands. Some of statistical methods for the selection of bands are as follows :

Variance-Covariance Matrix

After applying the variance-covariance matrix of a given data, only those bands are to be selected whose sum of the variance is maximum, because variance is the measure of information content of the data set (Sheffield, 1985). For example. Landsat Thematic Mapper has seven bands. If we choose any three bands out of these seven, then the total possible combinations will be 7C3 or 35. The subsets of the bands (e.g. triplet here) that provides the maximum determinants of 3 × 3

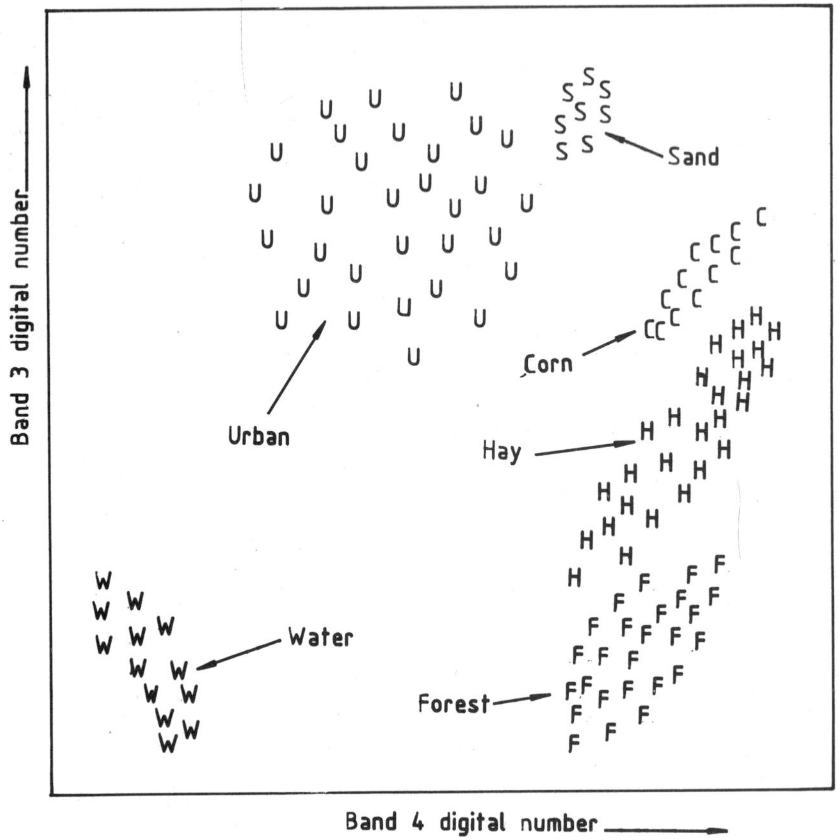

Fig. 10.3: Pixel observations from selected training sites plotted on scattered diagram

principal sub-matrix are to be selected.

Correlation Matrix

The analyst should choose those bands that have maximum interband correlation and the sum of variance is maximum. The optimum index factor (OIF) proposed by Chavet *et al.*, (1982 & 1984) is applicable to any multi-spectral remote sensing data set and is based on the amount of total variance and correlation and between various combination. The algorithm used is as follows:

$$\text{OIF} = \frac{\Sigma \text{ SD}_k}{\Sigma \text{ abs } (r_{ij})} \qquad \ldots\ldots \quad (1)$$

where

SD_k = Standard deviation for the band 'k' (i, 2, ...n)

r_{ij} = The absolute value of correlation coefficient between any two bands (e.g. i and j)

Any subset of bands having maximum OIF value must have the most information as measured by variance with least amount of duplication (as measured by correlation). For thematic mapper data, Jenson (1986) showed that bands 1, 4 and 5 is the best three band combination. Further, Sheffield (1985) also indicated similar results. However, Saha *et al.* (1990) found that four bands combination (2, 3, 4 & 5) of Landsat TM is best for wasteland mapping.

Statistical Separability

Unlike the earlier approaches, this method emphasizes on the distance or separability measure between two training classes in the pattern space instead of information context of the scene. The separability is the indirect measure of the predictor of classifier performance. Figure 10.5 shows pairs of normal probability density functions having various degrees of overlap. As mentioned earlier, the pattern classes (training classes) in question have equal a priori probabilities in which case the decision boundary is located at the central crossover of the probability functions. Divergence is widely used as a measure of spectral separability between two classes in remote sensing measurement. The divergence is nothing but the likelihood ratio of the two classes (Swain & Davis, 1978; Haralick & Fu, 1983). The divergence is defined by the following equation :

$$D_{ij} = 0.5 \text{ tr } [(\Sigma_i - \Sigma_J)(\Sigma_J^{-1} - \Sigma_i^{-1})] + 0.5 \text{ tr } [(\Sigma_i^{-1} + \Sigma_J^{-1})(\mu_i - \mu_J)(\mu_i - \mu_J)^T] \quad \ldots (2)$$

Where tr[A] denotes the trace of matrix A, i.e. the sum of the diagonal elements of the matrix. Σ_i denotes the variance-covariance matrix for the training class 'i', Σ_J is the variance-covariance matrix of training class 'j'.

Σ_i^{-1} and Σ_J^{-1} represent the inverse of the matrix Σ_i and Σ_J respectively, μ_i and μ_J are the mean vectors for the class 'i' and 'j' respectively. 'T' is for the transpose of the matrix, i.e. interchange of row into column.

In this method, the subset of bands should be so chosen that the divergence value for the two bands is maximum. But, for the multiple thematic classes, we should select the band subset which have maximum average divergence.

Brightness Value Overlapping Index

Another approach for selection of bands is based on the Bright Value Overlapping Index (BVOI) suggested by Ma and Olson (1989). It can be expressed as follows:

$$F_{jk} = \sum_{i=1}^{N_{jk}} f(X_{i,k}) \quad \ldots\ldots (3)$$

$$F_{aj} = \frac{1}{m} \sum_{k=1}^{M} F_{i,k} \qquad \text{..... (4)}$$

$$F_{ta} = \sum_{j=1}^{N} F_{a,j} \qquad \text{..... (5)}$$

Where $X_{i,k}$ = (i)th brightness value within a class of channel (k)

$f(X_{i,k})$ = number of pixels with brightness value X_i, k

$N_{j,k}$ = the range of brightness values within class (i) of channel (k)

$F_{j,k}$ = accumulated frequency for class (i) of channel (k)

M = number of spectral bands

$F_{a,j}$ = average accumulative frequency over all channel of class (i)

N = number of classes in the data set, and

$F_{t,a}$ = total average accumulative frequency over all classes

If overlap does not exist among any classes of channel (k), then

$$F_{t,k} = \sum_{j=1}^{N} F_{J,k} = F_0 = 100 \text{ per cent} \qquad \text{..... (6)}$$

where, F_0 has been defined as the cumulative frequency for the whole data set of a single channel, which always equal to 100 per cent; F_{tk} is the total accumulative frequencies of over all classes of channel (k).

If overlay exists among any classes of channel (k), then

$$F_{t,k} = \sum_{j=1}^{N} F_{j,k} > F_0 = 100 \text{ per cent} \qquad \text{..... (7)}$$

The degree of overlap among classes can be determined as

[BVOI = $F_{t,k}/F_{t,a}$ for channel (k)]

[BVOI = $F_{t,a}/N$ for the data set]

The steps involved for calculation of BVOI are as follows:

— Determination of the minimum and maximum brightness values of each land cover types from sample data of each spectral band.
— Determination of the accumulated percentages of pixels having brightness values ranging from the minimum to maximum for each land cover type, from

the histograms of the whole data set.

— Repetition of the above two steps for each cover types in each band.
— Computation of the average of the accumulative percentages of all bands for each land cover type.
— Summing the accumulative percentage for all land cover types in each band and sum the averages of all bands for each cover type.
— Thereafter the BVOI value of each band is to be calculated by dividing the total accumulative percentages for specified band to the sum of average of all bands for each cover type.
— The BVOI value of given whole data set is to be determined by dividing the sum of the averages of all bands for each cover type by the number of cover types.

In an experiment carried out by Kudrat and Saha (1993) for an area in Aligarh district in India, the range of brightness values of six land cover types for the five Landsat TM bands are shown in Table 10.1. The BVOI and OIF values for various land use and land cover classes were calculated. Table 10.2 reveals that the most appropriate three band combination out of the five Landsat TM bands is 2, 4 and 5 which has lowest BVOI values. Because the spectral bands with smaller BVOI values will experience less overlapping between classes, thereby they could be selected and combined for an efficient land use classification. If the BVOI value of a single band is larger than 1, the overlapping between classes is relatively large. However, if the BVOI value is less than 1, a relatively small overlap is likely to be found. Landsat TM band 3 had the highest and band 4 the lowest BVOI . Therefore, the TM band 4 having the less value of BVOI (0.96) gives least overlapping

TABLE 10.1: Range of Brightness values for various land cover types for different TM bands

Cover types	Range	TM Bands				
		1	2	3	4	5
Salt-affected land	Minimum	73	33	34	39	59
	Maximum	127	67	61	81	112
Water-logged land	Minimum	59	21	18	12	6
	Maximum	96	47	53	41	76
Crop 1	Minimum	60	24	19	46	34
	Maximum	80	38	39	93	77
Crop 2	Minimum	63	26	21	39	34
	Maximum	71	33	36	69	63
Built-up land	Minimum	62	25	21	32	27
	Maximum	80	36	40	70	76
Water body	Minimum	62	24	22	29	16
	Maximum	70	28	31	39	51

between land cover classes.

The second best three band combination (one each from visible, near-IR and mid-IR) is band 1, 4 and 5 which has second lowest BVOI values. The most unsuitable three band combination having highest values of BVOI is 1, 2 and 3. It is also necessary to consider correlation between bands in order to avoid band combinations in which both the bands have small BVOI and higher correlation.

TABLE 10.2: BVOI values for TM data of various land cover types

Cover types	Accumulative percentage of Brightness values distribution in the TM bands					
	1	2	3	4	5	Average
Salt-affected land	35	36	37	96	44	50
Water-logged land	98	98	97	9	82	77
Crop 1	84	86	76	82	81	82
Crop 2	58	71	77	93	65	73
Built-up area	83	81	81	97	82	85
Water body	55	16	53	7	36	33
Total	413	388	421	384	390	400
BVOI	1.03	0.97	1.05*	0.96	0.97	66.7

The values of OIF for five different TM band combinations are given in Table 10.3. The combination of bands 2, 4 and 5 out of five cases considered have higher OIF value and is the suitable band combination, which is in good agreement with the earlier observation in BVOI study. Data in this table also show that the bands 1, 2 and 3 and bands 1, 4 and 5 combinations are the worst and second best respectively.

TABLE 10.3: Values of OIF for different TM band combination for whole data set

Band combination			OIF
1	2	3	7.64
2	3	4	8.44
2	4	5	13.01
3	4	5	12.20
1	4	5	12.50

These results are also in agreement with the BVOI results. This phenomenon can be explained by the correlation coefficient between spectral bands (Table 10.4). The band combination of 1, 2 and 3 gives lower value of OIF and hence less suitable combination because the bands are highly correlated with each other and carry similar information for all classes. However, band combination of 2, 3 and 5 is most suitable because of higher value of IOF due to lesser correlation among the bands and more information content.

TABLE 10.4: Correlation matrix for five bands of TM data

Bands	1	2	3	4	5
1	1.000				
2	0.992	1.000			
3	0.960	0.981	1.000		
4	0.910	0.887	0.803	1.000	
5	0.936	0.952	0.959	0.817	1.000

The results of the above experiments do indicate that computational simple BVOI values can effectively be used for finding out the degree of spectral overlap between classes in a data set, and selecting small set of optimum bands from many spectral bands of remote sensing data presently available from Landsat TM or will be available in future from narrow band imaging spectrometer for more accurate and efficient classification. The results also show that selection of bands with BVOI and OIF must involve consideration of correlation between spectral bands.

Classification Scheme

The objective of the classification scheme is to generate meaningful thematic maps with the help of the remote sensing data, i.e. primarily the DN values in different bands. The thematic maps are of broad variety, such as land use or land cover maps, soil map, forest type maps and the like. Before analyzing the remote sensing imagery the categories of interest must be carefully be selected and defined. In order to achieve this, the analyst must understand the fundamental difference between the information class and the spectral class of the data (Jensen, 1983; Cambell, 1983). Information classes are the thematic classes that are achieved after the retrieval and processing the remote sensing data for particular use (Tomlinson, 1972). Conversely, the spectral classes are the inherent clusters of remote sensing measurements in the multi-dimensional space (Figs. 10.3 & 10.4) and must be identified and labelled by the interpreter for the information class so that the information class is the meaningful spectral class. Further, an analyst should thus realize that the spectral and the spatial characteristics of the sensor system and should be able to relate these parameters to the types and proportions of materials found within the scene and the pixel. Once this relationship is well apprehended, spectral classes can be thoughtfully relabeled as thematic classes. According to Robinove (1981), the classification schemes described are rigid and it is based on the a priori knowledge. In India, some classification schemes such as the nationwide land use mapping, wasteland classes, urban land use classes and forest type mapping are in practice.

There are two methods for multi-spectral classification of the remote sensing data : (a) supervised, and (b) unsupervised.

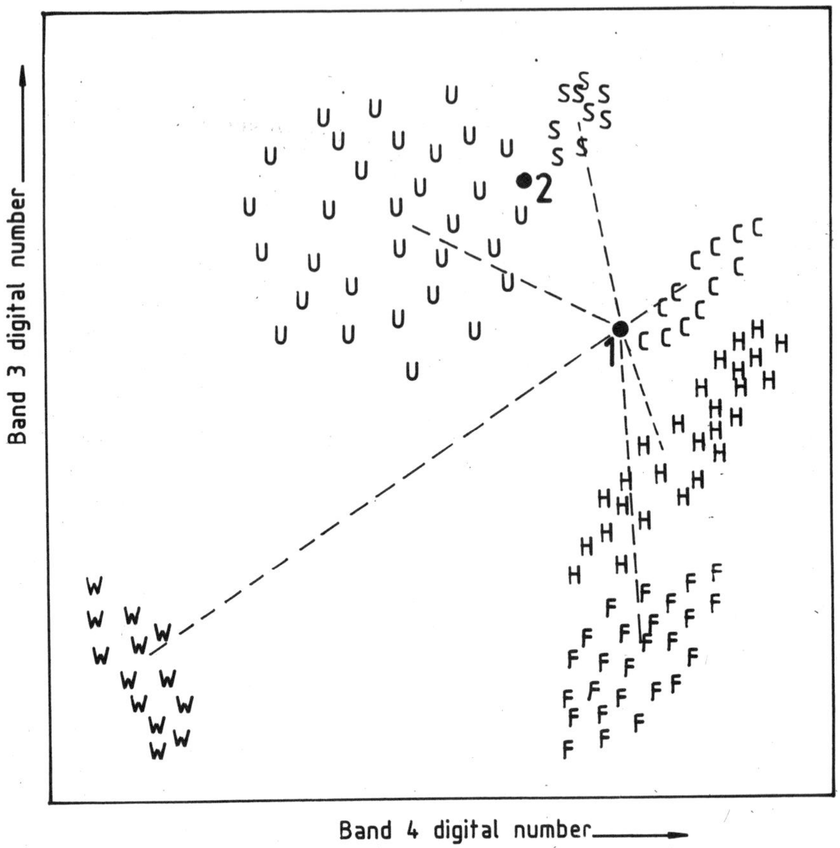

Fig. 10.4: Minimum Distance to Mean Classification Strategy

SUPERVISED CLASSIFICATION

In this method the analyst must supervise the pixel categorization process by specifying the computer algorithm of the statistical parameters, viz. mean vector, and variance-covariance matrix of various thematic classes present in the scene. In order to achieve this objective, representative sample site of known thematic classes, known as 'training sets', are to be computed in the above parameters. They are required for compiling the discriminant function or interpretation key which describe the spectral attributes for each thematic class of interest. Any known pixel in the data sets is then compared numerically to each category in the interpretation key and is assigned to specific class (Andrew, 1972; Swain & Davis, 1978; & Lillesand & Kiefer, 1987).

Training Sites Selection and Statistical Extraction

Multi-spectral supervised classification usually make use of the discriminant functions associated with the pattern class. However, the discriminant functions (e.g. probability density as in the case of maximum likelihood classifier) are generally unknown and are derived from a set of training sites (i.e. DN values in various bands) of known identity which are assumed to be the representative of the class of interest. In order to estimate the probability density function, certain parameters like mean, variance-covariance of each class need to be estimated. The acquisition of training sets is an art, and it needs a close interaction between the analyst and the image data. It also requires sufficient referencing information and a thorough knowledge of the area that is covered by the scene. Another vital point is the purity of the training samples which determines the success of the classification. To achieve a satisfactory classification result, the training samples must be both representative and complete. For example, agricultural class might contain several crop types and each crop type might be represented by several spectral classes. These spectral variation may be due to different stages of growth, soil conditions, soil moisture conditions, crop management practices, crop varieties, geomorphology and relief, and other associated environmental conditions (Swain & Davis, 1978). Thus the analyst should note that all the spectral classes present in the scene, constituting each information class must be adequately represented in the training sets and they must be spread all over the scene (Fig. 10.5).

The overall objective of the training process is to generate the sets of statistics that describe the spectral signature for each land cover class and the discriminant function or decision rule is derived from the statistics.

The training samples are acquired in an image normally by displaying a window of size 512 × 512 pixels. The first window will be the extreme bottom right hand corner. The training sample data are obtained by delineating the polygon for each site using cursor. As soon as the acquisition of all training sets covering all the spectral classes is over for the displayed window, the analyst can proceed further for another window and the above process is to be repeated. In the same manner the analyst will have to cover the entire scene. The training sets or polygons are to be carefully located and delineated to avoid the pixels located along the edges between two land cover classes. Spreading of training samples throughout the scene increases the chances that the training data will be representative of all the variations in the cover type (Lillesand & Kiefer, 1987).

Maximum number of pixels required for each training class must be n + 1, if the number of bands are 'n'. But, in practice, a minimum of more than 10n is used since the estimates of the mean vectors, and variance-covariance matrix improve as the number of pixels in the training sets increases (Jenson, 1983). These samples are then used for the computation of statistics of each class.

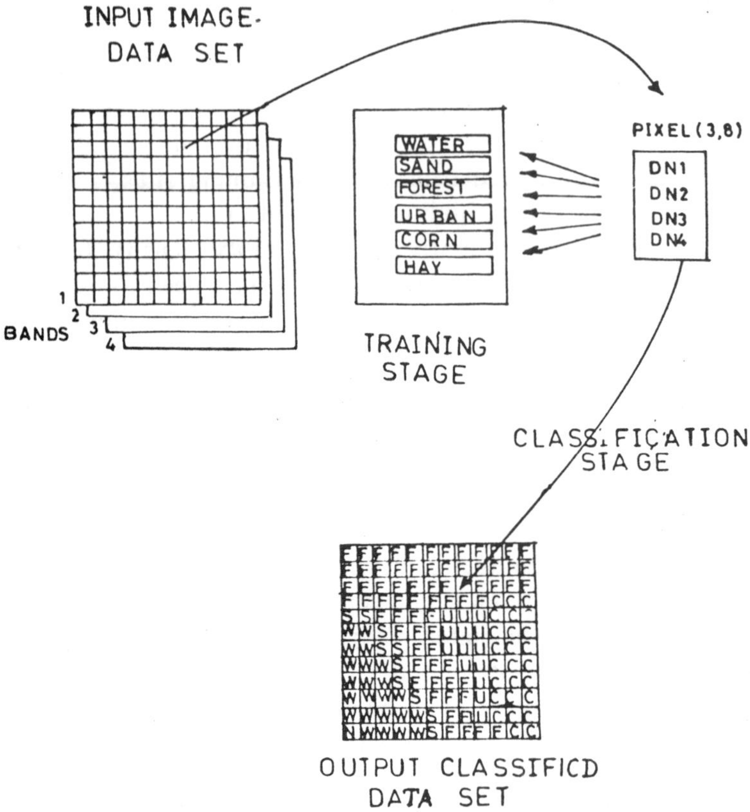

Fig. 10.5: Basic Steps in Supervised Classification

Discriminant Functions

Some statistical parameters which determine the property of any measurement vector (i.e. DN values of different bands of a pixel) decides the class to which it belongs. The discriminant function corresponding to any class (say i^{th}) has a value larger or smaller than any other class in the set at every point in that of the feature space corresponding to the ith pattern class (Swain & Davis, 1978; Duda & Hurt, 1973). These are probability density, Euclidean distance, Mahalanobis distance etc., and are widely used in remote sensing pattern recognition. For probability density function, any pixel is assigned to any class say ith for which the probability density is higher than any other class. But, in the case of distances, any class (i^{th}) will be assigned if interested pixel has the low distance value between the pixel of interest and the assigned class.

Probability Density Function or Maximum Likelihood Classifier

This is defined by the following equation as suggested by Fu (1983):

$$g(x) = \log \rho(wi) \cdot \rho(x/wi) \quad \ldots\ldots \quad (8)$$

where $g(x)$ = Probability density
$\rho(wi)$ = a priority probability
$\rho(x/wi)$ = probability of 'x' for falling in class i
i = 1, 2, 3, ...n

For equal a priority probability with Gaussian distribution, we have,

$$gi(x) = \log \rho(wi) - \frac{N}{2} \log 2\pi - \log |\Sigma_i| - \frac{1}{2} [(x-\mu_i)^t \ \Sigma_i^{-1} (x-\mu_i)] \quad \ldots \quad (9)$$

or simply,

$$gi(x) = -\log |\Sigma_i| - \frac{1}{2} \ \log [(x-\mu_i)^t \ \Sigma^{-1} (x-\mu_i)] \quad \ldots\ldots \quad (10)$$

where
$|\Sigma_i|$ = determinant of variance-covariance matrix of class i
Σ^{-1} = inverse of variance-covariance matrix
x = measurement vector, i.e. DN values of any pixel for all the channels
μ_i = mean vector for the ith class
t = transpose

Any measurement vector 'x' i.e. any pixel may be classified into i^{th} class if $g_i(x) \geq g_j(x)$ for all $i \neq j$.

Euclidian Distance

This distance can be defined as follows :

$$\text{Distance } D = [\Sigma(X_{ijk} - \mu_c)]^{1/2} \quad \ldots\ldots \quad (11)$$

where
X_{ij} = any measurement vector or DN values of all the bands at the pixel at the location (i, j)
μ_c = mean vector for c^{th} class
k = number of bands (1, 2 ......n)
c = 1, 2 ....m

Mahalanobis Distance

For multi-spectral remote sensing data Mahalanobis distance (MD) is defined by the formula given by Duda and Hart (1973):

$$MD_i = (x-\mu_i)^t \ \Sigma^{-1} \ (x-\mu_i) \qquad \ldots.. \qquad (12)$$

where

μ_i = mean vector for i^{th} class

x = any measurement vector i.e. pixel DN values

Σ^{-1} = inverse of variance-covariance matrix

t = transpose.

UNSUPERVISED CLASSIFICATION

In this digital technique, there is no need to train the computer algorithm for decision rule or interpretation key. The image data are first classified by aggregating them into natural spectral cluster present in the scene. Then the analyst has to determine the thematic class corresponding to each spectral cluster by comparing the classified image data to the ground reference information.

In an inaccessible area, collection of sufficient training samples is not possible, which is inevitable for the generation of decision rule or interpretation key to train the classifier. Maximum likelihood classifier, one of the most popular supervised classification technique, utilizes the probability density function for discrimination of classes. This is based on the assumption that the classes must have multivariate normal distribution. This implies that the measurement vectors must have a tendency to cluster around a central value (mean or mode). In agricultural scene, normal distribution assumption is violated. Because of the different dates for sowing, variation of crop management practices, and heterogeneous topographic and soil conditions, the probability distribution functions are multimodal instead of unimodal (Swain & Davis, 1978). This aspect is also common in the case of water classes or wetland environments due to the pollution, silt contents and the presence of various types and amount of aquatic vegetation. In order to overcome the above mentioned problem, it is imperative to find out an alternative solution for classification of the remote sensing data. The unsupervised classification is based on the clustering techniques and is being particularly used in studying vegetation in a big way (Sokal & Sneath, 1963; Duda & Hart, 1973). The unsupervised classification algorithm examines the unknown pixels in an image and aggregate them into a number of spectral classes based on the inherent groupings or clusters present in the remote sensing data of the concerned scene. The main logic behind this technique is that a given cover type should form a single cluster (Figs. 10.3 & 10.4 for a two dimensional feature space) in a n-dimensional feature space, whereas cluster corresponding to different classes are well separated (Lillesand & Kiefer, 1987). The unsupervised classification does not require any *a priori* knowledge for the analysis of the data (Fu, 1976). Once the classification is over, the user has to identify each spectral class to information class with the help of a post priori knowledge of the generation of the thematic map. This indicates that the spectral classes must be identified by ground verification or some other referencing data source.

Numerous clustering algorithms that can be used for unsupervised classification are available in remote sensing and pattern recognition literature (Duda & Hart, 1973; Fu 1976; Haralick & Fu, 1983). But the clustering algorithm used for unsupervised classification of remote sensing data vary according to the efficiency and objective of the analysis (Jenson, 1986). Different criteria of efficiency leads to different approaches.

Here we will only emphasize on two techniques for unsupervised classification that are (a) hierarchical or tree classification, and (b) hyper cube or cluster grouping following the minimum euclidean distance between the clusters (mean vectors).

Hierarchical or Tree Classification

This classification technique involves following steps:

- Decoding in data or formation of Hash Table.
- Generation of statistics and computation of multi dimensional histogram.
- Grouping the clusters following hierarchical ascending classification tree (HAC).

Hashing is a process by which a large amount of data can be stored in a small area in the computer memory without any loss of useful information (SEP, 1986). In order to store two dimensional histogram or two band histogram in computer memory, $256 \times 256 = 2^{16}$ memory locations are required. Similarly for storing seven bands histogram of Landsat TM data 2^{56} memory locations are required. Since most of the image processing computer has 32 bit addresses, hence a maximum of only 2^{32} addresses can be handled. Due to this reason, the scope of analyst for using the multiband data is restricted. To overcome this problem a code conversion is to be done by any one of the following operations:

User's Specified Bands

In this option the gray values of each band is to be partitioned by the user and for each band he has to specify some value. This means the upper and lower limits of a range in the input scene together with its value is specified by the user. For example, 20-30 : 4, i.e. a value of 4 is assigned in the conversion look-up table for the gray value range 20 to 30. This is to be repeated for all the bands covering the entire dynamic range of gray values (DN).

Histogram Partition

In histogram partition, the user has to display the histogram first, then interactively divide it in narrow dynamic range and for each partitioned range he has to specify some number (e.g. 20-30 : 5) as shown in the earlier option. The user should be careful while making partitions of the histogram so that the each partition range must be narrow to get maximum natural grouping or cluster of pixels.

Automatic

In this mode the user's role is very limited. Only the user has to specify the dynamic range of his interest by examining the histogram. This indicates that instead of 0-255 (i.e. whole range of DN values) the user can select any range, say 50-200, of his interest. After deciding this, the concerned range will be divided by the computer into 16 equal gray range (DN value). It obviously implies that each of these sub-range will have equal difference between the minimum and maximum bounds.

Equipopulation

This option is further simple. The computer divides the decided range of gray values into 16 parts so that each part has equal number of pixel population.

After code conversion of data from 8 bit to 4 bit, 7 × 4 = 28 bit address is needed to store all the entries in 7 dimensional histogram of Landsat TM data instead of 56 bit addresses. The next step is concatenation which is a process of arranging the gray values side by side for all the pixels.

Band 1	Band 2	Band 3	Band 4
G_1	G_2	G_3	G_4

Sometimes, the concentration code is so obtained that it may go beyond the maximum address which is 50,000. This may vary from computer to computer. In order to circumvent this problem, the next step is to find out the address for the concatenated code. Following prime number technique, it is being carried out. Since the largest prime number less than 50,000 that is 49,999, so it is used to calculate the hash table address for a concatenation code.

Address = e modulo (P1)

where e = concatenated code

P1 = represents the prime number.

Further, a case may arise when two different concatenated codes get the same number, e.g. 5. This situation is termed as collision. In this situation, one concatenated code should get new address and another hashing function may be used. Hue the next largest prime number is 49,991 and is less than 49,999, so 49,991 is used to find the second address.

Address = e modulo (P2)

where (P2) = next higher prime number

However, again, the collision may arise even the computation of second address. Then they are kept in new collision table. The final hash table is of the form as given below (SEP, 1986).

TABLE 10.5: Hash Table

Serial Number	Address	Concatenated code	Histogram n-dimension	New address if collision	Class

Hue histogram entry represents the number of pixels having the same concatenated code. The last entry is determined after the formation of the hierarchical ascending classification tree.

Formation of Hierarchical Ascending Classification Tree

From the hash table entry, i.e. the natural cluster pairs are further combined into more generalized clusters that form a hierarchical ascending classification tree (Fig. 10.6). The tree formation starts from the bottom (terminal node) to trunk (Sokal & Sneath, 1963; Fukunga, 1972). Each node is associated with a parent node to which its clusters also belongs. In order to achieve this, the analyst should concerned with the representative pairwise similarity or correspondence among the clusters present in the hash table entry. Hence we must consider some distance between the cluster centres in terms of the classification tree alone. Hue will consider the Mahalanobis distance (Duda & Hart, 1973) as distance or similarity between the clusters. According to Fukunaga (1972), hierarchical ascending tree formation comprises the following steps :

Step 1: Assign each cluster to a terminal node of level '0' and calculate the Mahalanobis distance between each and every entry pair.

Step 2: Add a new node as the parent node of two nodes corresponding to the closest pair of pixels (clusters) in the hash table. This is nothing but the linking of two pair that have minimum Mahalanobis distance. The level of this node is the distance between the corresponding clusters.

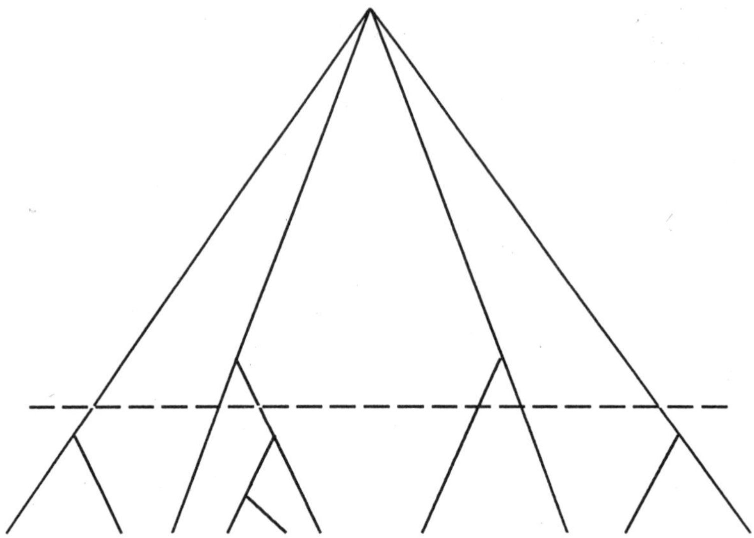

Number of classes = 10 Cut off level=6

Fig. 10.6: Hierarchical Ascending Classification

Step 3: Compute the distance between the cluster represented by the new node and the clusters corresponding to the remaining terminal nodes.

Step 4: Consider the new node to be a terminal node and delete the two nodes which were merged to form it. If there are two or more terminal nodes left, return to step 2 and repeat, otherwise algorithm is complete.

Step 3 is the most troublesome part of the above digital procedure. The resulting tree can heavily depend upon the way distances between the clusters is defined. After the generation of final hierarchical tree using the hash table, the concatenated code is generated for each pixel (DN) and the concatenated code is searched in the hash table. Each pixel is then assigned the class number following the class number of the hash table. As soon as the processing is complete for all the pixel, the number of classes, population of each class in terms of pixel and the percentage of pixels for each class will be recorded in the output file.

With the help of ground information and other reference sources, the analyst has to determine the information class corresponding to each class in the output image. For different ground cover class, separate colour is to be assigned to generate the colour coded thematic map. Here the ground information and the related referenced information are used as a posterior knowledge.

Hyper Cube Grouping

The hyper cube grouping operates in following three phases:

(a) Conversion of codes
(b) Formation of hash table
(c) Grouping of the hyper cube or clusters to have the user's specified number of clusters or spectral classes.

Phases (a) and (b) are very much similar to the hierarchical ascending tree classification as discussed earlier. In the phase (b), using the concatenated code for each and every pixel, it will form maximum possible clusters. Here the user has to specify the total number of classes or hyper cubes desired. Hash table in this algorithm has two parts (Table 10.5) : part one contains all the hyper cubes, pixels under each hyper cube and class numbers; second part comprised the final number of class after grouping the other clusters based on the minimum euclidean distance (Jenson, 1986). Immediately after the users interaction, the algorithm will determine the user's specified number of dominant clusters and other clusters will be merged or regrouped following the minimum distance criteria between the clusters. Classified unsupervised classification output will thus be generated. It is always better to have as many as reasonable number of classes that should be tackled by the user for identification with the help of a posterior knowledge and ancillary information for determining thematic classes. Many times two or more above spectral classes may belong to one single thematic class. Immediately after the identification of desired thematic class, separate colours are to be assigned for each class to generate the colour coded map. Plate 10.1 is an example of unsupervised classified output of IRS-1A, LISS 2 data around Shimla in India.

Conclusion

There has been a lot of initiatives to find out methods for reducing the required time for classification techniques. Several hardware and software solutions are likely to be brought out in near future. This has also become essential because the digital processing of remote sensing data is becoming more and more PC based. Fast computation, parallel processing and use of artificial intelligence are being considered for this purpose. Neural network are applied for parallel processing and stimulate the functionality and decision-making process of the human brain. Neural network features corresponding to the synapses, neurons, and axons of the brain are input weights, processing elements, and output paths (Civco, 1991; Sheldon, 1990; Benediktsson, Swaiin & Esory, 1990). Attempts are also made to have artificial neural network using minimum training sets in conventional supervised classification (Hepner, Logan, Rittner & Bryant, 1990).

Further, in a PC environment, transputors as a hardware platform for image classification are being considered (Crane, Weiler & Scarpace, 1991). Use of these boards are studied as a method to increase the speed of maximum likelihood classification in a desktop environment. This method of classification require huge computation and hence computer time. However, there can be two basic approaches to this problem: (a) faster algorithm (software), and (b) faster processing (hardware). Supercomputers can significantly reduce the time required. However, in the PC environment, the other method is to use transputer based parallel processing boards in an IBM PC/AT. Further, there is a possibility to reduce processing time with the help of appropriate algorithm, particularly in the case of maximum likelihood classification (Bolstad & Lillesand, 1991). The software strategies can be (a) image parallelism and (b) task parallelism. Image parallelism requires each processor to perform the same operation of different areas of the image (input data) simultaneously. The task parallelism is where different operations are performed on each processor. Task parallelism has been useful for an assembly or production line operation (Crane, Weiler & Scarpace, 1991; Healey & Desa, 1989).

REFERENCES

Anderson, R.R. & F.J. Wobber (1973), Wetlands mapping in New Jersey, *Photogrammetric Engineering*, vol. 39, No. 4, pp. 353-8.

Andrews, H.C. (1970), *Computer Techniques in Image Processing*, Academic Press, New York & London.

Andrews, Harry (1972), *Introduction to Mathematical Techniques in Pattern Recognition*, Printice-Hall, New Jersey.

Benediktsson, J.A., P.H. Swain & O.K. Esroy (1990), Neural network approaches versus statistical methods in classification of multisource remote sensing data, *IEEE Transcations on Geogscience and Remote Sensing*, vol. 28 (4), 540-52.

Bolstad, P.V. and T.M. Lillesand (1991), Rapid maximum likelihood classification, *Photogrammetric Engineering and Remote Sensing,* vol. 57 (1), 67-74.

Cambell, J.B. (1981), Spatial correlation effects upon accuracy of superimposed classification of land cover, *Photogrammetric Engineering and Remote Sensing*, vol. 47, 355-63.

Chaves, P.S., G.L. Berlin and L.B. Sowers (1982), Statistical methods for selecting Landsat MSS ratios, *Journal of Applied Photographic Engineering*, vol. 8, 23-30.

Chavez, P.S., S.C. Guptill and J.A. Bowell (1984), Image processing techniques for Thematic Mapper data, *Technical Papers, 50th Annual Meeting of the American Society of Photogrammetry*, vol. 2, pp. 728-42.

Civco, Daniel L. (1991), Landsat TM image classification with an artificial neural network, *1991 ASPRS/ACSM Annual Convention*, Baltimore, vol. 3, pp. 67-77.

Crana, Aaron, I, Peter R. Weiler and Frank L. Scarpace (1991), The use of transputers to reduce maximum likelihood classification time, *1991 ASPRS/ACSM Technical Papers*, vol. 3, pp. 93-102.

Duda, R.O. and Hart, P.E. (1973), *Pattern Classification and Scene Analysis*, John Wiley & Sons, New York.

Fu, K.S. (1976), Pattern recognition in remote sensing of the Earth resources, *IEEE Transaction on Geoscience Electronics*, vol. 14, pp. 10.11.

Fukunga, Keinosuke (1972), *Introduction to Statistical Pattern Recognition*, Academic Press, New York.

Gurney, Charlotte and J.R.G. Townshend (1983), The use of contextual information in the classification of remotely sensed data, *Photogrammetric Engineering and Remote Sensing*, vol. 49(1), pp. 55-64.

Haralick, R.M. and K.S. Fu (1983), Pattern recognition and classification, in R.N. Colwell (ed), *Manual of Remote Sensing*, American Society of Photogrammetry and Remote Sensing, Falls Church, vol. 1, Chapter 18, pp. 793-805.

Healey, R.G. and G.B. Deas (1989), Transputer based parallel processing for GIS analysis: Problems and potentialities, *AUTO-CARTO 9 Proceedings*, April 1989, pp. 90-9.

Hepner, G.F., T. Logan, N. Rittner and N. Bryant (1990), Artificial neural network classification using a minimal training sets: Comparison to conventional supervised classification, *Phptogrammetric Engineering and Remote Sensing*, vol. 56(4), pp. 469-73

James, M. (1985), *Classification Algorithms*, John Wiley and Sons, New York.

Jenson, J.R. (1983), Biophysical remote sensing, *Annals of the Association of American Geographers*, vol. 73, pp. 127-47.

Jenson, J.R. (1986), *Introductory Digital Image Processing: A Remote Sensing Perspective*, Prentice-Hall, Englewood Cliff, New Jersy.

Kudrat, M. and S.K. Saha (1993), Land productivity assessment and mapping through integration of satellite and terrain slope data, *Photonirvachaka*, vol. 21, No. 3, pp. 157-66.

Lillesand, T.M. and R.W. Kiefer (1987). *Remote Sensing and Image Interpretation*, John Wiley & Sons, New York.

Ma, Z. and C.E. Olson Jr. (1989), A measurement of spectral overlap among cover types, *Photogrammetric Engineering and Remote Sensing*, vol. 55 (10), pp. 1441-4.

Oza, M.P. (1989), Selection of band combination of IRS data, *Photonirvachak*, *Journal of the Indian Society of Remote Sensing*, vol. 17(2), pp. 23-6.

Robinove, C.J. (1977), The logic of multispectral classification and mapping of land, *Remote Sensing of Environment*, vol. 11, pp. 231-44.

Robinove, C.J. (1981), The logic of multispectral classification and mapping of land, *Remote Sensing of Environment*, vol. 11, pp. 231-44.

Robinove, C.J. (1982), Computation with physical values from Landsat digital data, *Photogrammetric Engineering and Remote Sensing*, vol. 48, pp. 781-4.

Saha, S.K. and M. Kudrat (1992), Software options for thematic interpretation and mapping using remotely sensed digital data, in P. Nag (ed), *Thematic Cartography and Remote Sensing*, Concept Publishing Company, New Delhi, pp. 149-67.

SEP (1986), *Manual of VIPS-32 Software*, RRSSC, Dehra Dun.

Sheffield, (1985), Selecting band combinations from multi-spectral data, *Photogrammetric Engineering and Remote Sensing*, vol. 51 (6), pp. 681-7.

Sheldon, R.A. (1990), Satellite image analysis using neural networks, in *Proceedings of the 1990 Goddard Conference on Space Applications of Artificial Intelligence*, NASA Goddard Space Flight Center, Greenbelt, pp. 349-55.

Sokal, R.R. and P.H.A. Sneath (1963), *Principles of Numerical Taxonomy*, Freeman, McGraw-Hill Inc., New York.

Swain, P.H. and S.M. Davis (1978), *Remote Sensing : The Quantitative Approach*, McGraw-Hill Inc, New York.

Swain, P.H., H.J. Siegel and B.W. Smith (1980), Contextual classification of multispectral remote sensing data using a multiprocessor system, *IEEE Transactions on Geoscience and Remote Sensing*, GE-18, No. 2, pp. 197-203.

Tomlinson, R.F. (ed.) (1972), *Geographical Data Handling*, UNESCO/IGU Second Symposium in Geographical Information Systems, Ottawa.

Weiler, P.R. and F.L. Scarpace (1990), Using architecture of the INTEL 80287/80387 to optimize maximum likelihood computations, in *1990 ACSM/ASPRS Annual Convention Technical Papers*, vol. 4, pp. 475-82.

ELEVEN

GEOGRAPHICAL INFORMATION SYSTEM

GEOGRAPHICAL Information System or GIS has become a slogan in the present day scientific literature. It has attained a status of multi-discipline. The thrust area basically meant for geographers has been taken over by resource scientists, physicists, computer scientists, electronic engineers, surveyors and statisticians, apart from the remote sensing application scientists. All have a say in this matter. Nevertheless, due to such interactions, this multi-disciplinary thrust area has developed very fast. Everybody would like to plot the information in order to get a spatial pattern of a particular theme. The computer companies, electronics industries, programmers have accepted this idea as a challenge.

With the possibility of using vector and raster information for GIS, there has been an impetus to use the remote sensing data at least as one of the inputs. GIS does provide a right opportunity for much desired integration of remote sensing data with the collateral data. The remote sensing companies would like to push the concept that the GIS application is only a step after processing the remote sensing data. There are computer companies which will plead for a total solution, i.e. integration of aerial photography, digital mapping, digital image processing, data base management system (DBMS), and of course field information. Many books and research papers on this subject are being added every day.

Basic Principles

GIS is expected to capture the information into digital form. Each information set is said to be a layer. GIS is a computer based system to capture, display, manipulate and edit a geographically referred information. It should be able to superimpose such layers for finding out additional information required for development, planning or monitoring of resources etc. There are four basic components of the GIS:

- — Geographic data base
- — Hardware
- — Software
- — Users

The geographic data base has two basic components : spatial data and attributional data. The former one is for the location which gives shape and spatial relationship with other attributes. The attributional data pertains to the characteristics of given location. These characteristics can be in the form of numerical data, words, symbol, text, shades or digital number (DN value in remote sensing). There are several models to describe the GIS model. The most common one is host of inputs feeding the data base which makes it possible for the users to interact and bring out required themes. The output can be in different forms as well, such as map, imagery, image maps, statistical table, text, CCTs. floppies, CD-ROMs, optical disks etc. (Fig. 11.1).

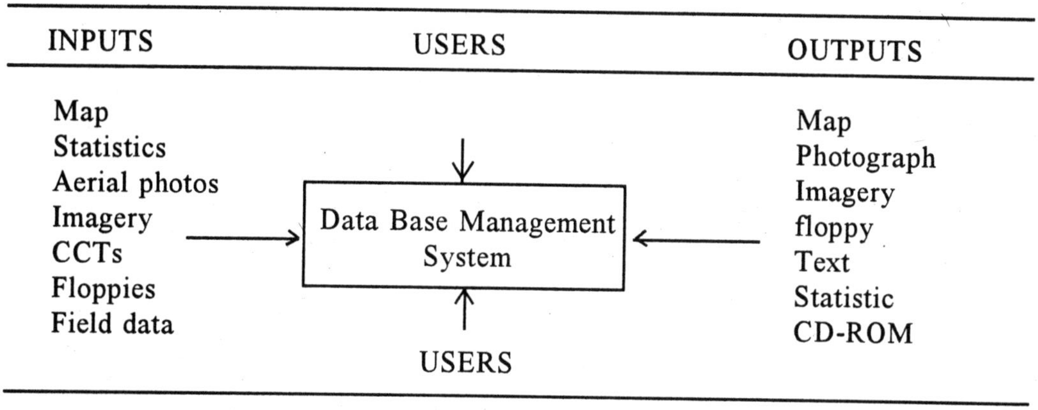

Fig. 11.1: General Model of GIS

The above model is too simplified pictorial view of the whole system. In fact for each type of input or even output, there has to be compatible input/output (I/O) devices. Further there should be compatible software as well which can link different I/O devices.

GIS Hardware

In order to capture different input and bring out desired outputs, a minimum configuration can be as follows:

— Host computer with tape drive, memory and storage capacity
— Digitizer
— Graphic terminals
— Plotter

At present a wide range of choice is available for the host computer starting from mainframe to super minis, workstations and powerful personal computers. Similar to the host computer for digital image processing, it is very important that the speed, number of users. storage capacity, operating systems, inter connectivity etc are determined as per requirements. A co-processor is also helpful for some of the operations. Further, digitizer are of different size, for example A3 or A0 size. Larger the size, more expensive will be the item. In some digital image processing system, digitizers are being used for the rectification of the image. The resolution, accuracy, functions, mouse, bit-pad back lighting and provision of inclination are important aspects for this subsystem.

Similar to digitizers, the plotters are also available for various size which also determines the cost. Plotters are of a different variety, such as pen plotters, electrostatic, ink jet, graphic printers and recently the thermal plotters. The price varies according to the speed, resolution (in terms of dots per inch), and colours. The plotters are cylindrical and flat bed. The former requires less space. However, the latter is preferred for cartographic accuracy. The speed of electrostatic plotter is generally high. In addition to above considerations, it is also important to find out the details of the terminals. Here, resolution, screen size, colour card, local memory and zooming facilities are important because it is used for interacting and viewing the data. For better functioning of the GIS, workstation is preferred though it is expensive (Chapter 5). It has following characteristics:

— Networked single user stations
— Resources shared across the network
— Powerful CPU
— Storage device may be different than CPU
— File server/client configuration
— Generally UNIX operating system
— Fast graphics due to integrated bitmap display
— Window and graphics possible.

At present there are different large capacity storage are available. Some of the examples are CCTs, CD-ROM reader and optical disks. Additional hardware is required for this purpose. With the improvement of the technology. All such hardwares can be connected through a single or more servers.

Raster and Vector Formats

At the outset, one must be able to understand the format which is compatible for the geographical information system. Conventionally, data has been stored in the form of hardcopy, such as paper or film. But for the GIS development such information has to be converted to a data format which is acceptable by a GIS programme. The digital data can be either in raster or in vector form (Table 11.1). The raster data comes in the form of individual, bit, without intelligence. It is good for viewing the information without changing, analysing or editing. On the other hand, the vector data has been widely used for CAD and GIS activities because the information can be manipulated, edited, analysed—some of the basic function for GIS. Vector data are stored as a series of individual points and lines that are geometrically and mathematically associated. This association means that the information can be attributed to individual points and lines to create an "intelligent" data set (Ogata, 1992). While remote sensing data is in raster format, the requirement of vector data for GIS application is almost essential.

TABLE 11.1: Raster vs. vector

Features	Raster	Vector
Data capture	Fast	Slow
Data volumes	Large	Small
Graphics	Medium	Good
Data structure	Simple	Complex
Geometric accuracy	Low	High
Linear network analysis	Poor	Good
Area/polygon analysis	Good	Poor
Combining data layers	Good	Poor
Generalization	Simple	Complex

Source: Dale and McLaughlin, 1988.

The typical raster formats are RLE, RLC, and TIFF. RLE stands for run-length encoding. It provides scanning parameters such as resolution; and supports gray scale up to 256 levels. Similarly, RLC is also run-length encoding format. It does not provide ancillary information and supports binary images. TIFF is a popular raster format due to enormous data compression facilities. It is almost considered as an industry standard for data format.

The vector formats are known as DXF, DWG, IGES, DGN and HPGL. DXF format is considered as the industry standard for vector data. Originally it was developed for AutoCAD applications. It is now widely accepted. It is well documented and character based thereby ensuring portability. DWG is the propriety of Autodesk format. Initial Graphics Exchange Specification or IGES is an standard for graphics files. DGN is a proprietary of Intergraph company. Similarly, HPGL is proprietary for Hewlett-Packard company.

Data Conversion Methods

Though remote sensing data is available in raster format, vector data is mostly used for GIS development. However, recently, raster based GIS solutions are being proposed which obviously have limited use. After carrying out the digital processing of the remote sensing data or other scanned data it is therefore important to convert the information to vectors. There are several methods now available to convert a hard copy document to a data format compatible with GIS. The advantages and disadvantages of the five methods of conversion are indicated in Table 11.2.

TABLE 11.2: Comparison of Conversion Technologies

Conversion technologies	Advantages	Disadvantages	Best used with
Point to point	Specialized knowledge not essential	Relies on GIS for conversion. Time consuming. Accuracy depends on operator's skill. Difficult to draw curves etc.	Dense intersection. Damaged or fragile drawings which cannot be scanned.
Table	Established technology. High accuracy. Large and un-usual size drawing possible	Time consuming. Required large space. Accuracy depends on operator's skill and physical condition.	As above
Heads-up	Raster and vector files can be viewed simultaneously. Can run under specific GIS menu.	As above	Simple drawing with few lines, curves and irregular polygons. Drawings that require only a portion of the raster to be converted to vector.
Automatic	Efficient conversion. No operator during conversion. Can run in batch mode. Supports pattern matching and rule bases for recognition of geometry, text and symbol.	Enhancement requires trained personnel. Limited symbol and font recognition. Character and symbol recognition not exact. Post-processing and pre-processing programming are time consuming. Difficult to convert selected portion of raster data. Mixed graphics increase editing time. Depends on the accuracy of raster file.	Large number of similar drawings. Drawing with primarily line data, and repeated symbology. Good drawings create good raster file.
Intelligent	Allows repetitive conver-. sions. Manual mode allows to vectorize close or unclear lines. No post-editing required. Symbol, Text and line can be supported. Attributing and layering done during conversion. Vectorising of selected data possible.	Dependent on the accuracy of the raster file. Accuracy subjected to operator's skill and physical condition.	Hybrid raster/vector applications and databases. Drawing that must be partially digitized. Complex drawings operator's discretion. Drawing requiring attributing or layering. Drawing in good condition.

Source: Modified from Ogata 1992.

GIS Software

The choice of software generally user oriented. There are standard packages now available of the shelf. The criteria of a GIS software is more-so-ever standardized. The detail criteria is as follows:

A. *Data entry*: digitizing, scanning, automated data capture, interface with existing digital files, key board entry.
B. *Analysis*: map overlay, proximity analysis, mathematical analysis, enclosure, buffer, measurements, attribute analysis and interpolation.
C. *Surface modelling*: 3-D surfaces, slope analysis, and draping.

Network modelling: route selection. allocate demand to network and facility siting.
Manipulation: image merging, projection, clip, window, update, generalize and aggregate.
Query: spatial and attribute.
Output: tabular list, map display, digital outputs.

From the above description it is obvious that the GIS soft has to be versatile (for further detail see Appendix-II). Considering the basic features of GIS software available globally or indigenously, we can have their assessment. One has to look into the particular requirements and the capabilities or the facilities offered.

Remote Sensing and GIS

From the recent trends it appears that all the mapping related technologies are coming closer and merging. The digital image processing and the GIS can now be integrated. The hardware requirements have much in common and the vendors are coming forward with interface solutions for both the programmes. If the basic GIS work is done in raster (or scan) format it is more the easy to connect with the remote sensing data. Contrary to the scenario in the near past, the GIS is having more and more raster data. This is also due to this reason that scanning is becoming common now a days. On the other hand, the vector digitization takes a long time.

There is certainly a possibility to include the remote sensing data as one of the inputs. The satellite data cannot be exactly be superimposed on a map of specific project. Apart from having different projections, the remote sensing data has its own inherent problems. There are standard procedures to reorient this information according to a geographic base. In Chapter 6 the image to map and image to image registration have been discussed. Same methods can be used here as well. Once the data is reoriented as per given projection, it becomes one of the inputs for GIS (as shown in Fig 11.1). This information can be further manipulated for bringing out several themes.

For the development of remote sensing based GIS, whether the data is to be rectified first and then manipulated for different themes, or vice versa is a debatable issue. Both have their own advantages and disadvantages. The rectified remote sensing data can readily be used for bringing out a variety of themes. However, image processed remote sensing data is more clear hence the Ground Control Points (GCPs) can easily be identified which in turn contributes to develop better mathematical model between the map and image, or image to image. There can be a mid way as well. The remote sensing data can be enhanced by basic image processing techniques, such as contrast stretching or filtering, and then rectification. The rectified data can now be used for more serious image processing initiatives, such as band ratioing or principal component analysis (Chapter 9).

The processed remote sensing data can be a valuable input in the GIS. The relevant data can be extracted and form one or more layers of information. Some GIS/image processing software do have provision to bring out limit lines if the gray values changes beyond a specific or given limits. These limit lines have to be analyzed with other information or ground realities so that they become meaningful. However, such options provide a wide scope for remote sensing data manipulation in order to fit in a GIS.

Application of Remote Sensing based GIS

The combination of image processing and GIS technologies are enormous. As indicated earlier, they both are coming closer and it is now becoming extremely difficult to differentiate each other. The hardware was almost similar, and now software are being integrated. The problems of transfer of data is also being circumvented. The private companies are combining their efforts in this direction (Parker, 1991).

The most common application of this combination is the landcover mapping. In addition to areal photographs, it also provides a synoptic view of the surface features. Due to its repeatability and scope of manipulation, it has a edge over the photographs. Further, such a combination is preferred where the integration of spatial data with current information is required, such as Land information System (LIS). Recently, GIS packages are also supported by Digital Terrain Models (DTMs) which has further enlarged the applicability. For example, in case of the Canadian GIS, "the widespread distribution of Landsat and SPOT imagery, and the availability of digital elevation models and street files in many countries have certainly lead to applications well beyond those used to justify the data's compilation" (Goodchild, 1992). With better resolution and with always improving software, the topographical mapping requirements are being met by GIS remote sensing combination. It has almost become unavoidable source for cross checking or updating in digital surveying. Furthermore, the GIS software can now accept the Global Positioning System (GPS) information in their programme, which is an additional advantage. For example, the ARC/INFO GIS software now accept GPS data through its GEOLINK module.

Experiments have been carried out for various applications through such efforts. It is not always traditional land based information which is generated, but unconventional themes have also been tackled. Urban and regional planning are obvious offshoots. Environmental issues related to ground cover such as forests, water bodies or wastelands are simple examples as well.

Even with above mentioned trends, much remains to be developed. More research work should be oriented towards the development of application oriented remote sensing software which can be easily integrated with the GIS. A recent develoment in this field has been integration of raster based GIS modules in conventional vector based GIS software like ARC/INFO. Apart from raster/vector conversion facilities in the GIS packages, vector overlays on remote sensing or other raster data is now possible. This has, to certain extent, facilitated the remote sensing data integration with GIS. Nevertheless, following functions would further help in such integration :

— Livelink
— Direct usage of compressed data in GIS
— Automatic registration of GIS and remote sensing data
— Projection support
— Preparation of ortho-images and image maps
— Solution based functions for specific applications.

The forthcoming GIS or image processing softwares are likely to include at least some of the above functions. However, integration of GIS and remote sensing data is being carried out in different institutions and laboratories with different ratio of manual/digital involvement.

Digital Topographical Data Standards

In September 1991, the Union Department of Science and Technology constitute a Committee of Experts to finalise the Standard Character Format of the topographical data to be produced by the Survey of India in digital form. The digital data was considered useful for the production of cartographic products such as base maps; and structured to support the GIS functions as well. Such data was to be collected, archived and distributed in vector formats in following three levels :

Level - 1

In this level, the topographic data was to be coded in prescribed standards in the simplest form without any topological linkage. The scanned digital data after vectorisation was to be supplied quickly to the potential users.

Level - 2

In this level, the full topology is not provided. However, it may not be possible to derive topographical data in DVD-3 structure from already digitised data if this structure has not been kept in view of digitizing. In the first few years Survey of India would be supplying DVD-1 and DVD-2 data only.

Level - 3

DVD-3 is a topologically structured data file which is composed of basic three separate but related elements, i.e. nodes, lines and areas. It contains text data as well. All lines, areas and texts can be assigned unique object identification numbers through which the lines/areas may be associated with text and/or non-graphic data outside the data file (SOI, 1992).

It was projected that in the first few years, Survey of India will only provide level-1 and level-3 data to the potential users. A detailed exercise was carried out to indentify the Major Data Categories and Feature Description. In addition, the data elements have attribute features or codes are based symbols used in the topographical map series as per the MRB (Map Review Board) recommendations. The whole operation is commonly known as the "National Standard Exchange Format for Digital Vector Data". However, it is expected that the digital data will soon be available to the users for GIS applications.

Geographical Data Bases Available in India

Organising data for a lerge country such as India is not an easy task. The 32 states and union territories do not use similar bases for the collection of data. As a result, the cartographic base and statistical contents vary considerably. In contrast, the federal agencies for mapping and data collection follow the same policies for the entire country; examples of such sets are the population, housing and livestock censuses, survey of industries, topographical sheets, district planning maps and thematic maps. India has also advanced in space technology and remotely sensed data is now available from the Indian Remote Sensing Satellite, IRS. Some places have attracted special attention, for which large scale maps, photographs and detailed information are available. All these informations are now available for development of a geographical information system for the whole country or its part.

Possible bases for GIS

After assessing the sources of information, data bases, experiments on information

TABLE 11.3: Geographical Bases

Number	Geographical Bases	Scale
1	NATMO thematic maps	1:1,000,000 or 1:2,000,000
2	Remote sensing data (FCC)	1:250,000 or 1:1,000,000
3	Topographical sheets	1:25,000, 1:50,000 or 1:250,000
4	PS/tahsil/taluk/majmuli	1:253,440
5	District Planning Maps	1:250,000 (inset 1:1,000,000)
6	Aerial photographs	1:15,000-1:60,000
7	Cadastral maps	1:1,014 and 1:2,028

Source: Modified from Nag (1987).

systems and possible new sources of new information, it is worth considering different types of data sets within its federal structure, great flexibility is required. Five levels (A-E) of spatial information can be identified according to the availability of data, the potential sources and the mapping work already undertaken.

Level A - 1: 1,000,000

At present maximum information is available at the 1:1,000,000 scale. The LANDSAT MSS and IRS LISS-1 data are useful at this scale. NATMO has developed a geographical information base at the same scale. The whole country is divided into 15 plates and this base has been used for the mapping of various aspects, such as administrative areas, physical aspects, population, transport and tourism, land use and cultural landscapes. Some of the maps series at 1:2,000,000 scale can be enlarged to this scale for additional information. These maps show slope, rock and minerals, rock types, rainfall, drainage, water resources, wildlife and wetlands, types of forest, density of rural population, crop regions and work-force. Some maps prepared by other institutions are based on this scale. NATMO has also developed a base where boundaries at sub-district level have been matched with the geographical features of the country. On the NATMO 1:1,000,000 base other information files or data bases can be superimposed and corrected. It can provide a useful framework to initiate a GIS for India. It can be enlarged to the extent that detailed data are available for specific uses.

Level B - 1: 1,250,000

From Table 11.3 it may be gathered that topographical sheets are available on this scale. In addition, the maps being prepared under the District Planning Map Series (DPMS) by the Survey of India and NATMO are also available at this scale. These DPMS have about eight inset maps on different themes at 1 : 1,000,000 scale.

National Remote Sensing Agency (NRSA) has also prepared a series of maps at the same scale on land use/land cover and wastelands. IRS LISS-2 and LANDSAT TM data can be meaningfully used. At this scale, location of villages can be shown but the mosaic of village boundaries would be too detailed. The greatest advantage of this scale is the availability of the PS map series.

Level C - 1: 50,000

The most important map series available at this scale is the topographical sheet prepared by the Survey of India. However, some of the maps of this series fall in the restricted category. NATMO also prepared land use maps of selected development blocks of the country at this scale. At this scale, village boundaries can be related to geographical features. Some air photographs are available at this scale. It is thought that remote sensing data can also be used at this level. Its advantage is the availability of uniform and accurate survey sheets, and great deal of research and survey work is being undertaken at this level.

Level D - 1: 25,000

A fresh topographical sheet series is now being prepared at this scale, primarily for planning and for providing greater details for flood control etc. Detailed physical features are shown on such maps. Large scale photographs are also available at this scale for priority areas. SPOT Pancromatic data can be useful input at this level.

Level E - 1: 5,000

No source of information is available at this scale. The closest is the cadastral map series. Maps at this level would be useful for high population or property density areas, such as towns and cities. Urban area authorities would welcome such maps showing details of property lines for tax and planning purposes. The Russian remote sensing data at 2m resolution can be integrated at this scale.

Assessment of GIS Packages

Several GIS packages are now available off the shelf. Some of them are purely for GIS, while with others image processing is also possible. In a recent trend, live links between the GIS and image processing or digital mapping are possible. A summary of such popular packages are as follows in Table 11.4.

GIS and the Private Sector

The contribution by the private sector towards the development of GIS software and

TABLE 11.4: Popular GIS Packages

Package	Environment	System	Data	DBMS
ARC/INFO	DEC, PRIME, DG, IBM, PC-DOS	GIS	V, TIN	INFO, Oracle, INGRES, Dbase
ERDAS	PCs/DOS, SUN, UNIX, VAX/VMS	GIS	R	Infor
IDRSI	PCs/DOS	GIS	R	Lotus, Quattro etc.
IMAGE	PCs/DOS	GIS,IP	V	Lotus, Dbase, etc.
LaserScan	DEC, VAX/VMS	GIS	V, R	RDB
PAMAP	VAX, PC-DOS, SUN	GIS	R, V	Any Relational DBMS
SPANS	PC-DOS, UNIX	GIS	Q	Dbase, RDB
ISROGIS	PC, UNIX	GIS	PM-Q	Any SLQ
GRAM	PC-DOS	GIS	R	Dbase
THEMAPS	PC-DOS	GIS	V	Dbase
UltiMap	Apollo, AEGIS	GIS, AM	V, R	Oracle, Informix, Ingress, IMS, etc.

Note: R= Raster, V= Vector, P.M.= Polygonal Map; AEGIS= Apollo Enbedded GIS; AM= Automated Mapping; IMS= Information Management System.

Source: Mukund Rao *et al.* (1994).

hardware has been enormous. Every day new advancements are brought out which makes almost imperative to keep track of the private sector. However, most is funded by governments, especially from defense budgets, much of the actual work is done by private contractors. Beyond the early R & D stage, the influence of the private sector is even stronger as entrepreneurial minds imagine new ways of using the technology developed for government (Parker, 1991). The software and hardware development in image processing of the remote sensing data has similar track record.

A user is inclined to gather impression of the path of development of different companies in private sector before accepting their proposals. There are following issues which one would like to know in this regard.

* Technical status of the product.
* User support.
* Quality of documentation.
* Training programme.
* Long term commitment to R & D.
* Scope for integration with other geoprocessing systems either available in the users' establishments or elsewhere.

* Record of technological breakthroughs.
* Market share and financial stability.
* Market reputation.

GIS is widely used in many industries, application centres and universities. It has become more integrated and less modular. With the fast changing scenario in the private sector, it is becoming possible to bring different applications in one integrated system. The integration of GIS, image processing and GPS data is a ready example. The applications of GIS will be endless.

REFERENCES

Blakemore, M. (1984), Generalization and error in spatial data bases, *Cartographia*, vol. 21, pp. 131-9.

Bolstad, Paul and Lillesand, T.M. (1991), Automated GIS integration in land cover classification, in *Technical papers of the 1991 ACSM/ASPRS Annual Convention*, vol. 3, pp. 23-32.

Burrough, P.A. (1986), *Principals of Geographical Information System for Land Resource Assessment*, Oxford University Press.

Burrough, P.A. (1992), Development of intelligent geographical information system, *International Journal of Geographical Information System*, vol. 6, No.1, pp. 1-12.

Carstensen, Jr,. Laurence W. and Campbell, James B. (1991), Desktop scanning for cartographic digitization and spatial analysis, *Photogrammetric Engineering and Remote Sensing*, vol. 57, No.11, pp. 1437-46.

Cadoux-Hudson, J. and Heywood, I. (1991) (eds), *Geographic Information 1991*, Taylor and Francis, London.

Chou, Y.H. (1992), Management of wildfires with a geographical information system, *International Journal on Geographical Information Systems*, vol. 6, No. 2, pp. 123-40.

Cosetino, Brian L. and Lillesand, Thomas M. (1991), Towards automated statewise land cover mapping in Wisconsin using satellite remote sensing and GIS *Techniques,1991 ACSM/ASPRS Technical Papers*, vol. 3, pp. 93-102.

Dale, Peter F. and John D. McLaughlin (1988), *Land Information Management*, Clarendon Press, Oxford.

De Bruijn, Cornelis, A., Sherif Amer and Desmond Dougall (1996), GIS and geointegration, *ITC Journal*, vol. 1996-1, pp. 71-6.

Dillworth, Mary E. (1991), Geographic windows in remote sensing: does window size matter ? *1991 ACSM/ASPRS Technical Papers*, vol. 3, pp. 122-8.

Egenhofer, M.J. (1992), Why not SQL !, *International Journal of Geographical Information Systems*, vol. 6, No. 2, pp. 71-85.

Frank, Andrew U., Egenhofer, Max J. and Kuhn, Werner (1992), A perspective on GIS Technology in the nineties, *Photogrammetric Engineering and Remote Sensing*, vol. 57, No.11, pp. 1431-6.

Franck, U.F. and M.J. Egrnhofer (1992), Computer cartography for GIS: An object oriented view on the display transformation, *Computer and Geosciences*, vol. 18, No. 8, pp. 975-87.

Goodchild, M.F. (1991), Geographical information science, *International Journal on Geographical Information System*, vol. 6, No. 1, pp. 31-45.

Goodchild, M.F. (1992), Geographical data modelling. *Computers and Geosciences*, vol. 18, No. 4, pp. 401-8.

Green, N.P., Finch, S. and Wiggins, J. (1986), The state of art in geographic information system,

Area, vol. 17.4, pp. 295-301.

Healey, R.G. and Desa, G.B. (1989), Transputer based parallel processing for GIS analysis: problems and potentialities, Auto-Carto, 9 Proceedings, April, pp. 90-9.

Hellen, M. and R.F. Tomlinson (eds) (1988), *Building Databases for Global Sciences*, International Geographical Union, Taylor and Francis, London.

Knapp, W.G.M. van der (1992), The vector to raster conversion: (mis)use in geographical information systems, *International Journal on Geographical Information Systems*, vol. 6, No. 2, pp. 159-70.

Lillesand, J.M. (1990), Satellite remote sensing : Its evolution and synergism with GIS technology, *Government Information Quarterly*, vol. 7:3, pp. 307-27.

Kay, Simon (1988), Geographic information systems: An integrated database for Badiyah, *The Journal of Oman Studies*, Special Report No. 3 (The Scientific results of the Royal Geographical Society's Oman Wahiba Sands Project 1985-1987), pp. 47-54.

Loveland, Thomas R., James W. Merchant, Donald O. Ohlen and Jesslyn F. Brown (1991), Development of a land-cover characteristics database for the conterminous U.S., *Photogrammetric Engineering and Remote Sensing*, vol. 57, No. 11, pp. 1453-63.

Maguire, D.J., Goodchild, M.F. and Rhind D., (1991), *Geographic Information System*, Longman, London.

Mounsey, H.M. and Tomlinson, R.F.,(1988), *Building Data Bases for Global Science*, Taylor and Francis, London.

Mark, David M. and Gould, Michael D. (1991), Interacting with Geographic information: A commentary, *Photogrammetric Engineering and Remote Sensing*, vol. 57, No.11, pp. 1427-30.

Nag, P. (1987), A proposed geographical information system for India, *International Journal on Geographical Information System*, vol. l, No.2, pp. 181-7.

Ogata, Al (1992), Making the most of today's data conversion methods, *GIS World*, April, pp. 82-6.

Parker, H. Dennison (1991), The role of the private sector in GIS, *Photogrammetric Engineering and Remote Sensing*, vol. 57, No. 11, pp. 1465-6.

Rao, Mukund, V. Jayaraman and M.G. Chandrasekhar (1994), *Organising Spatial Information Systems around a GIS Core*, Indian Space Research Organisation, Special Publication, ISRO-NNRMS-SP-7-94, Bangalore.

Rao, Mukund and V. Jayaraman (1995), *Guidelines for GIS Standardisation*, Indian Space Research Organisation, Technical Report ISRO-NNRMS-TR-105-95, Bangalore.

Rhind, D. (1992), Data access, charging and copyright and their implications for geographic information systems, *International Journal on Geographical Information System*, vol. 6, No. 1, pp. 13-30.

Scholten, Henk J. and Stillwell, John C.H. (1990) (eds), *Geographical Information System for Urban and Regional Planning*, Kluwer Academic Publishers, Dordrechht.

Survey of India (1992), *National Standard Exchange Format for Digital Vector Data*, Department of Science and Technology, Dehra Dun

Worrall, Les (1990) (ed), *Geographic Information System:Development and Applications*, Taylor and Francis, London.

TWELVE

APPLICATION TRENDS

THE TERM remote sensing is restricted to the methods that employ electromagnetic radiation or energy as the means of detecting and measuring target characteristics (Sabins, 1978). Remote sensing techniques can also provide information about the location, availability and changing conditions of natural resources in specific areas. Information is essential for establishing appropriate priorities and effective plans for natural resource management. Such techniques and data have been increasingly used in identifying the problems in the management of natural resources (Deekshatulu and Hebber, 1991; Jensen *et al.*, 1977; Kudrat *et al.*, 1990; Westins and Frazee, 1976; Rao, 1991; and Bhan *et al.*, 1991).

As discussed earlier, the spectral reflectance and emittance which are dependent on the nature and properties of target objects present in the earth surface is unique behaviour of each feature. These characteristics are being recorded in the remote sensing recorder and will provide clue for identifying, mapping and inventory of various natural resources (Hunt, 1980). The nature of remote sensing itself has changed over the past three decades. It has moved from relatively qualitative 'art' relying on inference for information to quantitative 'science' capable of measuring system states in the field of agriculture, soil (Mulders, 1987; Eugman, 1991; Westin & Frazer, 1976), hydrology (Engman and Gurney, 1991), and forestry (Trucker, 1979; Nelson *et al.*, 1984). In the present chapter, we are highlighting broad trends of remote sensing application in various field, particularly in India. These application areas are as follows:

AGRICULTURE

Application in agriculture includes different features, such as crop area, crop yield, crop growth conditions and the like. The information of above agricultural themes are needed for effective distribution of agricultural and marketing of agricultural commodities. The agricultural crop resources information are generally collected in the country by conven-tional large scale sample survey methods by several agencies viz. Directorate of Economics and Statistics, National Sample Survey Organisation, State Bureaus of Economics and Statistics, departments of agriculture and revenue.

However, the conventional system of crop production forecasting have several limitations. These limitations are in terms of time period, suitability for sampling, variations, accessibility and quick retrieval.

Different agricultural aspects viz. crop, soil and land showed different spectral be-haviour depending on the crop type and their vigour and stages of growth, and physio-chemical behaviour of soils throughout the electromagnetic radiation. Since space borne remote sensing data possesses extra advantage over the conventional data (i.e. aerial photography) and it can provide multispectral multidata and multisensor information, this data is very much useful in the following agricultural fields:

— Agricultural/crop landuse
— Crop area estimating
— Yield modelling
— Monitoring the agricultural hazards, viz. floods and droughts
— Soil mapping
— Crop stress monitoring -- water/nutrients.

LAND USE/LAND COVER

Land is the most important natural resource which embodies soil, water and associated flora and fauna involving total ecosystem. Comprehensive information on the spatial distribution of land use/land cover categories and the pattern of their change is a prerequisite for planning, utilization and management of the land resources of the country. Land use or land cover inventories are assessed in increasing importance in various sectors like agricultural planning, settlement and cadastral surveys, environmental studies and operational planning based on aero-climatic zones. Information on land use or land cover allows a better understanding of the land utilization aspects like cropping patterns, fallow lands, forests, pasture lands, wastelands and surface water-bodies which are vital for development planning.

Application methodology for this purpose can be accomplished either by visual interpretation based on the photointerpretation elements and actual field knowledge, or even digital classification based on spectral behaviour of different land cover categories in the multispectral remotely sensed data. Here a brief methodology for each technique has been highlighted.

Visual Interpretation

The visual methodology was developed by the National Remote Sensing Agency (1989) for a nationwide land use mapping from multispectral satellite data comprising the following six major steps as shown in Fig. 12.1:

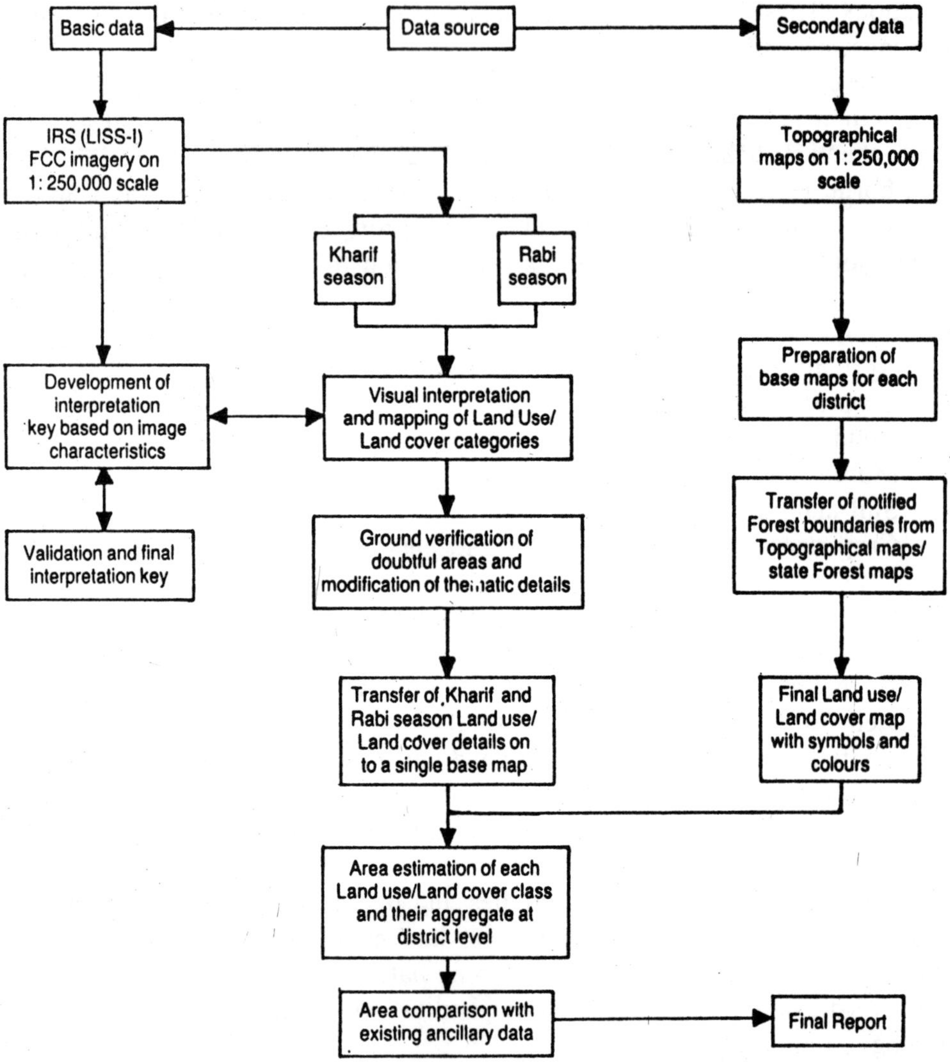

Fig. 12.1: Methodology for land use/land cover mapping using visual interpretation

— *Selection and acquisition of data*: Standard FCC imagery of IRS-TM of *kharif* and *rabi* seasons.

— *Prefield interpretation*: FCC imagery, for both the major cropping season in India, are interpreted individually based on the interpretation key. The boundaries of land use or land cover classes are being delineated following tone, texture, shadow, pattern etc. The boundaries of various categories are plotted onto a transparent overlay such as artisan on polyester sheets.

— *Ground data collection and verification*: Following the prefield interpretation scheme (map), traverse plan are to be formulated with the traverse plan and ground truth information is collected so that 10 per cent of the mapping area is being covered. The areas of doubtful preliminary interpreted site are particularly verified.

— *Postfield interpretation and modification*: Based on the ground truth information, modification are effected and classes as well as their boundaries are to be verified.

— *Computation of area*: Area under different classes are estimated by planimetric measurement instruments in order to complete mapping area land use statistics.

— *Final cartographic map preparation and reproduction*: Fair drawing of originals are made as per pre-designed specification and cartographic symbols.

Digital Image Processing

Operational methodology of land use mapping was developed by the Regional Remote Sensing Service Centre (1990). This methodology consists of following seven steps (Fig. 12.2):

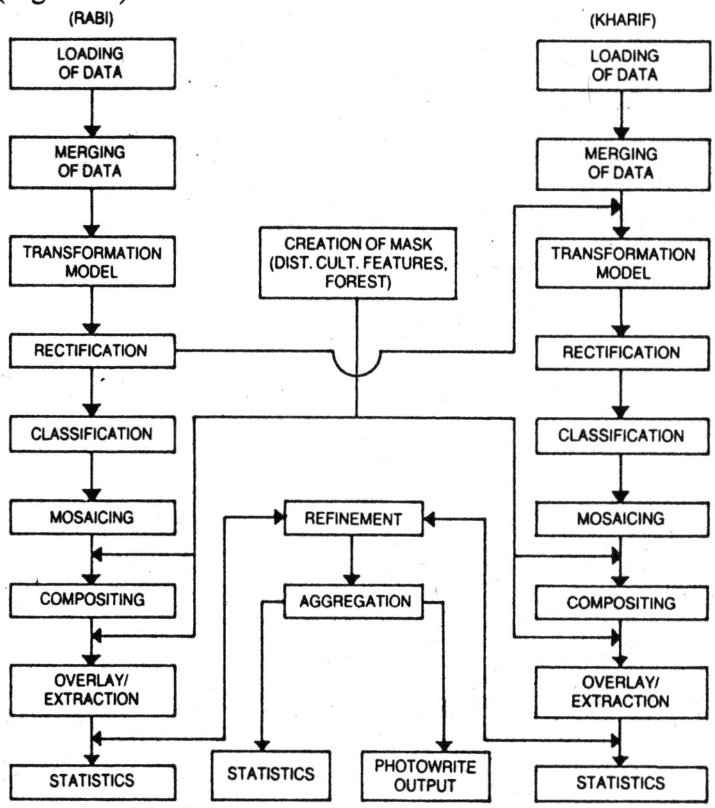

Fig. 12.2: Methodology for land use/land cover mapping using digital techniques

— Data acquisition, downloading, merging and georeferencing
— Demarcation of the concerned mapping area (area of interest)
— Ground truth collection
— Stratified classification of the two season data
— Refinement
— Aggregation of two or three season classification
— Statistics generation (main computation of area) and final output generation.

Supervised classification is carried out in stratified manner for *rabi* and *kharif* season with maximum likelihood classifier. Stratification is effected by digitisation of forest boundaries from the ancillary data source (i.e. map) and creation of work scene. The classification proceeds through selection of the training sets, calculation of statistics of the training sets and decision boundary of maximum probability based on mean vector, variance-covariance and correlation matrix of the pixels.

Normalized difference vegetation index (NDVI) or separate classifier is adapted for classification of forest areas. Gaussain post normalization is being operated for smoo-thening the scene. Further, a look up table (LUT) is prepared for depicting various classes by unique colour code.

The methodology allows human logic and intuition to refine the classified output. This operation takes care and modifies the discrepant classification arising out of spectral similarity of two or more classes in a non forest stratum. In referential refinement, classification of a pixel in one cropping season is evaluated and, if necessary, corrected with reference to its classification in other season.

Rabi and *Kharif* seasons classified scenes are reformatted to a single output through the process of pixel aggregation. Finally, the land use statistics and the colour coded land use or land cover classified map are being generated for the concerned area.

SOIL MAPPING

Research and development efforts of the past few decades have established potential application of remote sensing techniques for soil mapping with a high degree of reliability, efficiency and cost effectiveness (Karate *et al.*, 1991; Kudrat *et al.*, 1992; Myers, 1975). Innumerable studies have been reported on soil mapping with visual interpretation of remote sensing data. The methodology is very much similar to land use or land cover classification. Here we will mention about the digital image processing. Procedures for such mapping in digital environment rely primarily on statistical pattern recognition techniques for different soil classes. Implicit in this approach is the premise that the different soil classes manifest different combinations of digital numbers (DN values) based on the inherent spectral reflectance properties, attributable to their physical and chemical characteristics. Notable among these are soil colour, soil texture, organic matter, moisture content, mineralogy, iron, manganese, $CaCO_3$ and free silica.

Soil mapping is tedious and tricky since soil is a very heterogeneous and complex system. Soil reflectance is subject to the noise caused by several extraneous factors including vegetation. Soil mapping is based on the hypothesis that soil reflectance or emittance is dependent on the nature and properties of soil, terrain, and land cover. These parameters controls the soil variability. So the spectral classes must represent the soil classes. Of course it is not that simple. Also, it is not unlikely that two or more soil classes in the interested area are seemingly alike in their spectral profile. So careful consideration of the approach and procedure to be followed is therefore vital to the reliability of soil mapping, particularly in areas of diverse soil scape resulting from complex geology on intricate nature of other pedogenic factors.

Methodology

Soil mapping is accomplished in four stages. The detail methodology has been discussed by Kudrat *et al.* (1990 & 1992) and summarized in Fig. 12.3. The stages are as follows:

— Using FCC of Landsat-TM or IRS or even SPOT data and concerned topographical sheets, different physiographic units namely piedmont, valley, crust, river terrace alluvial plain, mountains etc. are to be identified considering the spectral variations with respect to the land use or land cover, soil erosion, and physiographic units, the accurate definition of the training classes are to be performed. Further, following the guidelines of Satterwhite (1984) training site selection and identification should be performed. At this stage, spectral attention is to be given so that entire spectral variability is to be covered, otherwise, it will lead to inaccuracies.

— Sample strip and soil profile site is to be planned considering the above aspect. Each and every physiographic unit must have 2 to 3 training sets and soil profiles. Representative soil profile data is to be gathered in the context of morphology, and physio-chemical characteristics. For this soil sample, all the representative profiles are to be collected for laboratory analysis. By integrating soil morphology and physio-chemical properties, soils have to be classified.

— A ground truth mask scene is to be created by integration spectral information and actual ground information. The ground truth mask scene consisting of various training classes must be generated by defining these training classes covering all the physiographic units and spectral variabilities. The training sets should be spread all over the scene, otherwise reject class (unclassified pixel) will be more.

— Generation of spectral statistics for various classes is inevitable, because this forms the basis for digital classification. Spectral statistics refer to mean vectors, standard deviations, variance-covariance matrix, and spectral separabilities between the classes. This has already been discussed in the chapter on 'Classification'. Modification of training classes is to be performed by critically examining the statistics list of each and every class.

— As soon as the training class statistic generated in the whole area of interest, classification will be performed by suitable classification scheme. Generally maximum likelihood classification algorithm is being practiced because this maximizes the proportion of correctly assigned observation when the data can be represented by multivariate normal distribution.

— Classified scene is then made smooth by adapting post classification filtering.

— Finally physiography-cum-soil colour coded map can be generated by assigning separate colour for each class.

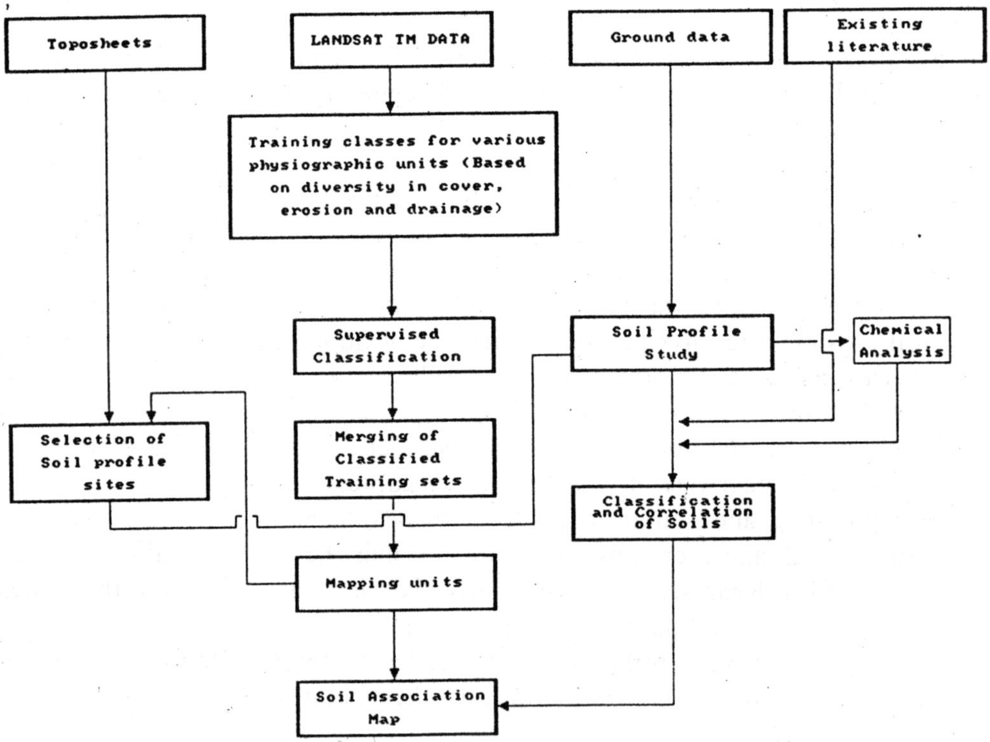

Fig. 12.3: Flow chart of methodology for soil mapping (digital)

Soil mapping is not very simple like land use or land cover mapping. Interpreter must have a thorough understanding of spectral behaviour of soils and he must have to be familiar with the concerned area. Nevertheless, many application scientists have mentioned different views regarding the soil mapping. Imhoff *et al.* (1982) and subsequently Lee *et al.* (1988) claimed that if principal component images can be used in addition to normal remote sensing data, landform delineation becomes more comfortable. Fukuhara *et al.* (1978) used ratio image for vegetated area for soil mapping. Further, several scientists (Weismiller & Kaminsky, 1978; Dipaolo & Hall, 1982; and Roudabush *et al.*, 1985) pointed out that soil map units discrimination only from remotely sensed data is problematic. However, Kudrat *et al.* (1990, 1991 & 1993) and Karate *et al.* (1991) claimed that soil mapping through digital analysis can be successfully carried out.

CROP INVENTORY

Accurate and timely information on the type of crop grown and their acreage, crop production and crop growth conditions are essential for strengthening country's food security and distribution system. Pre-harvest estimation of crop production are needed for guiding the decision makers in formulating optimal strategies for planning, distribution, marketing, transportation and storage of the essential agricultural produce.

It was the traditional practice to collect agricultural crop information in a country by conventional large scale sample survey method. In sampling approach, villages are being randomly selected following sampling procedure. Complete enumeration for land use and crops is done in these villages. This will provide the acreage estimation before the harvest of the crop. The final yield is being compiled through crop cutting experiment, carried on a number of randomly selected fields in the sample village in a district. The estimate is available only after the harvest of crops. Though the approach is statistically accurate, this procedure of crop production forecasting have several limitations. This fails to provide timely information, variation of statistical figures, lack of availability of data for minor crops like horticultural crops and cash crops.

Remote sensing techniques, visual and digital, can help in pre-harvest and post-harvest crop inventory.

CROP PRODUCTION FORECASTING

Three experiments were conducted in order to achieve the present agricultural assessment programme. These were Corn Blight Watch Experiment (CBWE), Crop Identification Technology Assessment for Remote Sensing (CITARS), and Large Area Inventory Experiment (LACIE). The objectives of above experiments were to map and assess corn blight effect in 1971 and demonstration of operational feasibility of using satellite data to make reasonably accurate crop estimate on a regional, national, and global basis.

LACIE was the first worldwide experiment to demonstrate the operational capability of satellite based remote sensing technology or crop production forecasting (MacDonald, 1980). It was jointly carried out by three U.S. government agencies, viz. National Aeronautics and Space Administration (NASA), the United States Department of Agriculture ((USDA), and the National Oceanic and Atmospheric Administration (NOAA). Landsat MSS data was 'used for wheat acreage estimation, and aero-meteorological models were employed for yield estimation'. The prediction of wheat production for U.S.S.R. was very accurate. However, results for Canada, India, China, Australia, Brazil and Argentina, though encouraging, were far from being accurate. Following the LACIE approach, a project on Crop Acreage and Production Estimate (CAPE) is now operational in India under the aegis of the Remote Sensing Application Mission (RSAM) (Anonymous, 1986; Navalgund *et al.*, 1991). Further, crop production forecasting of a region comprises identification of crop cover, acreage estimation, forecasting yield of the crop and assessment of crop condition (stress due to fertilizer or nutrient or water) during growth cycle.

CROP ACREAGE ESTIMATION

For the crop acreage estimation spectral reflectance characteristics of the crop must be understood. This is very much similar to the vegetation curve and was explained in the earlier chapter. However, it is discussed here once again in a summarized form. The typical pattern of crop or vegetation are divided into following three zones (Fig. 12.4):

> Zone I (0.4-0.7 μm: visible): Strong absorption by the leaf pigments which are namely chlorophyll 'a' and 'b', carotenes and xanthophyll. The portion of EMR absorbed by the plant leaves in this region are referred as photo-synthetically active radiation, and is used for the photosynthetic activity.
> Zone II (0.7-1.3 μm: near infrared): The high reflectance of the vegetation is due to the internal structure of spongy mesophilic cells and canopy geometry.
> Zone III (1.3-2.6 μm: SWIR or middle infrared): Absorption by leaf tissue water and characteristics of water absorption peak are at 1.45, 1.95 and 2.6 μm.

The spectral characteristics of crop canopy is influenced by leaf area index (LAI) and per cent crop cover, growth stages, agronomic practices, stress conditions, solar elevation and solar azimuth angles. Background soil, moisture content of the soil have significant influence on the vegetation reflection or absorption during the early stages of crop growth. Since every crop has its own unique leaf arrangement, growing period, and unique internal leaf structure, spectral signature or characteristics will enable the user to discriminate various crop cover through remote sensing techniques. Several times it has been observed that two crops shows similar spectral signature on a given data. This can be circumvented by using multidate data.

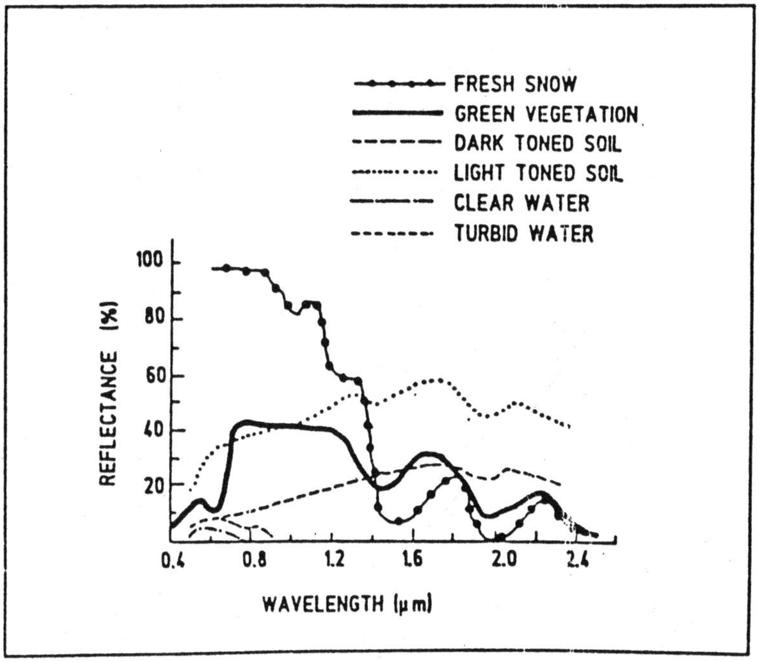

Fig. 12.4: Zones of cropping pattern

Crop area estimation can be accomplished by either visual or digital techniques. Here we will describe the digital analysis only, since visual interpretation is easy, subjective and very much similar to land use and land cover mapping.

Digital Analysis

This technique is non-biased and less subjective, and can detect a small variation in radiance values from land cover type. It utilizes the full dynamic range of the DN values (gray values) present in the scene in multi-spectral mode. This is cost and time effective. That is why digital technique is popular for crop acreage estimation, condition assessment and yield forecasting. For a large area, this is accomplished through sample segment approach to avoid spectral variation due to the variation of soil, management practices and other environmental factors. A brief methodology is presented in Fig. 12.5. Following steps are involved in crop acreage estimation:

- — Sampling plan
- — Sample segment extraction
- — Training signature generation

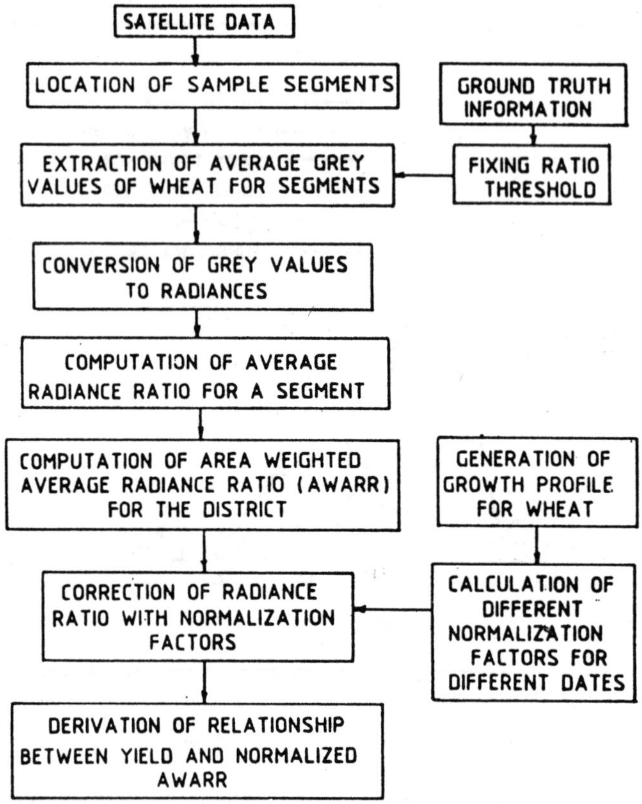

Fig. 12.5: CAPE Methodology

— Supervised classification
— Aggregation for acreage estimation

Sampling Plan and Stratification

Sampling plan consist of the stratification of the area, selection of sample size and selection of sampling fraction. Further, by stratification, the homogeneous yield potential zones of the study area is separated. In crop acreage estimation, stratified sampling techniques are commonly being practiced to overcome the problems of handling large volume of data hydrogenaty, and large efforts in ground truth data collection. There are a number of means to stratify an area. These are listed below:

— Aero-physical or agroclimatic zones
— Historical crop proportions and yield levels
— Vegetation density and crop vigour in the region based on the spectral variability on the satellite images.

Aero-physical stratification can be accomplished by spatial distribution of similar aero-physical zones (e.g. physiography, soil etc.) in the state. In case of aero-climatic stratification, each agroclimatic zone (i.e. homogeneous climatic conditions) in the state is treated as stratum (Fig. 12.6).

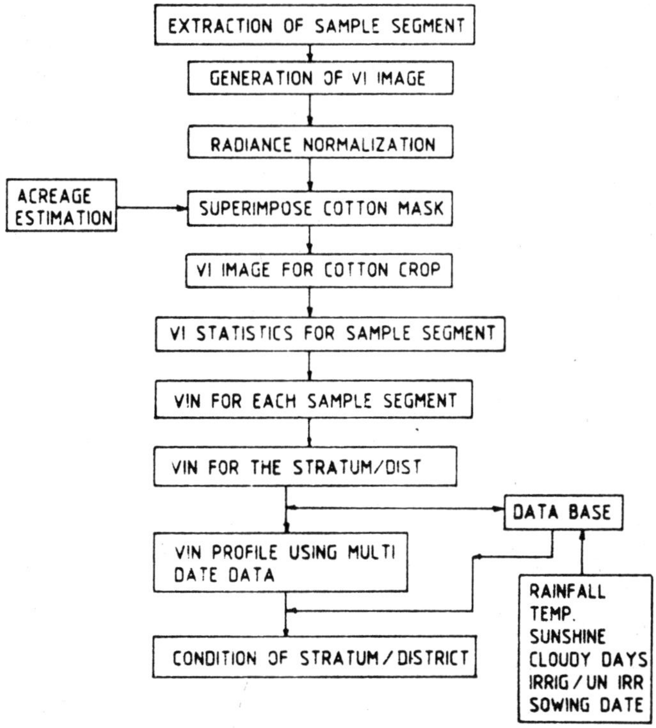

Fig. 12.6: Stratification

In historical crop statistics approach, crop acreage and yield data at district or *tehsil* level is prerequisite. For this, per cent geographic area under the crop of interest is calculated and is used for stratification for crop acreage estimation. This method has some limitations, it will follow only administrative boundaries. Many times this does not represent real time homogeneity.

The procedure for estimating vegetation density and crop vigour in the region based on spectral variability takes into account of standard false colour composite (FCC) hard copy at a scale 1:250,000 for the previous year. However, current year's FCC is preferable. The stratification is being practiced by visual interpretation of crop or vegetation density and crop vigour etc. based on the tonal variation. This kind of stratification is superior for crop yield forecasting and crop acreage estimation, and is especially suitable for areas which have well defined crop or vegetation strata due to variation in physiography, existing irrigation, command area network and the like.

Multi-stage Stratification

Any kind of stratification which considers more than one criteria is called multi-stage stratification. A combination of various criteria of stratification has been described earlier can be used for subsequent levels of stratification. In our country, three stage stratification is in practice for district level acreage estimation in addition to state level. Nevertheless, two stage stratification is suffice to predict district level crop estimates.

Each stratum is further subdivided into small square termed as sample segment grid. Size of the sample segment is largely determined by geographic extent and cropping system and practice of the area of interest. Most widely practiced sample segment sizes are 10 × 10 sq km and 5 × 5 sq km. For state level statistics generation 10 × 10 sq km is being practiced in India (Anonymous, 1995).

Selection of Sampling Fraction

Statistical sampling enable to predict large area estimate based few smaller estimates with minimum standard error. The choice of sampling fraction is governed by many parameters namely size of area, cropping pattern (mono cropped or multiple cropped), homogeneity and desired accuracy (i.e. optimum fraction to meet the 90/90 accuracy goals for crop acreage estimation). There is no thumb rule that a particular sampling fraction will provide best estimation. According to the area and its cropping pattern, specific sampling may be designed. For crop acreage estimation and production estimation in India, 10 per cent level has been practiced. However, at district level, 10 per cent sampling fraction showed reasonable estimation with fair amount of standard error.

Sample Segment Extraction

Sample segment extraction involves the process of identification of the sample fragment in the image and extraction of the scene of the desired sample area. This operation is performed in following three steps:

- Selection and distribution of GCPs and development of map to image transformation model.
- Determination of the image co-ordinate of top left corners of the sample fragment.
- Split or extraction of sample from the scene.

Training Signature Generation

For this purpose first of all training sets are to be identified. As soon as the training sets identification and delineation is over, spectral statistics for the training classes will be generated, and this is already discussed earlier. This signature will be used for classification.

Classification

The sample segment is classified by using maximum likelihood classifier. Mean area for the interested crop of all the samples within each strata will be computed.

Aggregation for Acreage Estimation

Crop acreage within a stratum is computed as a product of stratum population and the corresponding mean obtained from the sample estimate. The state or district estimate will be obtained by aggregation of acreage for different strata using the formula for stratified sampling design (Cochram, 1977). The simplified formula for the whole of area of interest (A) estimation will be as follows:

$$A = \sum_{i=1}^{n} W_i X_i \qquad \ldots\ldots \quad (1)$$

Where X_i is the crop production estimate in the ith segment, n is the number of digitally processed segments and W_i is a known weight based on the total number of segments and historical data on crop acreage.

Accuracy Assessment of Remote Sensing Estimate of Crop Acreage

In two different ways accuracy estimation can be accomplished by (a) site specification (local), and (b) non-site specific (total areal extent) methods:

(a) *Site Specific*: This approach provides a more informative appraisal of digital classification. Classification accuracy in a broad sense refers to the correspondence between classification results and actual observation in the ground. It is determined by classification of pixels in the training windows which are used for training signature generation. The accuracy is judged on the basis of correctly classified pixels, omission and commission errors.

(b) *Non-site Specific*: This statistical accuracy estimation approach compares the total area estimation, determined from remote sensing data with that of corresponding field measurements without regard to location. Statistical accuracy considers two components, viz. precision and bias. Precision refers to reliability with which a certain estimate is made, while bias refers to the deviation of the estimate from the true value. Precision in crop production estimation is measured in terms of co-efficient of variation (CV) of remote sensing based estimate and bias is measured as relative deviation (RD) from a standard estimate, e.g. estimate made by the Directorate of Economics and Statistics. The relative deviation (RD) can be expressed by:

$$RD = \frac{RS - Ref.}{RS} \times 100 \quad \ldots\ldots \quad (2)$$

where RS = Estimates of crop acreage from remote sensing
Ref = Standard estimate or reference estimates.

The most commonly used accuracy for crop acreage estimation is termed as the '90/90 accuracy criterion'. This means that only 10 per cent of the observation can deviate from the true population. The 'RD' and 'EV' of the estimated acreage are used to assess whether the goal of 90/90 accuracy criterion has been achieved.

CROP YIELD FORECASTING

Crop yield modelling refers to predict production per unit area with the help of remote sensing data. Production per unit area is an important component in crop production forecasting system. The various types of crop yield models which are in practice for crop yield prediction utilizing remote sensing data, meteorological and other collateral data. These are grouped under four broad heads, viz. (i) spectral yield models, (ii) meteorological yield models, (iii) time series yield models, and (iv) integrated yield models. However, here we will mention the spectral yield model which is basically considers the remote sensing data as input.

Spectral Yield Models

Spectral information of the crop is an integrated manifestation of the effect of the crop growth condition and the environment. Vegetation indices (VI) are the indicative of plant parameters, such as biomass, leaf area index (LAI) or physiological processes like photo-synthesis and/or evapotranspiration. Further, absorption in red region and high reflectance in the near-infrared region are related to the vigour of the crop. The radiance measured in red and infrared region, and their ratio, normalized differences, linear combination etc. redefined as vegetation index, which has been discussed earlier, are the indicators of crop vigour and may thus be related to yield. Furthermore, two approaches are commonly used for yield model generation. They are single date data, and crop growth profile developed using multidate data.

Single date data: In this approach spectral data or derived parameters (such as vegetation index) are related to yield. This is basically a statistical technique which relates spectral VIs of the crop at a particular stage which is normally peak vegetation growth or maximizing LAI to its yield (historical) through regression. Measurement of vegetation index generated or derived from remote sensing data acquired at crop stages other than peak vegetative growth stage generally do not relate well to yield and should be avoided. The occurrence of peak vegetative stage even for the same crop may differ in different regions based on existing agroclimatic conditions and management practices. The local crop calendars and in season crop sowing information may be used to decide the period of remote sensing data acquisition. The detail procedure is shown in Fig. 12.7.

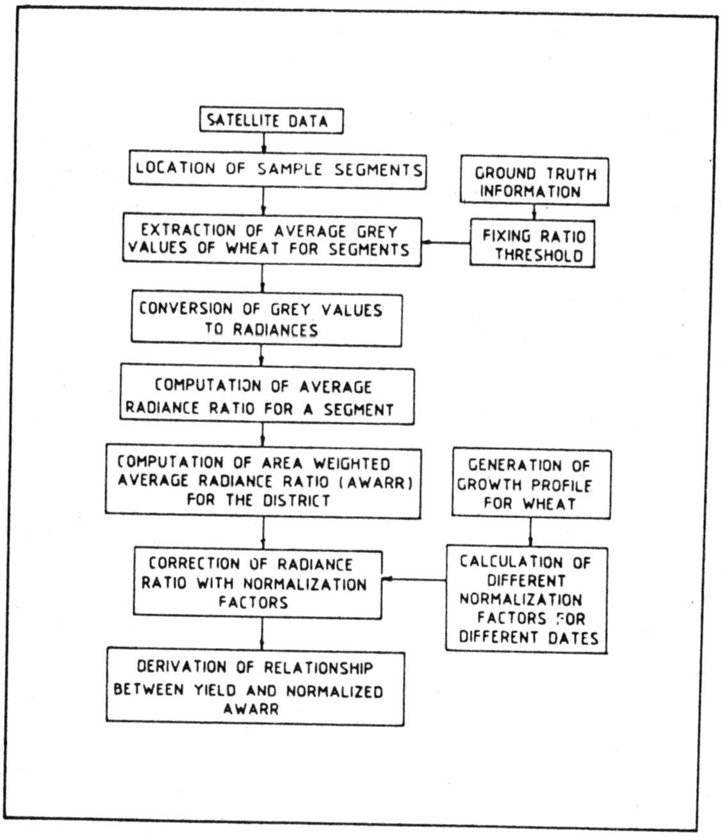

Fig. 12.7: Crop yield forecasting methodology

Using the classified image for sample segment a binary mask is generated to determine the crop of interest from other crop or land cover classes. The mask will have a value '1' for the pixels belonging to the crop of interest and '0' for land cover classes. The crop mask when superimposed on original four band data will allow only the pixels belonging to the concerned crop for further processing.

Conversion of Digital Numbers into Radiance

Spectral digital numbers is proportional to spectral radiance. The conversion of digital number into spectral radiance value will provide better yield model. This conversion requires sensor calibration constants, viz. bandwise offset and gain. A typical radiance conversion needs a coefficients A_i and B_i for converting digital number (DN) to radiance Y_i as per the following equation for band 'i':

$$Y_i = A_i DN_i + B_i \quad \ldots\ldots \quad (3)$$

where A_i is gain and B_i is the offset.

Sensor to Sensor Transformation

Due to the cloud and unsuitability of acquisition date, many times users may have to use different sensor data. Response characteristics of the sensors differ to the extent that simple radiance calibration is not sufficient for getting actual spectral indices while using data from more than one sensor. Bring homogeneity of the spectral data from different sensor is inevitable. By using common site location on different sensor data sets of the same date, spectral indices may be obtained (Markham and Baker, 1987; Suits *et al.*, 1988).

Regression analysis on radiance calibrated average spectral radiance in near infrared (NIR) and red bands from 10-15 sites of 3 × 3 windows for different surfaces collected through Landsat MSS and IRS LISS I sensors give the following relationship as shown in Fig. 12.8 (Patel *et al.*, 1991).

$$\text{IRS } B_4 = -0.45 + 1.84 \text{ MSS7} \quad (r^2 = 0.99) \quad \ldots\ldots \quad (4)$$

$$\text{IRS } B_3 = -0.32 + 1.18 \text{ MSS5} \quad (r^2 = 0.99) \quad \ldots\ldots \quad (5)$$

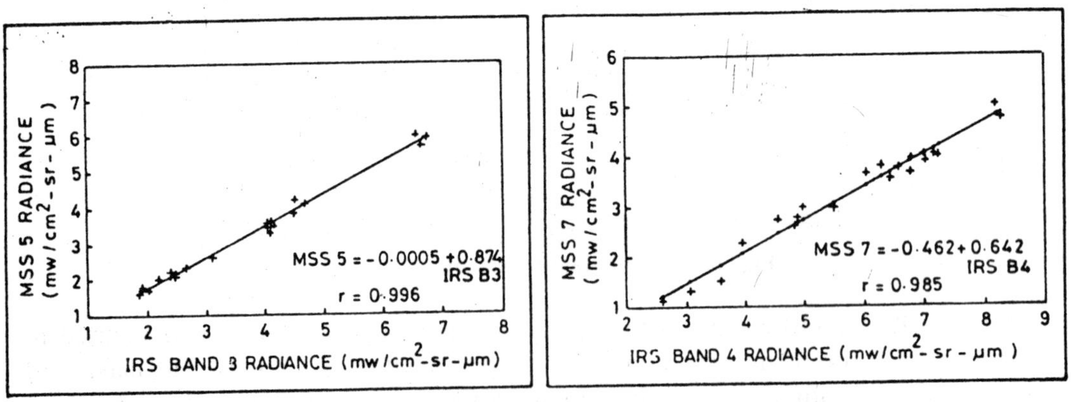

Fig. 12.8: Normalized factors calculation procedure

Normalization for Acquisition Data

Vegetation indices (VI) required for yield estimation should correspond to the peak vegetative growth phase of the crop. This signifies that acquisition of satellite data should be done within a narrow period or a 'bio-window' corresponding to heading or flowering stage of the crop. Spectral vegetation indices derived from satellite data acquired outside the specified period need to be normalized with respect to the peak value for the crop, using growth profile approach.

Statistical Analysis

Spectral indices after calibration and date normalization are weight average at the area of interest (district/*taluk*) and related to yields of the corresponding units using regression techniques. Since yield is not a direct observation from remote sensing data, an assumption based on functional form is to be developed. Normally, a linear relation is assumed for yield-spectral index data sets. In order to achieve consistency of relations in time and space, the exercise of generating vegetation indices (VI) should be repeated for three or more consecutive years. In practice, historical yield values of the concerned area obtained using crop-cutting experiments are used in developing regression equations.

Growth Profile based Model

Growth and decay of crop canopy can be characterized by temporal plot of spectral indices. A mathematical function fitted to this temporal spectral pattern is known as spectral crop growth profile. The peak of this profile corresponds to peak vegetative cover of the crop. Further, crop growth profile can also be used to derive information on the crop stage for a given remote sensing acquisition data, data normalization for data acquired at stages other than peak vegetative stage crop vigour and crop identification.

Crop growth profile can be generated using radiometric measurements, NOAA-AVHHR data or from data obtained from sensor of medium resolution. Badhwar (1982) has suggested the following empirical expression for crop growth profile (Fig. 12.9).

$$G_t = G_o\ (t/t_o)^{\alpha} \exp \beta\ (t_o^2 - t^2) \text{ for } t > t_o \quad \ldots\ldots \quad (6)$$

$$G_t = G_o \quad \text{for } t = t_o \quad \ldots\ldots \quad (7)$$

where, G_o = Soil base spectral index

t_o = Spectral emergence date

α, β = Growth parameters

g_t = Spectral index at time 't'

G_m, the peak VI is derived by substituting t, the time of peak occurrence in the

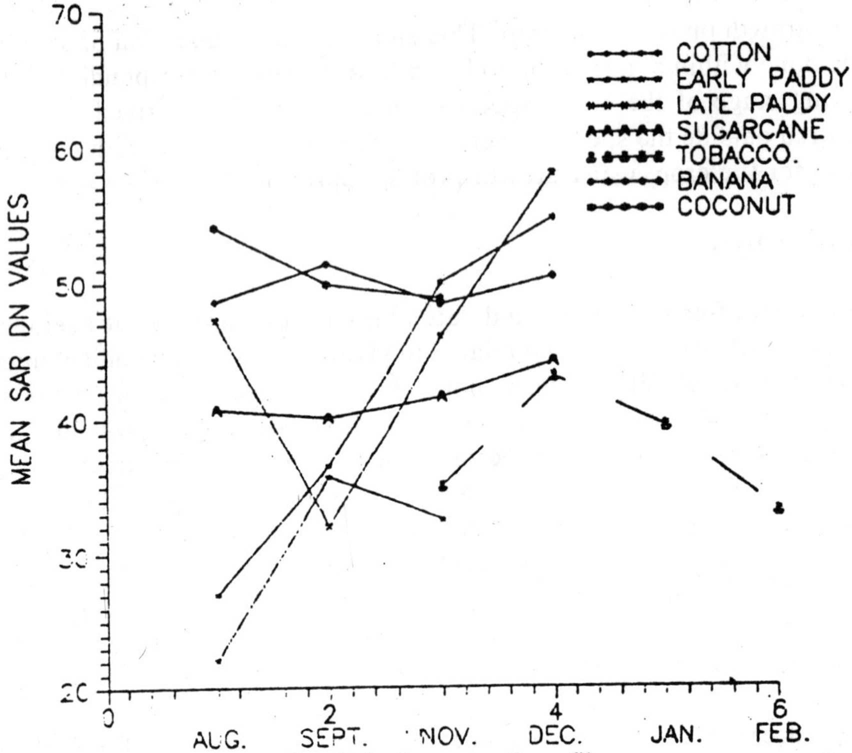

Fig. 12.9: Crop growth profile

growth profile equation:

$$G_m = C_t\,(t_m) \qquad \text{where } t_m = \frac{\sqrt{\alpha}}{2\beta} \qquad \ldots\ldots \quad (8)$$

Alternatively, a graphical plot of the profile can be used for generating approximate correction factor (C_t). The relationship between these parameters and crop yield can be served to develop a yield model. The parameter area under profile was found to be the best parameter for wheat crop.

Besides these a number of models are available to predict yield in the literature on the application of remote sensing in agriculture. However, their scope of discussion is limited in this book.

Change Detection

Change detection provides the information of natural resources changes and the trend of changes. This information is of immense use in formulating plans for development. Changes have occurred as a result of various activities including shifting cultivation, permanent cultivation, logging and clearing of forests, grazing and urbanization. Systematic investigation of such environmental dynamics should be undertaken to know the present status and trend of changes for sustainable ecological, agricultural and developmental strategies.

Change detection can be accomplished by visual interpretation as well as digital analysis of remote sensing data which is temporal in nature. Here we will outline the concept and possible approaches of digital change detection using remote sensing data.

Change Detection Techniques

Various procedures have been proposed in the literature (Crapper and Hyson, 1983; Singh, 1986; Colwell and Weber, 1981). The fundamental assumption is that any changes in land cover will result in changes in reflectance values which are sufficiently large to register despite other variables. The digital technique for change selection comprises many steps:

(a) Registration of temporal images,
(b) Sensor must be same but many times same sensor data may not be available. The two or many date data should be transformed with respect to one data in respect of radiometry and spectral bands. This operation is referred to 'normalization'.
(c) Generation of residual images are to be performed by taking the algebraic difference of any two remote sensing data (imagery) or transformed remote sensing data (NDVI or classified data). In this section we will discuss various approaches for change detection analysis. Some other techniques are also available.

Some of the popular techniques for change detection have been mentioned below:

Image Differences

In this method, first of all image is vital and then normalization of the images. Normalization refers to the transformation of different images into same radiometric condition by regression analysis.

Few areas are to be selected where there is no change. Then the gray values (digital numbers or DN) of the same area (site) were regressed to find out the regression equation. As soon as the regression equation is known, then by applying the linear band combination remote sensing digital data can be normalized. The resulting digital number for each pixel should be used to generate residual images.

By subtracting the two data images, we will be able to generate residual images. The resulting image will show the gray values for each pixel positive, negative or zero values. Both the positive and negative values represent the change areas, while the zero value will represent the 'no-change' areas. Pixels of no digital value change are distributed around the mean while pixels of radiance change (digital number) are distributed in the lower ends of the distribution (Stauffer and McKinney, 1979). Single band images or combination of different bands can be used in this technique.

Ratio Image Differences

With the help of two or more bands, images will be transformed into vegetation index images, e.g. NDVI, VI, GVI etc. which are dealt earlier in the chapter on Band Combination. The above vegetation index has been found to be strongly correlated with the amount of green biomass (Tucker, 1979) and is used here as a basis for comparing green biomass at different times. Image differencing were again used to detect changes between the ratioed images.

Principal Component Analysis

This multivariate technique is used to decorrelate and decompress the bands and accounts for most of the variance in the original multispectral images. Data from two or more dates are then compared as in the case of image differencing (Lodwick, 1979; Byrne *et al.*, 1980). Principal component analysis is separately performed on the whole of the data sets from each date. The transformation was investigated because it involves linear combination of data from all the input bands. The analysis uses the co-variance or correlation matrix. Principal component two (PC_2) from each date can be used in the image differencing procedure. PC_2 was chosen because it will emphasis the highly uncorrelated features and the bands. PC_2 is likely to be most suitable for the enhancement of vegetation features. On the other hand, PC_1 contains variability due to overall scene brightness, while the higher principal components (PC_3, PC_4....) accounts for very little variance in the original data set.

HYDROLOGY

Water is one of the most prevalent substances in the earth atmosphere system and it is one of the most fundamental to the existence of life. Life on the earth evolved in aquatic environment which depends on water for sustenance. However, distribution of this resource is highly uneven. The collection of basic data on water balance of the sea and river basins, continents and the planet as a whole is essential for rational utilization and protection of water resources and for understanding the evolution of our planet (Engman and Gurney, 1991).

For last few decades, remote sensing was considered as practical tool for mapping. With the advent of many sophisticated instruments in this field, more quantitative measure-ments have been gaining popularity in the disciplines like hydrology, oceanography, and solid-earth geophysics, where few fundamental data is available. In the recent years, the growth of hydrological sciences has now resulted in a much greater need for data in different areas including in remote areas. Hence demand for remotely sensed data has increased. There is an awareness among the hydrologists that more efforts can be put in developing the new models that can deal with many aspects of hydrology, taking the measurements of remote sensing data. Most enchanting aspects of remote sensing is its ability to measure spatial information as opposed to point data from which most of our hydrologic concepts and models have been developed (L. Vovich, 1979).

Remote sensing techniques provide unique information on the study and management of surface and ground water, spatial and temporal data on precipitation, soil moisture, transpiration, evaporation, and how drainage patterns lend themselves to the development of a wider range of hydrological models. In this section, we will discuss the exploration of ground water, distribution of surface water and their management and a few but related aspects of hydrological models.

Ground water: Ground water occurrence at any place on the earth surface is not a matter of chance but a consequence of the interaction of climate, geology, hydrology, geomorphology and ecology. The ground water occurs mainly in three types of geological formations:

(i) *Hard rocks*: These include crystalline rocks namely granite, gneiss, schists, basaltic rocks and compact sedimentary formations like sandstone and cuddapahs in India.

(ii) *Semi-consolidated rocks*: This type of rocks cover only 5 per cent of the total geographical area in India. The lithology includes primarily semi-consolidated sandstone formation of mesozoic and tertiary age.

(iii) *Unconsolidated rocks*: These rocks comprise the alluvial plain, fans and piedmont and river terrace. The lithology may consist of sand, gravel, pebbles and the like which are suitable for ground water recharge and can store large quantity of water.

Approach and Methodology for Ground Water Exploration

Owing to the synoptic coverage of the remote sensing data, it will provide quick and useful baseline information on the factor controlling the occurrence and movement of ground water like geology (lithology and structure) and geomorphology, soils, land use or land cover etc. Satellite imagery can be used in delineating and identifying linear features such as fractures or faults and paleo channels which are usually the zones of localization of ground water in hard rock and alluvial areas respectively. Above features are often good aquifers. The general methods and guidelines for detecting aquifers have been summarized by Moore (1977, 1988). These features relate to identification of springs. Seeps and phreatophytes indicate the presence of shallow water table conditions, differentiation of vegetation that are also closely related to depth and salinity of ground water, location and monitoring of ground water under stress. The detailed methodology has been presented in Fig. 12.10.

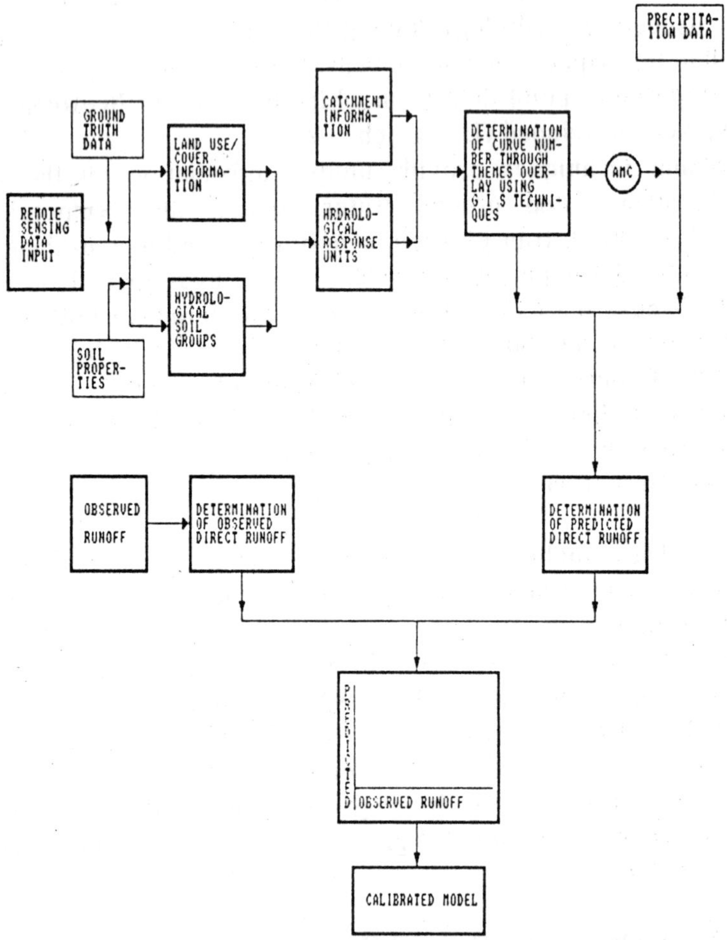

Fig. 12.10: Procedure of different parameters derivations

Considering the synoptic view, near orthographic, multispectral nature of space borne remote sensing data, Indian Remote Sensing Satellites has unique capability of providing regional level information on hydrogeomorphology with fair degree of accuracy and reliability. Though no detailed lithological interpretation was possible from the remote sensing imagery, differentiation of broad rock groups has been made based on factors like spectral characteristics of rock types in conjunction with associated structural, drainage and geomorphic features (Hunt, 1980). Fractures, faults and their intersection in the hard rock terrains as well as certain synclinal structures which at places are the loci for ground water occurrence in sedimentary rocks must be interpreted carefully from remote sensing data. There is much likelihood of missing the above features in the ground surveys. While certain land form like buried channels, valley fills, alluvial fans and the like having established relationship with ground water occurrence were readily mapped from the remote sensing imagery, the utility of the satellite data in studying in drainage pattern or paleo channels which are the bearing on the recharge conditions and localizations, have been gaining popularity. Some hydrogeomorphological maps have been prepared based on remote sensing data in India (DOS, 1988; Sahai *et al.*, 1985) (Fig. 12.11). The steps for preparation of such maps are as follows:

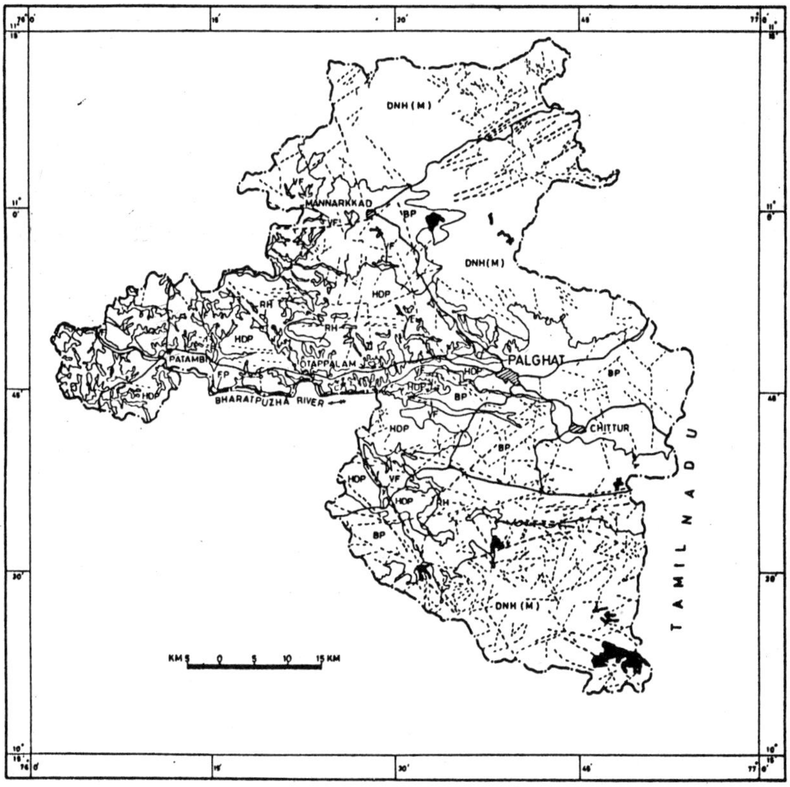

Fig. 12.11: Steps for hydrogeomorphological mapping

Base map preparation: In order to initiate this work, base map is to be prepared from topographical sheets showing usual features like drainage course, important localities, roads, rail lines and canals.

Prefield interpretation: Next step involves a systematic visual interpretation of satellite imagery to delineate various geomorphic features or land form and their depiction on the base map. These geomorphic features or land form were then evaluated critically in terms of major lithology associated structural feature, drainage patterns, broad soil types and their depth, the thickness of weathered mantle in the case of handrock areas and the land use or land cover classes.

Ground checks: Existing geological maps and other ancillary data were also used, supplemented by extensive field checks.

Post-field interpretation: By integration of ancillary data (geology, lithology, land use, geomorphology) and ground observations, hydrogeomorphological maps will be finalized.

Final map preparation: Following methodology for final map preparation by visual interpretation of land use or land use map, final hydrogeomorphological maps will be prepared (Fig. 12.12).

REMOTE SENSING STUDIES

↓

ACQUISITION OF SATELLITE DATA FCC PRINT ON 1:250,000 SCALE
or
1:1M FCC TRANSPARENCY

↓

BASE MAP PRELIMINARY INTEGRATION COLLATERAL PREPARATION OF ENLARGED IMAGE AT DATA 1:250,000 SCALE

↓

GROUND CHECKS

↓

FINAL INTERPRETATION

↓

PREPARATION OF FINAL HYDROGEOMORPHOLOGICAL MAPS

Fig. 12.12: Preparation of Hydrogeomorphological Maps using remote sensing data (*Source*: Sahai *et al.*, 1985).

SOIL MOISTURE

Soil moisture is the temporary storage of precipitation within the shallow layer of the earth (soil profile) that is generally limited to the zone of aeration. Soil moisture follows many paths and is a dynamic phenomenon. This can be returned to atmosphere through evapotranspiration, goes to ocean through runoff, and can percolate to a phreatic saturated zone as ground water recharge. Soil moisture is highly variable as the result of the inhomogeneity of soil properties, topography, land cover and the non-uniformity of input from precipitation.

Soil moisture is a very important parameter to the agriculturists as well because it is essential for seed germination for the growth and development of agricultural crops and many agricultural operations. Regional level estimation of soil moisture is an important input for crop yield modelling, water balance computations, irrigation scheduling, watershed modelling and the like (Idso *et al.*, 1975; Salomonson *et al.*, 1975; Kite, 1989; Draper and Rao, 1986; Bondelid *et al.*, 1982 and Price, 1982).

There is a need to develop ways to quantify the spatial variability of soil moisture to determine the scale of hydrologic units that characterize various hydrologic processes; and to develop criteria for delineating areas that can be treated as hydrologically uniform. Remote sensing data is advantageous over conventional methods for providing aerial estimates of soil moisture rather than point measurements which seldom yield information to characterize major hydrologic processes (Engman and Gurney, 1991). Here we will highlight some of the general approaches.

(i) *General approach of assessment of soil moisture using remote sensing*: Remote sensing of soil moisture can be accomplished by all regions of electromagnetic radiation (EMR). Successful measurement of soil moisture by remote sensing techniques depends on the type of reflected or emitted radiation.

(ii) Visible/Near-infrared remote sensing technique: It is established that moist or wet soil shows lower albedo or reflectance in wavelength region ranging from 0.3 to 1.5 μm than dry soil. Confusion factors such as organic matters, soil pads under ploughing condition (roughness), texture, angle of incidence etc., also effect the reflectance in the visible or near infrared region and make this approach impractical (Jackson *et al.*, 1978).
Infrared black and white and colour infrared photography are of particular value for sensing the spatial variation in surface soil moisture status quantitatively. Space borne remote sensing data having spectral bands in this region are also useful for large area, quantitatively assessing soil moisture conditions. The principal application of remote sensing in the visible or near infrared would appear to be qualitative rather than quantitative in respect of soil moisture. However, wetness index can be attempted in assessing for soil moisture (Crist and Cicone, 1984).

(iii) Thermal infrared remote sensing techniques: Surface temperature is primarily dependent upon thermal inertia of the soil. The thermal inertia, in turn, is dependent upon both the thermal conductivity and the heat capacity which increases with soil moisture according to the relationship as suggested

by Price (1982):

$$DT_s = T_s\ (PM) - T_s\ (AM) = f\ (I/D) \qquad \ldots\ldots \qquad (9)$$

where

DT_s = Diurnal temperature difference between afternoon surface temperature and early morning temperature.
T_s (PM) = Surface temperature in the afternoon
T_s (AM) = Surface temperature in the early morning
D = Diurnal thermal inertia = $w.Q_c k$
w = Day length
Q_c = Volumetric heat capacity
K = Thermal conductivity

'D' describes the ability of the soil to resist temperature change. For example, a dry sand has a relatively low value of 'D' compared with lower than that of clay and the volumetric heat capacity value for dry soil is lower than that for wet soils. Thus, by measuring the amplitude of the diurnal temperature change, we can develop a relationship between the temperature change and soil moisture. However, this relationship depends upon soil type, and is largely limited to bare soil conditions. Vande Griend *et al.* (1985) used thermal data from geostationary satellites to demonstrate the feasibility of measuring soil moisture.

(iv) Microwave remote sensing technique: Microwave remote sensing for measuring soil moisture and its spatial variation include both passive and active microwave approaches with several advantages. The theoretical basis for measuring soil moisture technique is based on the large contrast between dielectric properties of liquid water and dry soil. The large dielectric constant for water is the result of the water molecule's alignment of the electric dipole in respect to an applied electromagnetic field. The dielectric constant of water is approximately 80 as compared with dry soil which is of the order of 3 to 5. Thus, as the soil moisture increases, the dielectric constant can increase to a value of 20 or more (Schmugge, 1983). So the difference in dielectric properties between wet and dry soils permit the remote detection of moisture content in soil (Macdonald, 1980). For active microwave remote sensing, increase in soil moisture content increases the measured microwave back scattering by about 6-10 dB or even more. For passive microwave remote sensing, this change in moisture content decreases emissivity from 0.95 to 0.6 or less.

Here we will discuss each of the techniques separately.

Active microwave technique: In active microwave technique, the measured radar back scattering (St) is made up of backscatter from vegetation (Sv) and soil (Ss) and the attenuation caused by vegetation canopy (L). These terms can be related by simple model:

$$St = Sv + Ss/L \qquad \ldots\ldots \quad (10)$$

The soil backscatter, Ss, can be mathematically related directly to soil moisture content.

$$Ss = R \,.\, \alpha \,.\, Mv \qquad \ldots\ldots \quad (11)$$

Where, R is the surface roughness term, α is a soil moisture sensitivity term and Mv is the volumetric soil moisture content. Although 'R' and 'α' are known to vary with wavelength, polarisation and incident angle, there is no satisfactory theoretical model suitable for estimating these terms independently. Back scattering is dependent upon soil texture, roughness and overlaying vegetation. The relationship between measured backscattering and soil moisture requires an empirical relationship with ground data. Each of the above factors that influence the measurement of soil moisture has been discussed below:

Surface Roughness: Roughness is relative to the wavelength by the band used and are highly dependent upon angle of incidence of viewing. Fig. 12.13 shows how the microwave backscattering changes with the variations of angle of incidence. From the works of Ulaby *et al.* (1986), it can be recommended that angle of incidence less than 30 degree will minimise the effect of surface roughness. But for the most rough field, the back scattering is almost independent of angle of incidence.

Vegetation Effect: The effect of vegetation on active microwave remote sensing of soil moisture is greatly dependent upon the angle of incidence, frequency and polarisation. From the Fig. 12.13 it can be seen that the attenuation for the horizontal polarization is relatively weak, but the vertically polarised data are attenuated to a much greater degree because of their relationship to the canopy structure, which consists primarily of vertical stalks. From the researches of NASA (1988), it can be seen that depth of penetration increases with the increase in wavelength.

Measurement Depth: The depth of penetration of the microwave in soil is a function of moisture content. This is represented by the Fig. 12.14. In a series of field experiments with a C-band H polarization of the microwave, Bruckler *et al.* (1988) showed that the experimental results of penetration depth

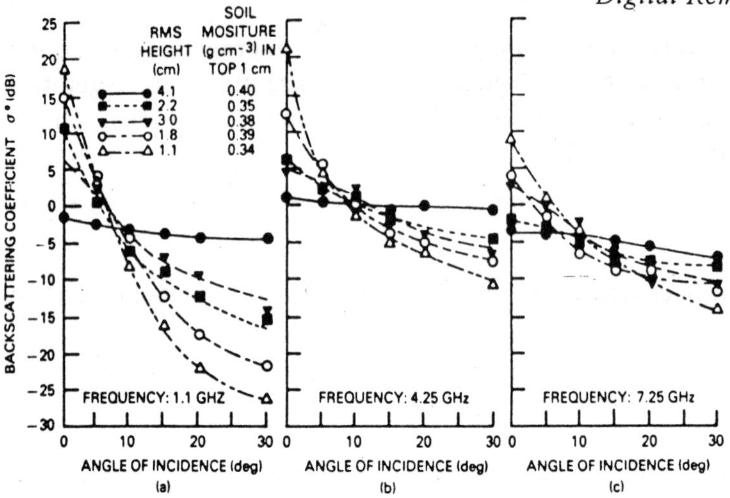

Fig. 12.13: Microwave back scattering changes with angle of incidence

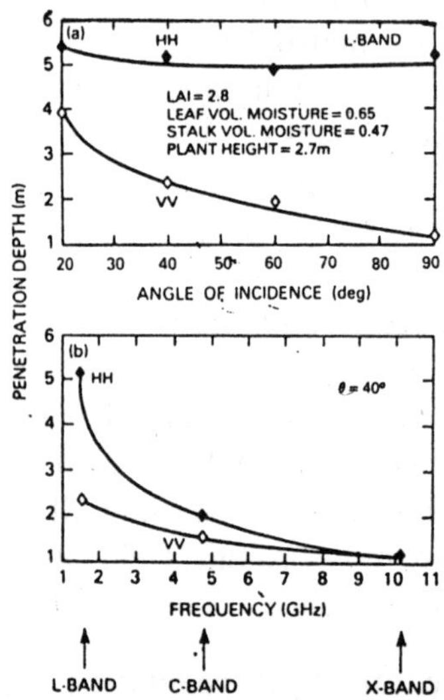

Fig. 12.14: Microwave back scattering changes with frequency

compared with soil moisture followed very closely the theoretical curve for a uniform profile.

Very few studies are carried out using satellite microwave data for assessing soil moisture conditions. Ulaby *et al.* (1983) compared a Seasat SHR image of an area near Waterloo, Iowa with rain gauge data. Soil Moisture experiment with shuttle imaging radar (SIR)-B were used to demonstrate the high resolution capability of SAR from space platform (Wang *et al.*, 1986).

C-band SAR data collected from ERS-I has been used to derive soil moisture information. The soil moisture estimation are still in experimental stage. Usually relationship between DN values of SAR data and soil moisture content in top 0-10 cm layer of soil has been attempted. Gravimetric measurement of soil moisture from the soil sample collected concurrent with the ERS pass were used for regression analysis (Mohan *et al.*, 1994; Venkatratnam *et al.*, 1993 and Saha *et al.*, 1995). With the help of above empirical relationship (regression equation) from the observation sites, soil moisture map of the concerned area could be generated. But in practice, vegetative cover limits the estimation of soil moisture. However, the soil moisture can be measured to a reasonable degree of accuracy. In order to account for the influence of vegetative cover, data from conventional remote sensing can be used successfully to delineate the various land cover or land use classes, and these classes can be used as mask image over the ERS data. Thus moisture determination for each pixel location can be carried out for the entire area of interest. The complete methodology for SAR ERS-I data analysis for soil moisture estimation is presented (Mohan *et al.*, 1994) in Fig. 12.15. But it is recommended that for vegetative area passive microwave approach is better than active microwave remote sensing.

Measurement of depth: The relationship between soil brightness temperature and soil moisture depends upon the dielectric constant across the soil air interface. This phenomenon results in some uncertainty as to how thick the soil layer is for determining the dielectric constant. According to Wilheit (1978) the optimum depth of soil would be of the order of a few tenth of a wavelength. Mo *et al.* (1980) determined that the radiometric sampling depth is between 0.06 and 0.1 times the wavelength. The measurement of depth is variable and is dependent upon the total amount of moisture present in the soil layer.

Surface roughness: Microwave emission from the soil is related to the emissivity (reflectivity) of the surface. Roughness or smoothness in the context of microwave is a relative term and is dependent on the wavelength. The effect of a rough surface is to increase the surface emissivity and thus to decrease the sensitivity to soil moisture (Newton and Rouse, 1980).

Passive Microwave Remote Sensing for Soil Moisture Estimation

Passive microwave is largely dependent on soil brightness temperature. This is the temperature of a black body radiating the same amount of energy per unit area at the wavelength under consideration as the observed body. Mathematically brightness temperature (Tb) can be defined as:

$$Tb = (1 - r)\, Tsoil = eTsoil \qquad \ldots\ldots \quad (12)$$

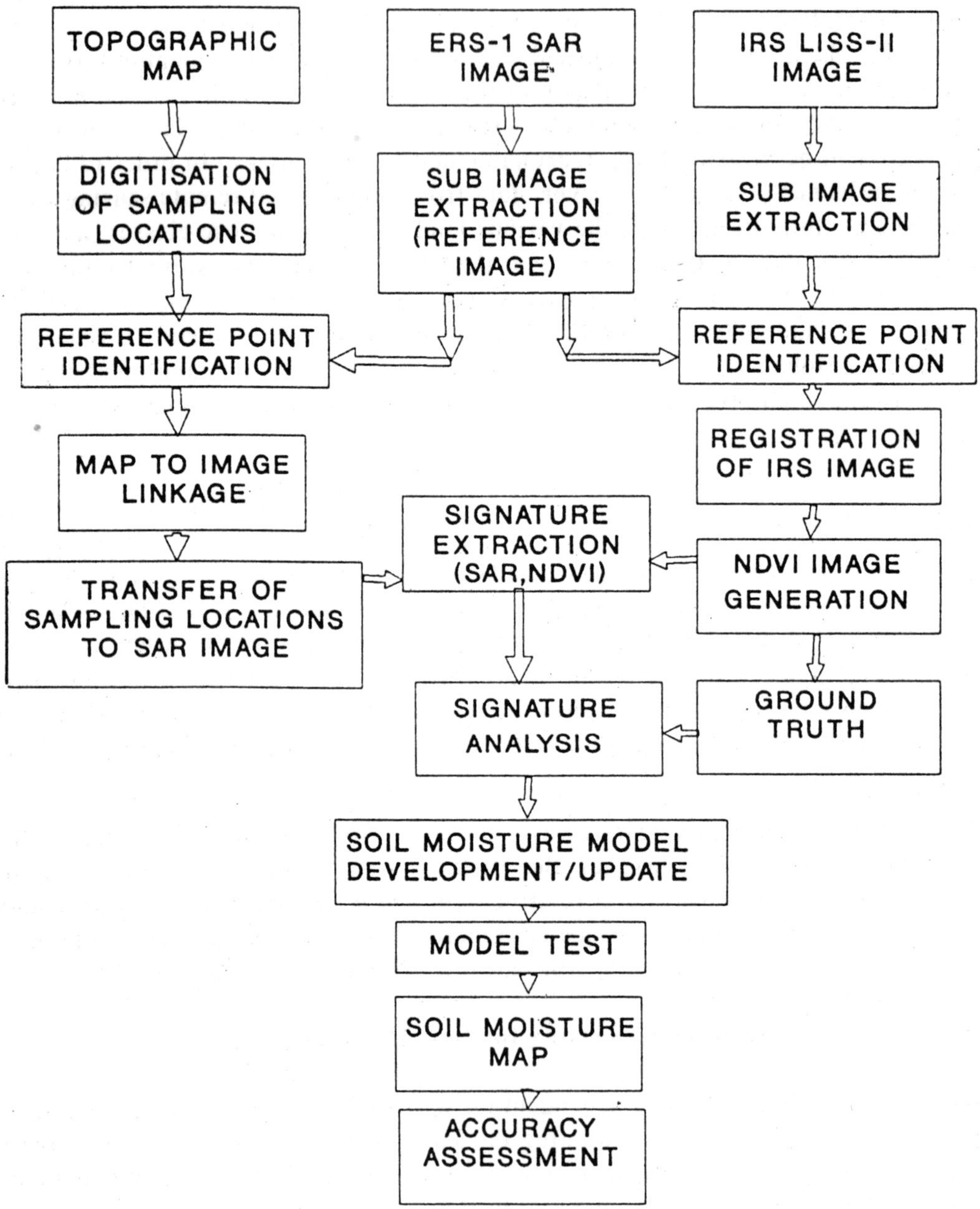

Fig. 12.15: Methodology for soil moisture estimation

where

Tsoil = soil temperature and

e = (1 - r) is the emissivity and r is the reflectivity of soil.

It is a common phenomenon that materials or soils with different dielectric constants would produce different brightness temperature (Tb). Since dielectric constant for soil are strongly dependent on their moisture content (Fig. 12.16), it should be possible to determine soil moisture content with brightness temperature observations. Fig. 12.17 shows how soil brightness temperature changes with the variation of soil moisture. These are the basic relationship for measuring soil moisture by passive microwave technique. Schmugge *et al.* (1978) showed that brightness temperature or soil temperature, i.e. emissivity, has a linear relationship with soil moisture content.

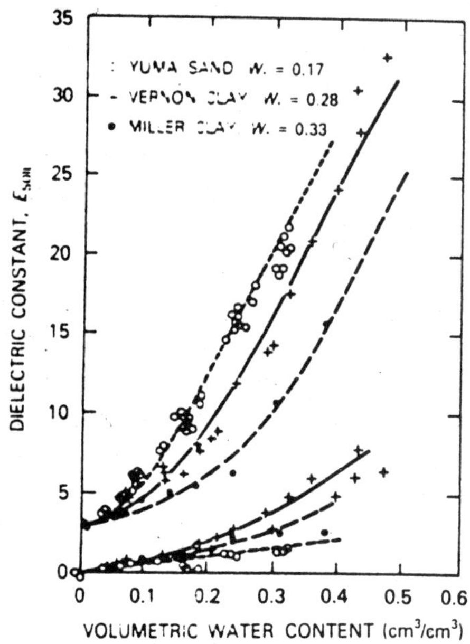

Fig. 12.16: Relationship between dielectric constant and volumetric soil moisture content

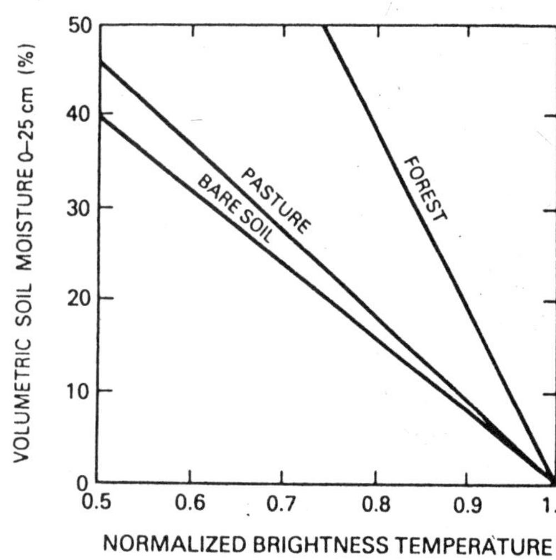

Fig. 12.17: Schematic representation of the SCS runoff model

Though the above relationship apparently looks reasonably good, there needs some additional considerations that will effect the soil moisture estimation by passive microwave. The various parameters to be considered are depth of measurements, surface roughness and vegetation effect. The following equation establishes the relationship between the soil moisture and water content.

$$Mv = 78.9 - 78.4\ [1 + (e-1)\ \exp\ (0.22W)] \qquad \ldots.. \quad (13)$$

where

Mv = volumetric soil moisture within 0-2.5 cm depth
e = emissivity
W = Water content of the vegetation (kg/m^{-2})

So far soil moisture estimation in presence of vegetation must be corrected following Jackson *et al.* (1978) approach. Theis *et al.* (1984) proposed that vegetation index or perpendicular vegetation index can be used to correct moisture estimation of the soil in presence of vegetation. They observed that for vegetation index below 4.3, satisfactory results can be achieved.

Theis *et al.* (1986) demonstrated the possibility of using a multi-sensor approach for improving the soil moisture estimates under field conditions. For this purpose, the effect of roughness were accounted for with scatterometer measurements. These were then used in a soil moisture equation which included terms related to the emissivity measurement by the radiometer and to the term improved the R^2 values from 0.22 to 0.65 for C-band and from 0.69 to 0.95 for L-band.

Vegetation effect: The vegetation attenuate the microwave emission from the soil as it also adds to the total radioactive flux with its own emission. Only in the bare field, there is strong relationship between microwave brightness temperature and soil moisture (Barton, 1978); Newton and Rouse (1980) observed that there is hardly any relationship between brightness temperature and soil moisture over a vegetation covered soils in the lower wavelength. But on the higher wavelength (21 cm), the radiometric data is sensitive to soil moisture even under tall sorghum. This was also observed by Barsharinav and Shetko (1978). Based on this approach, Jackson *et al.* (1976 & 1978) modelled the vegetation layer absorption that can be quantified in terms of the water content of the vegetation.

RUNOFF

Runoff is one of the important component in the hydrologic cycle. Accurate prediction or estimation of runoff volume is of immense use for water supply, forecasting, flood prediction and warnings, navigation and water resource management (Engman and Gurney, 1991). Conventionally runoff data are being collected from ground measurement for large river system and reservoirs where day to day management is a goal. But this process is time consuming and of course expensive.

Background

Land use is an important characteristic of the runoff process that influences infiltration, erosion, and evapotranspiration. Hence any physical hydrological model uses some form of land use information that can be successfully derived from remote sensing data. Distribution model (Jackson *et al.*, 1976) in particular, need land use and their location within basin. Hydrologic runoff model is nothing but soil conservation service (SCS) curve number model (USDA, 1986). The SCS models are the most widely popular to the hydrologists.

Methodology

The empirical SCS curve number models need two basic input parameters that are defined in terms of land use and soil types, and do not require any hydrologic data for combination. These types can successfully be derived from remote sensing data by adapting visual or digital techniques. By integration of land use data and hydrologic soil groups derived from the soil type can be used for the computation of surface runoff and peak discharge of any storm (USDA, 1986; Jackson *et al.*, 1977, Kudrat *et al.*, 1990).

Soil Conservation Service (USDA) Method

The curve number represents the runoff potential of the hydrological soil cover complexes. The SCS model involves relationships between land cover, hydrological soil class and curve number. Hydrological soil groups were determined by considering the soil physical properties and infiltration rate studies (USDA, 1986). The SCS runoff model for small watersheds is described below:

$$Q = \frac{(P - 0.25)^2}{P + 0.85} \qquad \ldots\ldots \quad (14)$$

where

$$S = \frac{1000}{CN} - 10 \qquad \ldots\ldots \quad (15)$$

However, equation (14) modified for Indian conditions (Handbook of Hydrology, 1972) is given by:

$$Q = \frac{(P - 0.35)^2}{P + 0.75} \qquad \ldots\ldots \quad (16)$$

$$Q = \frac{(P - 0.1S)^2}{P + 0.95} \qquad \ldots\ldots \quad (17)$$

and

$$CN = \frac{25400}{254 + S} \qquad \ldots\ldots \quad (18)$$

where

Q = Actual direct runoff (cm)
P = Total rainfall (cm)
S = Potential maximum retention (cm)
CN = S (soil, land, vegetal cover, antecedent moisture conditions etc.)

Determination of Curve Number from IRS-1A Digital Data

The land cover and soil information derived by the IRS digital data is more general than that specified in the standard SCS table. It is necessary, therefore to establish a curve number classification system compatible with the IRS data. This was accomplished with the help of the standard SCS table of runoff curve number, modified for Indian conditions. For example, a curve number for cultivated land is an arithmetic average of all curve number values under the headings of straight row, contoured, contoured and terraced and bunds. Antecedent moisture condition, hydrologic soil cover complexes and hydrologic soil groups are the other parameters involved.

The weighed curve number for each sub-basin can be calculated by using the following equation:

$$\text{Weighted CN} = \frac{CN_1 \times A_1 + CN_2 \times A_2 + \ldots\ldots\ldots\ldots CN_n \times A_n}{A_1 + A_2 + \ldots\ldots\ldots\ldots A_n} \quad \ldots\ldots \quad (19)$$

Where CN_1, CN_2 .... CN_n are the curve numbers for different land use and treatment, and hydrological soil groups present in the sub basins of the watershed; and A_1, A_2 .... A_n are its respective sub-basin areas.

The weighted CN for the whole basin can also be calculated by using the following equation:

$$\text{Weighted CN} = \frac{\text{CN of sub basin} \times \text{its area}}{\text{Total area of the watershed}} \quad \ldots\ldots \quad (20)$$

FOREST/VEGETATION

Forest and vegetation is one of the popular theme which has taken help of remote sensing, particularly for monitoring this natural resource in developing countries (Malingreau, 1991). Monitoring of forests is not only limited at national level but at global scale as well (Hudson, 1987; Horler & Ahren, 1986; Malingreau *et al.*, 1989; Singh, 1990). Such studies are not only concerning coverage but also forest types, density, species composition, stand height, crown density classes, health and disease damages, forest fire and the like. Landsat MSS and TM and AVHRR data have been used for this purpose. In India, IRS-1A has been widely used for vegetation type and crown closure. Though visual interpretation is popular, digital techniques have also been used. Roy *et al.* (1991) have applied normalize vegetation index (NVI) on IRS-1A LISS-II data by using the following formula:

$$\text{IRS NVI} = \frac{\text{Band}_4 - \text{Band}_3}{\text{Band}_4 + \text{Band}_3} \quad \ldots\ldots \quad (21)$$

Further, the NVI band was added for normal four band IRS-1A LISS-II data in order to create five band data. In an another investigation, the radiometrically calibrated data of B3 and B4 of WiFS were used to calculate NDVI image. The NDVI equation used for both WiFS and AVHRR is as follows (Thriuvengadavhari *et al.*, 1996):

$$\text{NDVI} = 222.5\text{–}350 \; \frac{\text{near infrared} - \text{red}}{\text{near infrared} + \text{red}} \quad \ldots\ldots \quad (22)$$

Furthermore, supervised classification was used to stratify on the basis of type and density. Other indexes have also been used for the identification of forest or vegetation types. Madhavan Unni (1983) applied supervised classification using the data of leaf fall season presence of under stormy bamboo in part of Nallamalai Hill Range. Vogelmann and Roac (1986), using Landsat TM simulator data, have used band ratios to discriminate low and high damage sites. They also ratio colour density slice images for accurate detecting, quantifying and mapping of forest declines. Further, pre-launch investigations of utility of LISS-III and WiFS simulated data was carried out by combining spectral information from SPOT XS for visible and NIR channels with MIR spectral band from Landsat TM over a previously investigated site in north Gujarat. Separability amongst various crop or vegetation classes as estimated from Bhattacharya distance increased when MIR band was also included in classification (Navalgund *et al.*, 1996). It is interesting to note that spatial mix of different digital processes is also possible. For example, for a IRS-1C LISS III scene covering Kaziranga National Park, NDVI has been used for the areas inside the park while maximum likelihood classification for the rest of the area (Rao *et al.*, 1996).

GEOLOGY

Remote sensing has proved to be a valuable tool for the regional geological investigations whether geomorphological studies, mineral exploration, hydrogeology, engineering geology, or even geological mapping. Such mapping primarily depicts rock types and associated minerals, and their structural relationships. It has helped many regional geological studies specifically in areas of poor rock exposure due to dense vegetation or in less accessible areas. Though geologists prefer to work on large scale maps, regional geological maps are considered to be essential basis for understanding the causes of magnetism, metamorphism and structure, regional controls of mineralization, tectonic evolution, and various possible geological hazards (Chetty, 1995). Earlier Landsat TM data was considered reasonable good for regional geological mapping at 1:250,000 scale or smaller. When the TM imagery is compared with the MSS imagery, it has been observed that the identification of number of lithological units remains more or less same, but due to high resolution of TM data, clarity of features is more which helps in better readability and quicker interpretation (Bhattacharya, 1988). However, the availability of IRS LISS III data at 5.8m resolution has enlarged the scope of mapping at larger scales as well.

Visual interpretation techniques on FCC is popular for geological mapping. Such techniques based on inductive and deductive analysis have been explained in Chapter 4. Digital techniques are also found to be useful particularly for tracing lineaments by spatial filtering techniques (Majumdar, 1995) (Chapter 8). Further, spatial integration of various data sets, through GIS technology (Chapter 11) or otherwise, has also been used widely. The information sets may consist of geological maps, remote sensing data, airborne magnetic data and gravity data and the like. The example of application of GIS technology in geological data integration for Vasundhara Project has been mentioned in the next chapter.

Following few examples will indicate the application of digital image processing techniques for geological mapping. Ravindran (1987) used IRS LISS II data for lithological mapping of Dehra Dun area. For this mapping work, he used enhancement techniques such as principal component analysis, correspondence analysis, brightness index and edge enhancement for generating colour composites. Bhan *et al.* (1989) applied Laplacian filtered bands and HSI transformed colour composites of IRS LISS I data for geological and structural interpretation of the hilly terrains of Shimla-Solan-Sarahan ranges and Nahan-Ponta areas of Himachal Pradesh.

In lineament studies, sun angle plays an important role. For the Himalayan region, the sun-angle on Landsat MSS/TM, SPOT and IRS images is found to vary from 30° in winter to over 70° in summer. Since the slopes are moderate to steep, it has been observed that the winter images with lower sun angle (30-45°) give sharper shadows which enables better interpretation of topography and related lineaments (Ravindran *et al.*, 1991). However, the digitally enhanced images using edge enhancement and directional filtering have brought out several lineaments which are not discernible in the raw images (Ravindran & Bhan, 1991).

The application of remote sensing data in geological studies has some limitations. The vegetative cover do provide a relatively good picture of some of the geological features. However, such information are cost effective. Nevertheless, the images have become popular in geological studies. Perhaps, the digital techniques will become more acceptable in the near future.

Other Applications

There are several themes where remote sensing data have been widely used. It varies from simple digital spectral analysis or resampling to principal component analysis. The images are brought to specific cartographic base by extracting ground control points or GCPs from topographical sheets and recently by geographical positioning system or GPS and by using image transformation models. For land use or land form mapping, multispectral classifi-cation by using maximum likelihood algorithm is common. Another approach for evaluating spectral separability of land features is the use of confusion matrix. Ensuring all non-diagonal elements of the confusion table as zero, misclassification can be minimized. The overall accuracy of the classification is finally assessed with reference to ground truth data (Rao *et al.*, 1996). But the visual techniques is still considered popular. Some of the emerging application areas are as follows:

— Geographical information system (GIS)
— Environment
— Wildlife
— Mineral exploration
— Digital elevation model
— Ground and surface water
— Flood and snow mapping
— Fisheries
— Natural hazards including cyclones
— Urban sprawl
— Coastal zone management, reclaimed land, mangroves, coral reef.

The above applications are only indicative. Studies can be made on different aspects of the above mentioned themes. With increasing facilities for digital data processing, better resolution and training infrastructure, it is likely that the digital techniques become essential for application of remote sensing data (Nag, 1996). It is now possible to take up dynamic themes, such as floods, by repetitive coverage of the same area at 6 days interval and taking into account of the data available from IRS-1B, P2 and Landsat 5 satellites. Even with such developments, the application scientists would have to go a long way to utilize the remote sensing data for topographical mapping or for land information system (De Bruijn, 1996).

REFERENCES

Anonymous (1986), Large area crop acreage and production estimation, Project Proposal, Remote Sensing Applications Mission Programme Plan, Report No. RSAM/MP/02/June/1996.

Anonymous (1995), *Manual for Crop Production Forecasting*, RSAM/SAC/CAPE-II/TN/46/96,

Space Applications Centre, Ahmedabad.

Barsharinav, A. Ye and H.M. Shutko (1978), Determination of moisture content of the earth's cover by superhigh frequency (microwave) radiometric methods: A review, *Radiockhmika i Electronika*, vol. 23, pp. 1778-91, NASA Technical Transaction TM 75411.

Bartolucci, L.A., B.F. Robinson and L.F. Silva (1977), Field measurements of the spectral response of natural water, *Photogrammetric Engineering and Remote Sensing*, vol. XLIII, pp. 595-8.

Barton, I.J. (1978), A case study of microwave radiometer measurements over bare and vegetated surface, *Journal of Geophysical Research*, vol. 83, pp. 3513-7.

Bhadwar, G.D. (1980), Crop emergence date determination from spectral data, *Photogrammetric Engineering and Remote Sensing*, vol. 46, pp. 369-77.

Bhadwar, G.D. (1982), Profile modelling for crop discrimination in *Machine Processing of Remote Sensing Data*, pp. 454-9.

Bhan, S.K., K.V. Ravindran, A.K. Tiwari, M. Kudrat, P.G. Diwakar, and T.P.S. Vohra (1989), Application potential of IRS-1A data for land resources mapping-case studies from Haryana and Chandigarh, Regional Workshops on IRS-1A Mission and its application potential at Chandigarh and Shimla.

Bhan, S.K., A. Bhattacharya, P.K. Guha and K.V. Ravindran (1991), IRS-1A applications in geology and mineral resources, *Current Science*, vol. 61, pp. 247-51.

Bhattacharya, A. (1988), Geology and geomorphology, in N.R.S.A., *Landsat Thematic Mapper Data Application--Illustrated Examples*, National Remote Sensing Agency, Hyderabad, pp. 17-20.

Bondelid, T.R., T.J. Jackson and R.H. Mecuen (1982), Estimating runoff curve numbers using remote data, *Proceedings of the International Symposium on Rainfall-Runoff Modelling Applied Modelling in Catchment Hydrology*, Water Resource Publication, Littleton Co., pp. 519-28.

Bruckler, L., H. Witono and P. Stengel (1989), Near surface soil moisture estimation from microwave measurement, *Remote Sensing of Environment*, vol. 26, pp. 101-21.

Byrne, G.F., P.F. Crapper and K.K. Mayo (1980), Monitoring land cover changes by principal component analysis in multitemporal Landsat data, *Remote Sensing of Environment*, vol. 10, pp. 175-85.

Chetty, T.R.K. (1995), Structural interpretation of high grade belts of South India using remotely sensed data, *NNRMS Bulletin*, September, pp. 34-7.

Cochran, G. (1977), *Sampling Techniques*, Third edition, Wiley Eastern Ltd., New Delhi.

Colwell, J.E. and F.P. Weber (1981), Forest change detection, *Proceedings of the 15th International Symposium on Remote Sensing of the Environment*, vol. 2, pp. 839-52.

Crapper, P.F. and K.C. Hyson (1983), Change detection using Landsat photographic imagery, *Remote Sensing of Environment*, vol. 13, pp. 291-30.

Crist, E.P. and R.C. Cicone (1984), Physically based transformation of thematic mapper data, VTM tassel cap, *IEEE Trans on Geoscience*, Remote Sensing GE 22, pp. 256-63.

De Bruijn, Cornelis A., Sherif Amer and Desmond Dougall (1996), GIS and geointegration, *ITC Journal*, vol. 1996-1, pp. 71-6.

Deekshatulu, B.L. and Hebber, K.J. (1991), Integration of Remote Sensing and Geographical information system. Book on Environmental Monitoring-Application of Remote Sensing and GIS R.B. Singh (ed). *Geo Carto. International Centre*, Hong Kong.

Dipaolo, W.D. and L.B. Hall (1982), Landsat data for soil investigations on federal lands, in C.J. Johannson and J.L. Sanders (eds.), *Remote Sensing for Resource Management*, Soil Conservation Society of America, Ankey, pp. 201-6.

DOS (1988), *Preparation of Hydrogeomorphological Maps of India on 1: 250,000 scale Using Satellite Imagery*, Project Report, Department of Space, Bangalore.

Draper, S.E. and S.G. Rao (1986), Runoff Prediction Using Remote Sensing Imagery, *Water*

Resource Bulletin 22, pp. 941-9.

Engman, E.T. and R.J. Gurney (1991), Soil moisture in Remote Sensing, in *Remote Sensing in Hydrology*, Chapman and Hall, New York, pp. 127-51.

Fukuhara, M., S. Itayashi, Y. Yasuda, I. Asanuma and J. Iisaka (1978), Extraction of soil information from vegetated area, in J.J. Cook (ed.), *Proceedings of the 12th International Symposium on Remote Sensing of the Environment*, Manila, Environment Research Institute of Michigan, Ann Arbor, pp. 1707-16.

Horler, D.N. and F.L. Ahren (1986), Forestry information content, Thematic Mapper data, *International Journal on Remote Sensing*, vol. 7(3), pp. 405-28.

Hudson, W.D. (1987), Evaluating Landsat classification accuracy from forest coverage-type maps, *Canadian Journal of Remote Sensing*, vol. 13(1), pp. 39-42.

Hunt, G.R. (1980), Electromagnetic radiation: The communication link in remote sensing, in B.S. Siegal and A.R. Gillespie (eds.), *Remote Sensing Geology*, John Wiley & Sons, New York, pp. 5-45.

Idso, S.B., T. Schmugge, R.D. Jackson and R.J. Reginate (1975), The utility of surface temperature measurements for remote sensing of surface soil water status, *Journal of Geophysical Research*, vol. 80, pp. 3044-9.

Imhoff, M.L., G. Peterson, S. Sykes and J. Irons (1982), Digital overlay of cartographic information on Landsat MSS data for soil survey, *Photogrammetric Engineering and Remote Sensing*, vol. 48, pp. 1337-47.

Jackson, R.D., R.J. Reginato and S.B. Idso (1977), Wheat canopy temperature: A practical tool for evaluating water requirements, *Water Resource Research*, vol. 13, pp. 651-6.

Jackson, R.D., J. Ahler, J.E. Estes, J.L.Heilman, A. Kakle, E.T. Kanemasu, J., Millard, J.C. Price and C. Wiegand (1978), Soil moisture using reflected solar and emitted thermal radiation, *Soil Moisture Workshop*, Chapter 7, NASA CP-2073.

Jackson, R.D., S.B. Idso, R.J. Reginato and P.J. Pinter, Jr. (1980), Remotely sensed crop temperatures and reflectance inputs to irrigation scheduling, in the *Proceedings of ASCE Speciality Conference*, Boise, Idaho.

Jackson, R.D., R.S. Reginato and S.B. Idso (1976), Timing of ground truth acquisition during remote sensing assessment of soil water content, *Remote Sensing of Environment*, vol. 4, pp. 249-56.

Jackson, T.J., R.M. Ragan and R.P. Shubinski (1976), Flood frequency studies or un-gauged urban watersheds using remotely sensed data, in the *Proceedings of National Symposium on Urban Hydrology, Hydraulics and Sediment Control*, University of Kentucky, Lexington, pp. 31-9.

Jensen, H., L.C. Grahman, L.J. Porcello and E.N. Leith (1977), Sidelooking airborne radar, *Scientific American*, vol. 237, pp. 84-95.

Karale, R.L., L. Venkataratnam, J.L. Sehgal and A.K. Sinha (1991), Soil mapping with IRS-1A data in areas of complex soil-scape, *Current Science*, vol. 61, pp. 198-203.

Kite, G. (1989), Using data for hydrologic modelling, in the *Proceedings of the IGRASS*, 1989, Vancouver, vol. 2, pp. 553-57.

Kudrat, M., S.K. Saha and A.K. Tiwari (1990), The potential of LISS-II digital data for soil mapping and land evaluation studies, *Asia-Pacific Remote Sensing Journal*, vol. 2, pp. 73-8.

Kudrat, M., S.K. Saha and A.K. Tiwari (1992), Soil resources mapping using IRS-1A LISS II digital data--A case study of Kandi area adjacent to Chandigarh, India, *International Journal on Remote Sensing*, vol. 4, No. 2, pp. 129-33.

Kudrat, M. and S.K. Saha (1993), Land productivity assessment and mapping through integration of satellite and terrain slope data, *Photonirvachak*, vol. 21, No. 3, pp. 157-66.

Kudrat, M., S.K. Saha and A.K. Tiwari (1990), Potential use of IRS-LISS II in soil and land use mapping and productivity assessment, *Proceedings of the National Symposium on Agricultural Applications of Remote Sensing*, New Delhi, pp. 25-34.

Kudrat, M., S.K. Saha and A.K. Tiwari (1990), The potential of IRS-LISS II digital data for soil mapping and land evolution studies, *Asia and Pacific Remote Sensing Journal*, vol. 2, No. 2, pp. 73-8.

Lee, K.S., G.B. Lee and E.J. Tyler (1988), Determination of soil characteristics from thematic mapper data of a cropped-organic-inorganic soil landscape, *Soil Science Society of America Journal*, vol. 52, pp. 1100-4.

Lodwick, G.D. (1979), Measuring ecological changes in multispectral Landsat data using principal components, in the *Proceedings of the 13th International Symposium on Remote Sensing of the Environment*, vol. 1, pp. 1-11.

L, Vovich, M.I. (1979), *World Water Resources and their Future*, American Geophysical Union, Litho Crafters, Chelseaa, M. 1 (Transl.).

MacDonald, H.C. (1980), Techniques and applications of imaging radars, in B.S. Siegal and A.R. Gillespie (eds.), *Remote Sensing in Geology*, John Wiley & Sons, New York.

Madhavan Unni, N.V. (1983), Application of remote sensing in forest mapping and management with special reference to air borne and multispectral techniques, *Proceedings of the NNRMS Seminar*, Hyderabad.

Malingreau, J.P., C. Tucker and N. La Porte (1989), AVHRR for monitoring global tropical deforestation, *International Journal on Remote Sensing*, vol. 10(4), pp. 855-67.

Malingreau, J.P. (1991), Remote Sensing for tropical forest monitoring: An overview, Alan S. Belward and Carlos R. Velenzuela (eds.), *Remote Sensing and Geographical Information Systems for Resource Management in Developing Countries*, Kluwer Academic Publishers, Dordecht, Boston and London, pp. 253-78.

Markham, B.L. and J.L. Baker (1987), Radiometric properties of US processed Landsat MSS data, *Remote Sensing of Environment*, vol. 22, pp. 39-71.

Mayers, V.I. (1975), Crops and soils, in R.G. Reeves (ed.), *Manual of Remote Sensing*, American Association of Photogrammetry, Falls Church, Virginia, pp. 1715-813.

Majumdar, Tapan J. (1995), Application of fast fourier transformation of a part of the Singhbhum shear zone for extraction of linear and anomalous features, *ITC Journal*, vol. 1995-3, pp. 241-5.

Mo, T., T.J. Schmugge and Choudhury (1980), *Calculation of the Spectral Nature of Microwave Emission from Soils*, NASA Technical Memo. 82002.

Mohan, S., N.S. Mehta, R.L. Mehta, P. Patel, H.S. Shrivastava, S. Sharma, C.M. Saxena and Sutradhar (1993), Soil moisture estimation using ERS-1 SAR data, in the *Proceedings of the ERS-1 Symposium: Space at the Service of our Environment*, Hamburg, Germany, ESA SP-361, pp. 875-8.

Mohan, S., P. Patel, H.S. Shrivastava, M. Yadav, B.S. Kundu, R.P. Agarwal and V.K. Phogat (1994), A methodology for soil moisture estimation using ERS-1 SAR data, in the *Proceedings of Microwave Remote Sensing Applications*, Space Applications Centre, Ahmedabad.

Mohan, S., S.K. Basu, A.M. Jha, T. Mishra, R.K. Rewalia, K.N. Padia, P. Patel and H.S. Shrivastava (1995), Calibration of ERS-1 SAR data, Scientific Note, SAC/RSA/ERS-1/SN/07/95, Ahmedabad.

Mintz, Y. (1984), The sensitivity of numerically simulated climate to land surface conditions in J.T. Houghton (ed.), *The Global Climate*, Cambridge University Press, New York, pp. 79-105.

Moore, G.K. (1977), Prospective for groundwater with Landsat Images, World Conference, Mar. Del. Plara, Argentina.

Moore, G.K. (1988), The role of remote sensing in groundwater exploration, in the *Proceedings of the Joint Indo-US Workshop*, Hyderabad.

Mulder, N.J. (1980), A view on digital image processing, *ITC Journal*, pp. 452-77.

Nag, P. (1996), A proposal for the preparation of satellite atlas of India, in the *Proceedings of the International Workshop on International Mapping from Space*, November 27-1, December, 1995, Madras, pp. 143-7.

Navalgund, R.R., J.S. Parihar, Ajai and P.P. Nageshwar Rao (1991), Crop inventory using remotely sensed data, *Current Science*, vol. 61, pp. 162-71.

Navalgund, R.R., J.S. Parihar, L. Venkataratnam, M.V. Krishna Rao, S. Panigarahy, M.C. Chakraborty, K.R. Hebbar, M.P. Oza, S.A. Sharma, N. Bagia and V.K. Dadhwal (1996), Early results from crop studies using IRS-1C data, *Current Science*, vol. 70, No. 7, pp. 568-74.

Nelson, R.F. (1983), Detecting forest canopy changes due to insect activity using Landsat MSS, *Photogrammetric Engineering and Remote Sensing*, vol. 49, No. 9, pp. 1303-14.

Nelson, R.F., R.S. Latty and G. Mott (1984), Classifying northern forests with Thematic Mapper simulator data, *Photogrammetric Engineering and Remote Sensing*, vol. 50(5), pp. 607-17.

Newton, R.W. and J.W. Rouse (1980), Microwave radiometer measurements of moisture contents, *IEEE Transaction Antennas Propagation*, AP-28, pp. 680-6.

Patel, N.K., N. Ravi, R.R. Navalgund, R.N. Dash, K.C. Das and S. Patnaik (1991), Estimation of rice yield using IRS-1A digital data in coastal tract of Orissa, *International Journal on Remote Sensing*, vol. 12, pp. 2259-66.

Price, J.C. (1982), Estimation of regional scale evapotranspiration through analysis of satellite thermal infrared data, *IEEE Trans. on Geosciences and Remote Sensing*, GE 20, pp. 286-92.

Rao, D.P., N.C. Gautam, R. Nagaraja and P. Ram Mohan (1996), IRS-1C application in land use mapping and planning, *Current Science*, vol. 70, No. 7, pp. 575-81.

Rao, U.R. (1991), Remote sensing for sustainable development, *Photonirvachaka*, vol. 19(4), pp. 217-35.

Ravindran, K.V. (1987), Study of geology and geomorphology of Doon Valley from Digitally enhanced IRS LISS II data, National Seminar on Indian Remote Sensing Satellite-1A, Mission and its Application Potentials, National Remote Sensing Agency, Hyderabad.

Ravindran, K.V. and S.K. Bhan (1991), Satellite remote sensing for geological studies in mountainous terrain--Potentials and constraints, in P.N. Gupta and A.K. Roy (eds.), *Mountain Resources Management and Remote Sensing*, Surya Publication, Dehra Dun, pp. 39-48.

Roudabush, R.D., R. Herriman, R. Barmore and Schellent (1985), Use of Landsat multispectral scanning data for soil survey of Arizona rangeland, *Journal of Soil Water Conservation*, vol. [illegible], pp. 242-45.

Roy, P.S., P.G. Diwakar, T.P.S. Vohra and S.K. Bhan (1991), Vegetation analysis in Aglar watershed (Garhwal Himalaya) using Indian Remote Sensing Satellite in P.N. Gupta and A.K. Roy (eds.), *Mountain Resource Management and Remote Sensing*, Surya Publishers, Dehra Dun, pp. 174-7.

Sabins, F.F. (1978), *Remote Sensing: Principles and Interpretation*, W.H. Freeman and G. San Francisco.

Saha, S.K., L.M. Pande and J. Prasad (1995), Evaluation of ERS-1 SAR data for soil moisture assessment and crop discrimination--A case study, *Proceedings of ISRS Silver Jubilee Year Seminar*, Dehra Dun.

Sahai, B., R.K. Sood and S.C. Sharma (1985), Ground water exploration, *International Journal on Remote Sensing*, vol. 3, pp. 443.

Salomonson, V.V., R. Ambaruch and J.P. Ormsby (1975), Remote sensing requirements as suggested by watershed model sensitivity analysis, in the *Proceedings of the Xth Symposium on Remote Sensing of Environment*, Environmental Research Institute of Michigan, Ann Arbor, pp. 1273-84.

Satterwhite, M.B. (1984), Discrimination vegetation and soils using Landsat MSS and Thematic bands and band ratio, *Technical Paper*, 50th Annual Meeting of the American Society of Photogrammetry, vol. 2, pp. 479-85.

Schmugge, T.J. (1983), Remote sensing of soil moisture: Recent advances, *IEE Transactions on Geosciences and Remote Sensing*, GE-21, pp. 336-44.

Schmugge, T.J., E.G. Njoku, E. Peck and F.T. Ulaby (1978), *Microwave and Gamma Radiation Observations of Soil Moisture*, Soil Moisture Workshop, NASA, CP-2073.

Singh, A. (1986), Change detection in tropical forest environment of northeastern India using Landsat, in M.J. Eden and J.T. Parry (eds.), *Remote Sensing and Tropical Land Management*, John Wiley & Sons, London.

Singh, A. (1987), Spectral separability of tropical forest cover classes, *International Journal on Remote Sensing*, vol. 8, pp. 97-9.

Singh, K.D. (1990), Design of a global tropical forest resources assessment, *Photogrammetric Engineering and Remote Sensing*, vol. 56(10), pp. 1353-54.

Stauffer, M.L. and R.L. Mckinney (1979), Landsat image-differencing as an automated Land cover change detection technique, Report Contract NASS-24356, General Electric Company, Beeltsvile, Maryland.

Suits, G., W. Malia and T. Willer (1988), Procedures for using signals from one sensor as suitable for signals of another, *Remote Sensing of Environment*, vol. 25, pp. 395-408.

Theis, S.W., B.J. Blanchard and R.W. Newton (1984), Utilization of vegetation indices to improve microwave soil moisture estimation over agricultural lands, *IEE Transactions on Geosciences and Remote Sensing*, GE-22, pp. 490-6.

Thiruvengadachari, S. Jonna, K.A. Hakeen, P.V. Raju, V. Bhanumurthy, G. Srinivasa Rao, P.R. Paul, E. Siva Shankar and A.T. Jeyaseelan (1996), Improved water management: The IRS-1C contribution, *Current Science*, vol. 70, No. 7, pp. 589-99.

Tiwari, A.K. (1992), Role of aerial photography and satellite remote sensing in thematic cartography: A case of forest mapping, in P. Nag (ed.), *Thematic Cartography and Remote Sensing*, Concept, New Delhi, pp. 137-48.

Tucker, C.J. (1979), Red and photographic infrared linear combination for monitoring vegetation, *Remote Sensing of Environment*, vol. 8, pp. 127-50.

Ulaby, F., B. Brisco and M.C. Dobson (1983), Improved spatial mapping of rainfall events with space borne SAR imagery, *IEE Transactions on Geosciences and Remote Sensing*, GE 21, pp. 118-21.

Ulaby, F.T., R.K. More and A.K. Fung (1986), *Microwave Remote Sensing: Active and Passive*, vol. III, Artec House, Dedham.

USDA (1986), *Soil Survey Manual*, Handbook No. 18, Department of Agriculture, Washington.

Weismiller, R.A. and S.A. Kaminsky (1978), Application of remote sensing technology to soil survey research, *Journal of Soil Water Conservation*, vol. 33, pp. 288-96.

Westin, F.C. and C.J. Frazee (1976), Landsat data, its use in a soil survey program, *Proceedings of Soil Science Society of America*, vol. 49, pp. 81-9.

Wilheit, T.T. (1978), Radioactive transfer in a plane stratified dielectric, *IEE Transactions on Geosciences and Remote Sensing*, GE-16, pp. 138-43.

Vande Griend, P.J. Camillo and P.J. Gurney (1985), Discrimination of soil physical parameters, thermal inertia and soil moisture from diurnal surface temperature fluctuation, *Water Resource Research*, vol. 21, pp. 997-1009.

Venkataratnam, L., P.V.N. Rao, B.R.M. Rao, K. Sreenivas, K.V. Ramana and R.S. Dwivedi (1993), Soil Moisture estimation using ERS-1 SAR data, Second ERS-1 Symposium: Space at the service of our Environment, Hamburg (Germany), October 11-14, ESA-301, pp. 855-60.

Vogelmann, J.E. and B.N. Roac (1986), Assessing forest decline in coniferous forests of Vermant using NS-001 Thematic Mapper simulator data, *International Journal on Remote Sensing*, vol. 7(10), pp. 1303-21.

Wang, J.R., E.T. Engman, J.C. Shiue, M. Ruzek and C. Steinmeier (1986), The SIR-B observations of microwave back scatter dependence on soil moisture, surface roughness and vegetation covers, *IEE Transactions on Geosciences and Remote Sensing*, GE-24, pp. 510-6.

THIRTEEN

CONCLUSION

IN THE FAST changing technological scenario in the field of remote sensing it has become imperative to keep abreast with the options available or likely to be available in near future. This has become more and more important because expectations from the space technology is increasing. On the other hand, the remote sensing or orbiting satellites are becoming application oriented. If the Indian example is an indication, it will reveal that the latest IRS satellites (1C, P1 and P2) and the planned ones (P3, P4 and P5) are having or will have sensors for very specific applications. The availability of a variety of data opens a panorama for the applications to study, monitor and map the features of current national, regional or local importance. Obviously the application scientists and engineers have to look into the options available to them in conjunction with the options which the industry has to offer in terms of software and hardware.

Remote Sensing Data Products

One of the most important aspect that an application scientist has to look into is the remote sensing data products which can be made available to him. The nature of data available from Landsat and SPOT have been discussed in details in chapter 3. But due to availability of IRS data, which is comparable if not better than any remote sensing data available internationally, the Indian scientists are inclined to use such data. Indian data is economic as well. Hence, the IRS data products have been discussed here in some details.

The basic source for the procurement of the IRS data is the National Remote Sensing Agency (NRSA) Data Centre located in Hyderabad. The IRS-1A and IRS-1B are identical and are providing data with a combined repetivity of 11 days. Various data products like browse, standard, precision and special products are generated regularly for meeting the demands of the users. The details are provided in Table 13.1.

The data products available from the NRSA for IRS, Landsat and SPOT is indicated in Table 13.2. The data products are available as (a) black and white photographs, (b) false colour composites, and (c) digital data on a magnetic media. The cost and other details are given in Appendix V. Further, AVHRR, TOVS, ERS-

TABLE 13.1: IRS Data Products

Type	Processing level	Format	Location Accuracy	Turn around time (days)
Quick look	Level-0	70mm	--	Real time
Browse	Level-1 (Earth Rotation & Radiometric corrections)	63 × 63mm	± 10 kms	2 days
Standard*	Level-2 (including geometric correction)	241mm	± 2.2 kms	7 days
Precision*	Level-3 (use of GCPs)	241mm	better than 100mm	15 days

*Computer compatible tapes (CCTs) of 1600 BPI/6250 BPI are available for these products.

TABLE 13.2: Data Products available from NRSA

Sensor	B&W Photo	FCC	Digital-BIL	S-MSDP	FSP
IRS 1A & 1B Standard DP	Y	Y	Y	Y	N
IRS 1A & 1B Geocoded DP	Y	Y	Y	Y	N
IRS 1A & 1B Precision DP	Y	Y	Y	N	N
Landsat MSS Standard DP	Y	Y	Y	N	N
Landsat TM Standard DP	Y with CCR	Y with CCR	Y with NNS	Y	Y
Landsat TM Geocoded DP	Y	Y	Y	Y	N
SPOT MLA	Y	Y	Y	Y	N
SPOT PLA	Y	N	Y with BIL	Y	N

Abbreviation used: B&W — Black and white; FCC — False colour composites; S-MSDP — Segment based miniature standard data product; FSP — Full scene products; DP — Data product; MLA — Multispectral linear array; PLA — Panchromatic linear array; Y — Yes; N — No; CCR — Cubic convolution resampling; NNS — Nearest neighbour resampling; BIL — Band interleaved by line.

SAR data products are also available from this agency.

The variety of IRS-1C data products is similar to the IRS 1A and 1B. The products are supplied to users on both photographic and digital media. Black and white and FCC photographic products are to be made available as films and paper prints. Digital products are to be supplied on various magnetic media, viz. CCTs, cartridges, exabyte cartridge and floppies. Further, IRS-1C mission has been planned for global coverage and it is expected that there would be simultaneous requirements from various users, particularly for the PAN data calling for appropriate tilting of the camera to cater to the needs of the user community. The types of data products for IRS-1C is as follows (Table 13.3).

TABLE 13.3: IRS-1C Data Products

Product	Sub-type	Sensor	Scene size
Standard	Path based, Fix/Float	LISS-III	141 × 141 km
			141 × 148 km for SWIR band
	Quadrant, Fix/Float	LISS-III	72 × 72 km
	Geocoded	LISS-III	Topographical sheet size
	Scene based	WiFS	810 × 810 km
	Path based, Fix/Float	PAN	23.9 × 23.9 km
			23.9 × 30.5 km at maximum tilt
	Geocoded (no mosaic)	PAN	Topographical sheet size
Stereo	Basic stereo pair	PAN	As per input
	Derived stereo pair	PAN	As per input
	Ortho image	PAN	As per input
Special System	Corrected raw data	All sensors	As per input
	Merged	LISS-III + PAN	23.9 × 23.9 km
	City geocoded	PAN	As per input
	Three strip mosaic	PAN	70 × 70 km (nadir)
			90 × 70 km (off nadir)
	District geocoded	LISS-III	Variable
	Vegetation index map	WiFS	Full size zonal

Source: ISRO (1995).

NRSA can supply current and archived data products from almost all existing satellites such as IRS-1A, 1B and P2, Landsat 5, NOAA series, ERS-1 and SPOT. Very soon, NRSA will commence supply of products from the IRS-1C satellite.

Trends of RS Data Application

The history of the application of remote sensing data in India is older than the successes in the satellite technology. Already established centres like the National Remote Sensing Agency (NRSA) in Hyderabad and the Space Applications Centre (SAC) in Ahmedabad were instrumental in promoting the application of remotely sensed data. The Indian Photo-interpretation Institute (IPI) in Dehra Dun geared itself for providing trained manpower and thereby changed its name to Indian Institute of Remote Sensing (IIRS). Further, the formulation of peers groups and standing committees and finally establishment of the National Natural Resources Management Systems (NNRMS) is more than a decade old. Following this the five Regional Remote Sensing Service Centres (RRSSCs) were established which in a way became centres for promoting applications based on remote sensing data. This infrastructure was ready to accept the IRS data for intensive applications.

Activities related to applications were not only limited to the Indian Space Research Organisation (ISRO) or the Department of Space (DOS), but in other centres as well. Following RRSSCs, the state remote sensing centres and universities also used remote sensing data for different applications. In some major national level projects handled or managed by ISRO/DOS, the involvement of the national and state level centres was desired. Table 13.4 will indicate a list of such projects. As a result, mapping on different themes was carried out at 1:50,000, 1:250,000 and 1:1,000,000 scales. Based on specific guidelines, themes like wastelands, forest, ground water, landuse, crops, droughts, floods and snow, urban sprawl, environmental issues and the like were taken up.

TABLE 13.4: DOS National Remote Sensing Application Projects

Project Name	Objectives	Achievements	Sponsoring/User Agency
National Wasteland Mapping	To identify and extent of wasteland at village level	Maps prepared on 1:1M scale for entire country and 1:50,000 scale for 237 selected districts	Department of Wasteland Development. District authorities
Forest cover mapping	Biennial forest maps on 1:250,000 scale for entire country	Forest cover maps for 1989-91 has been mapped	Dept. of Environment and Forests, FSI, State forest departments
Ground Water	Hydrogeomorphological maps (on 1:250,000 scale) to help in scientific source finding of drinking water	All the districts in the country have been mapped	Min. of Rural Development, state public health engineering departments and ground water departments
Oceanography/Fisheries	To generate routine information on ocean and coastal related parameters	SST charts are being generated and fishery potential zones identified and are being communicated to users	State fisheries departments, Fisheries Survey of India, Fisherman co-operative societies
Crop Acreage and Production Estimation (CAPE)	To estimate the crop acreage and forecast production	Project has been carried out for major crops like rice or wheat covering major parts of the country	Department of Agriculture and Cooperation, State agricultural departments, Ministry of Textiles
Landuse/Landcover mapping	Mapping landcover for agroclimatic zones at 1:250,000 scale	All districts in the country have been mapped	Planning Commission
Agricultural Drought Monitoring	To assess and monitor agricultural drought at district level	Crop condition reports for 10 drought prone states	Dept. of Agriculture and Coop, State agricultural departments, District authorities
Command Area Studies	Monitoring of crops/ irrigation status in command area	Bhadra, Tungabhadra, Indra Gandhi Nehar Project command areas are monitored. Nationwide command area	Command Area Development Authorities, Central Water Commission

TABLE 13.4: *(Contd.)*

Project Name	Objectives	Achievements	Sponsoring/User Agency
		development programme through remote sensing is being taken up	
Mapping of Saline/ Alkaline Soils	Mapping of saline/alkaline soils and estimation of areas of different types of salt affected lands of whole country on 1:250,000 scale	Kerala, Karnataka, Tamil Nadu, Gujarat, West Bengal, Punjab, Goa and Haryana are covered	Central and state departments of agriculture
Flood Mapping	Near realtime mapping of flooded areas and estimation of damage	Major floods in Ganga, Brahamaputra, Godavari rivers etc. are covered	Flood control boards, Central Water Commission, State irrigation departments
Snow Mapping	Delineation of snow cover area and estimation of snow melt run-off	Run-off models were made and predicted snow melt with accuracy of about ±6%	Bhakra-Beas Board and Central Water Commission
Geology and Mineral Resources	Regional Mineral targeting using multithematic data for area below 16 degree North of India	GIS based data base has been created and analysed	Geological Survey of India
Urban Studies	To study the urban landuse/ landcover, sprawl, growth of all major cities	Urban sprawl maps for Madras, Hyderabad, Nagpur, Calcutta, Mumbai have been prepared	Concerned urban development authorities, Ministry of Urban Development
Environmental impact of mining and thermal power complexes	To develop a suitable methodology for monitoring environmental parameters	Pilot scale studies have been carried out	Deptt. of Environment and Forests, National Thermal Power Corp., Coal India Ltd.
Coral Reef Mapping	Coral reef mapping on 1:50,000 scale	Gulf of Kutchch, Lakshadweep, Gulf of Manner, Andaman and Nicobar Islands are being mapped	Department of Environment and Forests

Source: K. Kasturirangan (1994).

If one would like to survey the trends of different applications of remote sensing, he has to go through the (a) publications brought out the Indian Space Research Organisation, including *Space India Quarterly,* (b) Bulletin of the National Natural Resources Management System (NNRMS), (c) *SAC Courier* of SAC Ahmedabad, (d) *Interface* which is a bulletin from the NRSA Data Centre, and (e) *Photonirvachak*, a journal of the Indian Society of Remote Sensing. There are other journals and books which have a good number of contributions on the applications of remote sensing data. Such publications are from the disciplines of geography, geology, soils, environmental science, forestry, water resources and the like.

In fact, the survey of research trends as being revealed from the published sources, indicate that the applications of remote sensing technology in institutions outside the ISRO/DOS have been on traditional lines and traditional themes basically depending on visual techniques. Nevertheless, due to their share of participation in the national projects as mentioned in Table 13.4, such departments and institutions do get exposure of the digital technology. As a result, several small centres are now having atleast PC based digital image processing facilities.

TABLE 13.5: Some latest example of image processing techniques used in India

Applications	Area/RS data	Digital technique	Reference
Environmental study	Karnataka coast IRS LISS II	Transformation model for rectification and Maximum likelihood algorithm	P. Manavalan & S. Adiga in *NNRMS Bulletin* September 1995
Forest encroachment	Sanjay Gandhi National Park IRS LISS II & SPOT	Visual	R.N. Jadhav in *NNRMS Bulletin* September 1995
Structural interpretation	Peninsular India: Eastern Ghats Landsat TM	Visual	T.R.K. Chetty in *NNRMS Bulletin* September 1995
Identification and mapping of sugar cane types	Ahmadnagar District IRS LISS-II	Visual	Sapna P. Dhoptey & R.S. Ayyangar in *NNRMS Bulletin* March 1994
Growth conditions of forests	Middle Andaman forests and Central Tarai forest (UP) Landsat TM, IRS LISS II & SAR *x* Band Airborne	Spectral values of data recorded during peak growth state correlated using linear and stepwise regression	P.S. Roy in *NNRMS Bulletin* March 1994
Crop monitoring	Bardhman district simulated IRS WiFS data & Landsat TM	Normalized Difference Vegetation Index, Maximum likelihood classification, Synchronous Leaf Area Index	S. Panighahy *et al.* in *Photonirvachak*, Vol. 23, No. 4, 1995
Soil Conservation Service Runoff Curve Number	Kaliaghat river basin (Midnapore district) IRS-1A LISS II	Supervised classification using maximum likelihood algorithm	Pramod Kumar, K.N. Tiwari & D.K. Pal in *Photonirvachak*, Vol. 19, No. 4, 1991.
Forest land use planning	West Garo Hills, Meghalaya IRS LISS II	Image to image registration and hybrid classification (visual & digital) using Combine software	S.P.S. Kushwaha *et al.*, in *Photonirvachak*, Vol. 23, No. 3, 1995
Identification of Wastelands	Sultanpur district (UP) Landsat TM	Soil Brightness Index and Principal Component Analysis	M. Kudrat, P. Nag & S.K. Bhan in Mishra (ed.), 1989
Wheat yield model and historical trends	Punjab Landsat MSS & IRS LISS I	NDVI, Linear regression analysis, Multiple regression analysis	T.T. Medhavy *et al.*, in *Photonirvachak*, Vol. 23, No. 1, 1995

The digital image processing techniques have been mentioned in the preceding chapters. Some of these techniques are popular while others are rarely used in Indian situation though the facilities exist in any established centre. The digital techniques used for image processing are varied. In several cases, more than one technique has been used for specific applications. Though radiometric corrections should be done as a matter of rule, it is not always carried out. Further, digital rectification with some enhancement techniques required for specific applications are common. Most of the important projects are related to agriculture and it is expected to do so in this agricultural based country.

The above list is not exhaustive but only indicative of the research trends of application of remote sensing in India. One may gather that visual techniques are still popular through facilities for digital processing of data is becoming ubiquitous. Perhaps the cost of remote sensing data and its processing is coming in the way of the digital remote sensing. Nevertheless, in the national projects the digital processing of data is widely used. In several cases it appears that remote sensing techniques have been used to supplement the findings, but in some cases it has been added superfluously.

Another trend which is becoming obvious is the integration of remote sensing data with other information. This is possibly due to the advent of the geographical information system in the scene. After the major joint activity of ISRO and Geological Survey of India, known as Vasundhara Project, several projects have been taken by ISRO/DOS (ISRO, 1994). Some of such projects have been mentioned in Table 13.4. Further, the current two important projects are as follows:

NATIONAL (NATURAL) RESOURCES INFORMATION SYSTEM (NRIS)

This project was initiated as a part of the greater NNRMS. The task force appointed for this purpose submitted the report in February 1988 (NNRMS, 1988). An assessment of the availability of information on different themes was made. Recently, such a system is considered to be essential for the decision makers to ensure the efficient and optimum utilization of natural resources which is also environmentally sound. This information system is meant to provide periodic, updated and systematic information on land, water, forests, minerals, soils, oceans and the like. It is to be integrated with the socio-economic data, such as demographic data, financial allocations, and developmental targets. Under the NRIS, suitable computer system based on geographical information system (GIS) are being developed with capabilities for data integration and easy retrieval. Further, NRIS will have advanced capabilities and is proposed as an integrated information system with linkages with the other existing systems. Obviously, in order to achieve this objective, co-operation from different agencies would be essential as extra-remote sensing information, i.e. conventional information, would also be required.

INTEGRATED MISSION FOR SUSTAINABLE DEVELOPMENT (IMSD)

IMSD is another national level project which is based on GIS methodology. But in this case the objective is multi-dimensional. "Sustainable development of natural resources is based on maintaining the fragile balance between productivity functions and conservation practices through monitoring and identification of problem areas which require application of alternate agricultural practices, crop rotation, use of bio-fertilizers, energy efficient farming methods and reclamation of underutilized lands" (ISRO, 1995). This model has local and global objectives. The solutions at the local level, similar to NRIS, are aimed to arrive through the effective use of space-based remote sensing data merged with collateral socio-economic data. This project has been launched in the 157 problematic districts of the country which cover about 45 per cent area.

As per the approved methodology, this study involved generation of thematic maps showing current landuse/landcover, types of wastelands, forest cover/types, surface water resources, drainage pattern, potential ground water zones, landforms (geomorphology), geology (rock types, structural features, mineral occurrence), soil types etc. using data from IRS. Slope maps from topographical sheets and meteorological data were also used. With the help of above mentioned information, derived maps showing land capability, land irrigability, status of soil erosion, run-off potential, priority watersheds needing immediate treatment etc. are to be prepared. All this information is to be used to identify land units which are unique in terms of their resource potentials and problems. Specific development plans with the help of the concerned departments have to be initiated. National level institutions like NRSA and SAC, RRSSCs and state level remote sensing centres are participating in this project.

New Data Products and Possible Applications: Example of IRS-1C

According to evaluation of the Indian Space Research Organisation (ISRO), the unique combination of data products from IRS-1C will open new vistas for applications specifically those related to vegetation, terrain and micro-level planning. The variety of applications based on IRS-1C and even IRS-1D will significantly multiply with the LISS-III 23.5m visible and near-IR data being able to discriminate mixed-crops to a larger extent and 70.5m SWIR data being able to detect moisture content in crops--thus making it suitable for crop-stress studies and also for crop-growth analysis. Further, the WiFS data will be most suited for large-area crop and vegetation inventory and will provide frequent observations for trend analysis. In addition, the 5.8m panchromatic data will provide digital elevation models (DEM) and facilitate better contour mapping and also high-quality cartographic mapping of features. The data will also find greater utility in urban applications (ISRO, 1995). Possible applications based on IRS-1C data products have been mentioned in Table 13.6.

TABLE 13.6: Possible applications of IRS-1C data products

Applications	Features
Agriculture	Mixed-crop discrimination and inventory Large-area crop inventory Crop stress detection/monitoring
Forests	Types and species discrimination Forest stock mapping Bio-mass estimation
Urban	Town and city mapping Facility mapping
Landuse	Landuse/landcover level-III mapping Change detection
Soil	Soil mapping Erosion prone mapping
Geology	Geological mapping Geomorphological mapping
Terrain	Cadastral mapping
Water Resources	Surface water inventory Ground water targeting Command area management
Digital Elevation Models	Contours Slope/aspect Visualization/perspective
Environment	Environmental impact assessment/ Monitoring siting applications
Disasters	Flood damage assessment Flood relief measures

Source: ISRO (1995).

It is expected that the application of IRS-1C data integrated in a geographical information system environment and alongwith global positioning satellite data will enable the generation of detailed digital cartographic data base. This should lead to engineering solution to complex integrated resource management problems, specifically for micro-level development.

Emerging Indian Scenario

The emerging Indian scenario is the combination of the availability of data product, the facilities for processing of the digital data and the research trends. As indicated in several places in this book, the better resolution of the sensors are opening new vistas for applications. There will be pressing demands from the governments and the public to tap this vital source of information to solve the day to day problems of the country, like floods, droughts, drinking water and the like. In any case, environment has become an inevitable factors for all development initiatives. The utilization of resources has to be linked with it. The projects like Integrated Mission for Sustainable Development is one of the example of the current thought in this direction. Further, the finer resolution will allow to use the remote sensing data for cartographic applications. As planned, the future IRS satellites will be having sub-metre resolution which with on-board GPS facilities will come closer to the requirements of the cartographers.

There are several projects successfully completed or under operation which clearly indicate that the remote sensing has to be integrated with the conventional data. GIS and LIS technology has to be widely used for making the remote sensing technology more meaningful and closer to the problems of the people at large. Only by doing so, the attainments in space, computer and associated technologies could be used for solving the day to day problems of India which in any case are multiplying.

A strategy has been worked out for assessing the future perspectives of remote sensing in India which include forthcoming missions as well. Based on this evaluation and also the need for identifying low-cost, quick turn-around, application-specific solutions, candidate payloads for the continuation series have been defined. These candidate payloads cater to a majority of the application needs and specific gap-areas would be taken care of (Chandrasekhar *et al.*, 1996). A few of the measures considered to are (a) MSS data of 5-10m, (b) PAN data of 1m thereby replacing aerial photography for cartography and resource management applications, (c) ocean observation mission for physical and biological parameters, (d) observation of the atmospheric constituents, radiation budgets, clouds etc. -- much of which will be oriented towards global change applications, (e) advanced systems with active microwave instruments, and (f) continuity of the programmes. The thrusts of the IRS missions in the 2000s would be as follows:

— High accuracy resource management applications
— High accuracy terrain applications
— Global change applications
— All weather applications

The details of the IRS-P3 have been discussed in chapter 3. The forthcoming IRS-P series will be application-specific missions. The IRS-P4, IRS-P5 and IRS-P6 have been planned to be launched by PSLV-C1, C2 and C3 respectively. The mission profile for the IRS-P series is mentioned in Table 13.7.

Conclusion

The success in the space technology in India has yielded rich dividends in communication and meteorology which in turn has helped in achieving the technological unity. The nooks and corners have been brought into the main stream

TABLE 13.7: Mission profile for IRS-P series satellite

Mission	Payload-1	Payload-2	Payload-3
IRS-P4	Ocean colour monitor (OCM) in 9 bands	Multi-frequency scanning microwave radiometer	AO
IRS-P5	LISS-IV 3 band XS, steerable, better than 10m resolution 140 km swath, 6/7-bit	LISS-III 4 bands (VNIR and SWIR with 23m resolution	AO
IRS-P6	PAN 2.5m resolution, 10 km swath fore-aft stereo	AO	--
IRS-P7	OCEANSAT mission (Ku-scatterometer, Ku-altimeter, radiometer)	--	--
IRS-P8 (2000)	Environmental/atmospheric satellite mission	--	--

Source: Chandrasekhar *et al.*, 1996, *op. cit.*

Note: In the NRSA Data Centre Bulletin, *Interface*, vol. 7, No. 3, IRS-P5 is for CARTOSAT-1, IRS-P6 is the state-of-art satellite, mainly for agricultural purposes. It appears IRS-P5 and IRS-P6 have been interchanged.

of this federally administered country. Remote sensing is another gift of the progress made in space technology. It cut across all sorts of barriers and provide a synoptic and integrated view of the country which can help in planning and development as well. The planned application-specific missions will bring the benefits of this high technology closer to people and their aspirations.

REFERENCES

Chandrasekhar, M.G., V. Jayaraman and Mukund Rao (1996), Future perspectives of remote sensing, *Current Science*, vol. 70, No. 7, pp. 648-53.

ISRO (1944), Workshop on Project Vasundhara, in *Space India*, July-September, pp. 16-7.

ISRO (1995), *Indian Remote Sensing Satellite-1C*, Indian Space Research Organisation, Bangalore.

Kasturirangan, K. (1994), *Sustainable Development and Role of Geomatics (Remote Sensing and Conventional Data Integration)*, Indian Space Research Organisation, Special Publication, ISRO-SP-68-94, Bangalore.

Kumar, Pramod, K.N. Tiwari and D.K. Pal (1991), Establishing SCS runoff curve number from IRS digital data base, *Photonirvachak*, vol. 19, No. 4, pp. 245-51.

Kushwaha, S.P.S. and G. Oesten (1995), A rule-based system for forest land use planning, *Photonirvachak*, vol. 23, No. 3, pp. 115-24.

Kudrat, M., P. Nag and S.K. Bhan (1989), Identification of waste lands in Sultanpur district, Uttar Pradesh, in B.M. Mishra (ed.), *Land Utilization and Management in India*, Chugh Publication, Allahabad, pp. 23-30.

Medhavy, T.T., Ţara Sharma, R.P. Dubey, R.K. Dubey, R.K. Mahey and P.K. Sharma (1995), Development of a wheat yield model for Punjab using remotely sensed data and historical tends, *Photonirvachak*, vol. 23, No. 1, pp. 23.30.

Nag, P. (ed.) (1992), *Remote Sensing and Thematic Cartography*, Concept Publishing Company, New Delhi.

NNRMS (u.d.), *Satellite Remote Sensing*, Indian Space Research Organisation, Bangalore.

NNRMS (1988), Report of the Task Force on National (Natural) Resources Information System, No. NNRMS-SP-08-88, Bangalore.

NRSA (1996), *Interface*, NRSA Data Centre, Hyderabad, vol. 7, No. 3 : July-September.

Panigrahy, S., R.P. Singh, S.A. Sharma and M. Chakraborty (1995), Results on potential use of simulated IRS-1C WiFS data for crop monitoring, *Photonirvachak*, vol. 23, No. 4, pp. 175-83.

Raghavswamy, V., S.K. Patan, P. Ram Mohan, R.J. Bhandari and Padma Priya (1996), IRS-1C applications for urban planning and development, *Current Science*, vol. 70, No. 7, pp. 582-8.

Rao, D.P., N.C. Gautam, R. Nagaraja and P. Ram Mohan (1996), IRS-1C applications in land use mapping and planning, *Current Science*, vol. 70, No. 7, pp. 575-81.

Rao, D.P., R.R. Navalgund and Y.V.N. Krishna Murthy (1996), Cadastral applications using IRS-1C data--Some case studies, *Current Science*, vol. 70, No. 7, pp. 624-8.

Rao, U.R. (1991), Remote sensing for sustainable development, *Photonirvachak*, vol. 19, No. 4, pp. 217-35.

APPENDICES

LIST OF COMPANIES AND AVAILABLE INSTRUMENTS

Company and address	Instruments
ACE 50 West Wilmot Street Richmond Hill Ontario Canada l48 1M5	S/w for cartographic solution
Audre 10915 Technology Place San Diego California 92127 U.S.A.	Audre conversion s/w
Autodesk Inc. 206 Raheja Plaza 17 Commissariat Road Bangalore 560 025	Autodesk GIS solutions : Mapping & GIS s/w AutoCAD Release 13 CAD s/w AutoCAD Map Mapping and GIS s/w
Bentley Lot 5.01, 5th Floor Wisma HLA Jalan Raja Chulan Kuala Lumpur Malaysia 50200	Micro Station GeoGraphics : Mapping s/w
Carl Zeiss Geschaftsbereich Vermessung D-73446 Oberkochen Germany	PC P3 Planicomp Phocus P3 Planicomp Phodis ST Digital Stereoplotter Phodis Photogrammetric Image Processing System Visopret Image Intrepretation Instruments Cadmap/dgn Micro Station Driver Cadmap Photogrammetric Cad system

Company and address	Instruments
Cipher Data Products 9715 Businesspark Av. P.O. Box 85170 San Diego, CA 92138-9198, U.S.A.	Cipher M990 Cache Tape Drive
Computer Graphics Group 2411 West La Palma Av. Anaheim California 92801 U.S.A.	CalComp Pen Plotters
Cymbolic Sciences International 26072 Merit Circle Suite 113 Laguna Hill CA92653 U.S.A.	Fire 2240 Color Digital Film Recorder
Earth Resource Mapping Suite 900 4370 La Jolla Village Drive San Diego, CA 92122 U.S.A.	ER Mapper image processing s/w
GETCO Corporation 7125 Riverwood Dr. Columbia MD 21046 USA	AccuTab Digitizer
Graphic Data Systems 6200 South Syracuse Way, Suite 250 Englewood, CO 80111 U.S.A.	MicroGDS Professional CAD s/w
GRI GeoResearch, Inc. 2815 Montana Avenue Billings, MT 59101 U.S.A.	GeoLink s/w GPS Pathfinder Receivers

Company and address	Instruments
HCL Hewlett-Packard A-10-11 Sector III Noida 201 301 (U.P.)	HP Apollo Series 700 Model 735, 715/33, 715/50 and 715/75 workstation
Intergraph Corporation Huntsville Alabama 35894-0001 U.S.A.	GIS and mapping solutions (h/w + s/w)
LASICO 2451 Riverside Dr. Los Angeles, CA 90039 U.S.A.	Digitizer/Planimeter Compact computing planimeter
Laser-Scan Limited Cambridge Science Park Milton Road Cambridge CB4 4FY U.K.	LAMPS Automated Map Production System
Leica AG CH-9435 Heerbrugg Switzerland	Wild RC30 Aerial camera system Wild TSP1 Pocket mirror stereoscope SD Line analytical workstation PC-PRO 600 data collection and editing s/w MAPIT COLLECTOR mapping software SD 2000 photogrammetric workstation Wild APT2 Stereo Photointerpretation System
Mapinfo Corporation The Global View Troy New York 12180-8399 U.S.A.	MapInfo GIS s/w
PAMAP Technologies 200-6772 Oldfield Rd. Victoria, B.C. Canada V8X 3X1	PAMAP GIS s/w
PCI Inc. 50 West Wilmot Street Richmond Hill Ontario	EASI/PACE Remote Sensing Software ACE Works automated cartographic editing s/w HI-VIEW digital photogrammetry DEM s/w Image Works Image analysis s/w

Company and address	Instruments
Canada L4B 1M5	FLY Terrain visualization s/w RadarSoft s/w EASI/PACE ORTHO for Airphoto ortho and DEM, and satellite ortho and DEM s/w Truth field/portable remote sensing/GPS system
Polaroid Far East 32/F Windsor House 311 Gloucester Rd. Causeway Bay Hong Kong	CI-5000S Digital Palette Color Film Recorder
Seimens Nixdorf Informations-system AG P.O. Box 830951 Otto Hahn-Ring 6 W-8000 Munchen Germany	SICAD Hybrid geoinformation system s/w SICAD-CARIS GIS application s/w
Speck System Pvt. Ltd. B-49, EC Kushaiguda Hyderabad 500 762 India.	Fotorite 1010C digital film recorder Fotorite 1040C Large format photowrite system Speck Isrovision image analysis s/w SACIMAGE image processing s/w
Strategic Mapping Inc. 3135 Kifer Road Santa Clara, CA 95051 U.S.A.	Atlas GIS s/w
Sun Microsystem, Inc. 29/F, Shui On Centre 8 Harbour Road Wanchai Hong Kong	Sun Sparc work station and servers
SysScan Dyrmyrgt 35 P.O. Box 433 3601 Kongsberg Norway	Tellus GIS s/w

Company and address	Instruments
Universal Systems Ltd. 270 Rookwood Avenue P.O. Box 3391 Station B Fredericton, NB Canada E3A 5H2	CARIS GIS s/w SAMI raster to vector conversion s/w
Tydac 2 Gurdwara Road Suite 210 Nepean, Ontario Canada K2E 1A2	SPANS Explorer desktop mapping software
Vexcel Imaging Co. 3131 Indian Road Boulder, Colorado USA 80301	VX3000 Image Scanning System s/w IDAS Integrated digital aerotriangulation system

Note: H/W : Hardware
S/W : Software

APPENDIX-II

FUNCTIONAL SPECIFICATIONS REQUIRED FOR GIS

1. Functional Capabilities

(i) *Data Input*

— Methods for vector and raster data input
— Registration procedures (tic marks, scale, projection)
— Cleaning methods
— Creation of topology and polygon building
— Support for meta data bases

(ii) *Nonspatial data bases*

— Methods to create and handle nonspatial data
— Types of spatial queries linking nonspatial data bases with spatial ones

(iii) *Data Import/Export*

— Data convertibility to standard vector and raster formats

(iv) *Data Model Conversion*

— Raster to Vector and Vector to Raster conversion
— Methods of data handling in case specific data structure is used

(v) *Coordinate Transformation*

— Transformation of coordinates between 2 projections
— Geometric Transformation/Geo Registration of 2 maps of different scales
— Edge matching and mosaicing or splitting a map into submaps

(vi) *Analysis*

— Map Algebra

- Overlay, Union or Intersect
- Buffer Generation
- Distance from point to point, point to line and point to polygon
- Visibility
- Area of Influence
- Flow analysis

(vii) *Spatial and Nonspatial Statistics*
(viii) *Terrain Modelling*

- Spatial Interpolation
- TIN Generation
- Compilation of slope and relief
- Prospective view
- Profile plotting
- Cut and Fill calculation

(ix) *Digital Image Analysis*
(x) *Output Generation*

- Display capabilities—Visualization (2D/3D)
- Plot/Print results

(xi) *Linkage with External Models through Language Interface*
(xii) *Limitation in Data Handling*

- No. of points per arc
- No. of arcs per polygon
- No. of arcs per node
- Total no. of arcs
- Total no. of label points
- Total no. of polygons
- Total no. of tics
- Size of symbol sets
- No. of fonts
- Plot size
- Points per polygon in polygon fill
- Limitations in nonspatial data bases

II. User Interface

- Availability of menus

— Explanation of menus and commands
— Help facility
— Error explanation
— Command language availability
— User Friendliness
— Manuals support

III. Training

— Time Requirement
— Availability of Demonstration kits
— Sample case studies with commands illustrations

IV. Select 3 Examples

— Spatial Analysis
— Spatial Query
— Vector data processing (for example Network)

Each example can be executed to know

— Time taken for each interaction
— Resources usage
— Accuracy Evaluation

Source: CSRE, IIT Mumbai.

APPENDIX-III

IRS-1C ORBIT, COVERAGE AND REFERENCING SCHEME

The primary objective is to provide systematic and repetitive acquisition of data of the earth's surface under nearly constant illumination conditions. The satellite operates in a circular, sun-synchronous, near polar orbit with an inclination of 98.69 deg. at an altitude of 817 Km. in the descending node. The satellite takes 101.35 minutes to complete one revolution around the earth and completes about 14 orbits per day. The entire earth is covered by 341 orbits during a 24 day cycle. The orbital parameters are summarised in Table III.1.

TABLE III.1: IRS-1C Orbit

Orbits/cycle	341	
Repetivity	24	days
Altitude	817	Km
Semi-major axis	7195.11	Km
Inclination	98.69	deg
Eccentricity	0.001	
Period	101.35	minutes
Distance between adjacent traces	117.5	Km
Distance between successive ground tracks	2820	Km
Ground track velocity	6.65	Km/sec

The mean equatorial crossing time in the descending node is 10.30 a.m. ± 5 minutes. The orbit adjust system is used to attain the required orbit initially and it is maintained throughout the mission period. The gound trace pattern is controlled within ± 5 Km of the reference ground trace pattern.

The sensors collect data with different swaths. The swath of LISS-III sensor in the visible bands is 141 Km while in SWIR band it is 148 Km. The swath of PAN and WiFS sensors are 70 Km and 810 Km respectively. Details of overlaps and sidelaps between scenes of a sensor are given in Table III.2. The successive orbits are shifted westward by 2820 Km at the equator. Fig. III.1 shows a typical Ground trace of the orbits. The entire globe is covered in 341 orbits between 81° North and

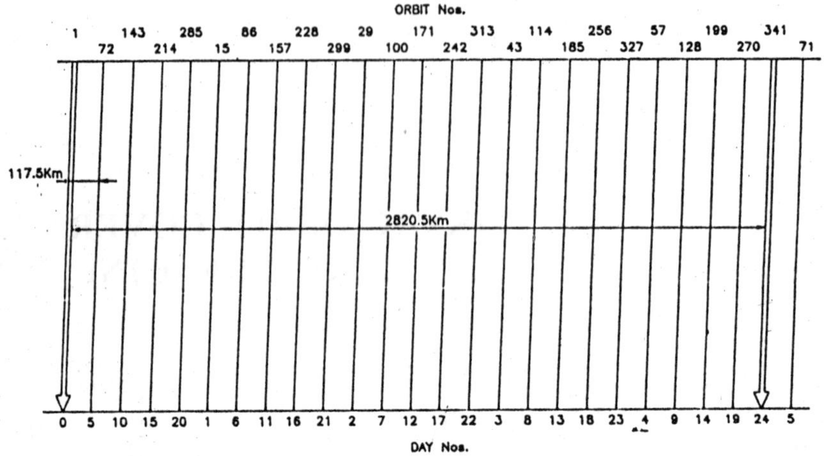

Fig. III.1: Ground Trace Pattern

TABLE III.2: Overlap and Sidelap between the Scenes

Payload	Resolution (meters)	Ground swath (Km)	Image size Km × Km	Overlap (Km)	Sidelap at equator (Km)
LISS III					
Visible	23.5	141	141 × 141	7	23.5
SWIR	70.5	148	141 × 148	7	30
PAN	5.8	70	70 × 70	2	~1 (Opt)
WiFS	188.3	810	810 × 810	~80%	~85%

81° South latitudes during the 24 day cycle.

Scene Layout: Fig. III.2 shows the scene layout of LISS-III (visible and NIR bands), LISS-III SWIR, band and PAN scenes within one WiFS scene. The corners are numbered as shown in the figure. Same pattern of marking the corners is followed for other images also. The LISS-III (SWIR) scenes are framed in such a way that their length is same as LISS-III (V, NIR) scene though its breadth is 7 Km more than a LISS-III (V, NIR) scene.

There is an overlap of 7 Km between adjacent scenes of LISS-III along a path. Also there is a sidelap of 23.5 Km between scenes of adjacent paths at equator. The sidelap is minimum at equator. As we go away from the equator, the sidelap increases because the paths come closer to each other as we move towards the pole. Typically, at 40° latitude the sidelap is around 40% of the swath and at 81° latitude it is 99%. The Figure depicts the sidelap and overlap in case of LISS-III (SWIR)

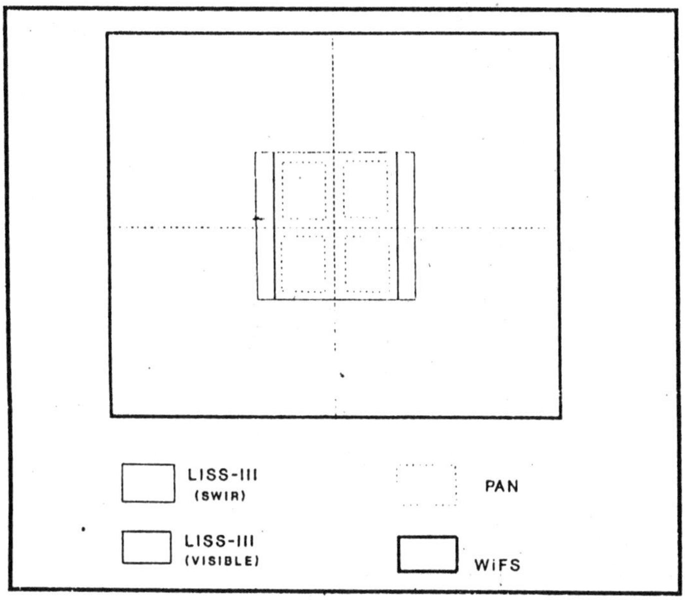

Fig. III.2: Scene layout of PAN, LISS-III (V, NIR) and LISS-III (SWIR) within WiFS scene
Source: IRS-1C Data User's Handbook, NRSA.

scenes. It can be seen that the overlap is same as LISS-III (V, NIR) as the lengths of LISS-III (V, NIR) and LISS-III (SWIR) scenes are same. The sidelap is more in the case of LISS-III (SWIR) as the swath is also more when compared to LISS-III (V, NIR).

As the swath of WiFS is very large, there is a sidelap of about 85% between WiFS scenes of adjacent paths at equator. But, between the n-th and n+6th path, the sidelap is around 105 Km at equator. Also there is a overlap of around 80% between adjacent scenes in a path. But the overlap between m-th and m+5th scenes along a path is around 130 Km. Hence one out of every consecutive five scenes can be downloaded for data products generation.

Revisit capability of PAN: Because of PAN's tilting capacity, a given area can be viewed more than once within one cycle. This is known as revisit due to PAN's steerability. Figure III.3 shows a path with three adjacent paths on either side from equator, the tilt angle with which the central path can be viewed from adjacent paths and also the day number on which the adjacent paths occur relative to the central path. From the figure it can also be seen that the maximum wait period to view an area is 5 days only. The maximum tilt angle being ± 26°, PAN camera can see only three paths on either side at equator. As we go away from equator, paths become closer to each other. Hence, more number of paths can be viewed by PAN at high latitudes.

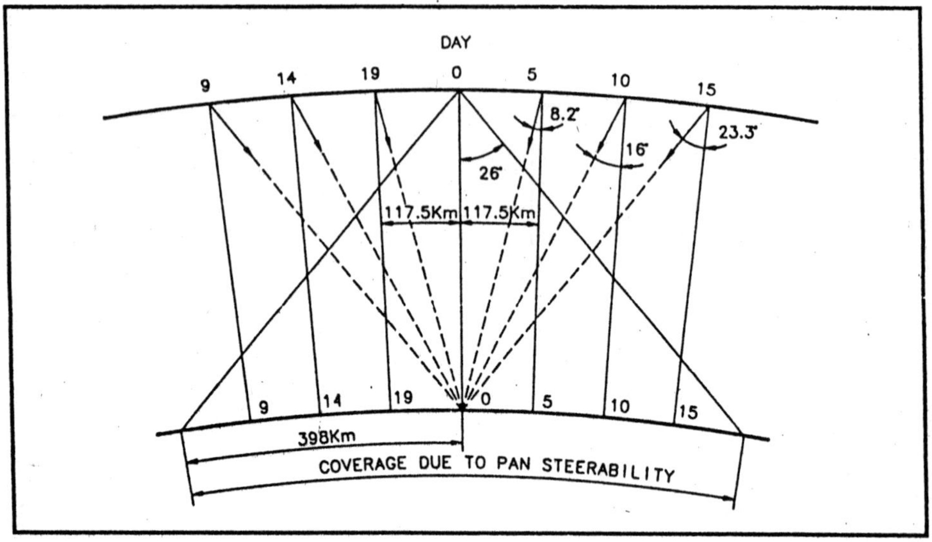

Fig. III.3: PAN off-nadir viewing capability

Source: IRS-1C Data User's Handbook, NRSA.

APPENDIX-IV

ESTIMATION OF THE CENTRE AND CORNER COORDINATES OF LISS-III AND PAN SCENES

From the ephemeris information, it is possible to compute geographical coordinates of LISS-III scene centres which lie on the ground track. However, this is not the case with PAN scene centres as they lie on either side of the ground track. The time of occurrence of any PAN scene center or any corner coordinate is obtained by using the information that the scanning is line by line at an interval of integration time of the respective cameras. Taking LISS-III scene center as the origin, the coordinates of any point is established in terms of lines and pixels.

In Figure IV.1 let a_1 be a point on the ground track with coordinates (ϕ_1, λ_1). Let P be a corner point of a scene.

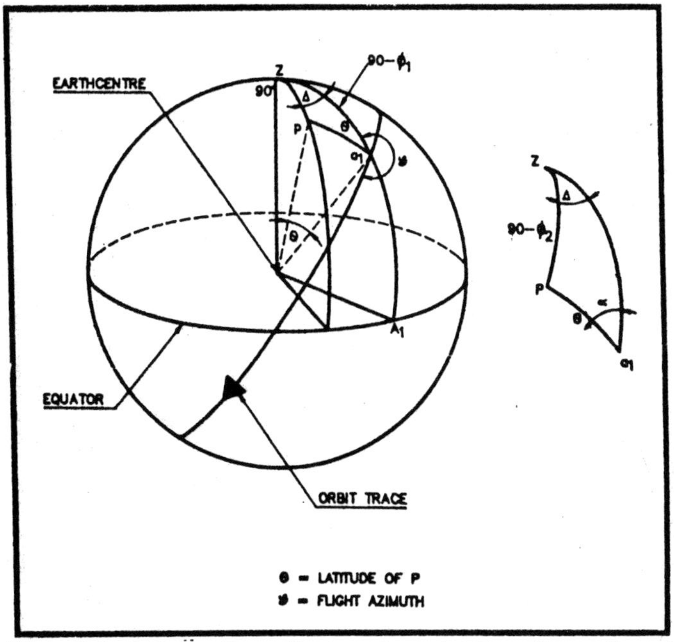

Fig. IV.1: Calculations of coordinates of a point P on the earth surface

The coordinates of P, say (ϕ_2, λ_2) can be calculated as follows.

$$\text{Sin}\ \phi_2 = \text{Cos}\theta\ \text{Sin}\phi_2 \pm \text{Sin}\theta\ \text{Cos}\phi_1\ \text{Cos}\alpha \qquad \ldots\ldots \qquad (1)$$

Where $\alpha = 2\pi - \zeta \pm \pi/2$
and θ, ϕ_1, ϕ_2 and ζ are the angles as shown in Figure IV.1.

$$\text{Cos }\theta = \text{Sin}\phi_1 \ \text{Sin}\phi_2 \pm \text{Cos}\phi_1 \ \text{Cos}\phi_2 \ \text{Cos}\Delta \qquad \ldots\ldots \quad (2)$$

Where Δ is longitudinal difference of λ_2 from λ_1 is the angle subtended by arc a_1P (Fig. IV.1) at the centre of the Earth.

Using equation (2) the expression for 'Cos Δ' can be derived as

$$\text{Cos}\Delta = \frac{\text{Cos }\theta - \text{Sin}\phi_1 \ \text{Sin}\phi}{\text{Cos}\phi_1 \ \ \text{Cos}\phi_2} \qquad \ldots\ldots \quad (3)$$

Appropriate sign is used to denote 'Δ' depending on whether the point P (ϕ_2, λ_2) is east or west of point a_1 (ϕ_1, λ_1)

The longitude λ_2, is, therefore, obtained as

$$\lambda_2 = \lambda_1 \pm \Delta \qquad \ldots\ldots \quad (4)$$

Thus, the geographical co-ordinates of any required point can be obtained either for any corner or the centre of the scene.

Source: IRS-1C Data User's Handbook, NRSA.

APPENDIX-V

ANNOUNCEMENT OF SATELLITE DATA PRODUCT PRICES FOR INDIAN USERS FOR THE PERIOD 1-4-1997 TO 31-3-1998

The prices of satellite data products being supplied by NRSA data centre (NDC) for the above period is given in the following Tables. However for orders for multiple copies at a time, a discount of 25% will be provided from the second copy onwards.

Priority services will be at (i) 50% additional charge over the product price for products supplied within three days and (ii) 100% additional charge over the product price for products supplied within one day. For such priority requests, orders must be received at NDC before 11 A.M. on a working day. The turn-around time (TAT) is calculated from the date of receipt of the confirmed order (or the date next to that of the day of acquisition of the desired data) till the date of despatch of products from NDC. In case the order is accepted but the product is despatched after the stipulated TAT, no additional charges will be levied, other than the products listed below, Raw data on CCTs, 2x enlargement paper prints and non-standard products can also be supplied.

INDIAN REMOTE SENSING SATELLITE (IRS-1A/B/P2) DATA PRODUCTS

Product description	Nominal scale LISS-I	LISS-II	Product code	Price in Rs.
STANDARD data products with polyconic projection and cubic convolution resampling				
1. Full scene based standard data products.				
1.1 Photo products				
1.1.1 Black & White (B&W) per band scene				
240mm Film negative	1:1,000,000	1:500,000	STPC01202	1500
240mm Film positive	1:1,000,000	1:500,000	STPC01212	1500
240mm Paper print	1:1,000,000	1:500,000	STPC01222	1500

NRSA Data Centre, National Remote Sensing Agency, Balanagar, Hyderabad-500037, A.P. (India).

	Product description	Nominal scale LISS-I	LISS-II	Product code	Price in Rs.
	480mm Paper print	1: 500,000	1:250,000	STPC01223	1800
	960mm Paper print	1: 250,000	1:125,000	STPC01224	2300
1.1.2	False Colour Composite (FCC)				
	240mm Film negative	1:1,000,000	1:500,000	STPC01232	4400
	240mm Film positive	1:1,000,000	1:500,000	STPC01242	4000
	240mm Paper print	1:1,000,000	1:500,000	STPC01252	3600
	480mm Paper print	1: 500,000	1:250,000	STPC01253	4000
	960mm Paper print	1: 250,000	1:125,000	STPC01254	5600
1.2	Digital products with Band Interleaved by Line (BIL) format				
	Computer Compatible Tape 1600 BPI		STPC00266	5300	
	Computer Compatible Tape 6250 BPI		STPC00267	5300	
	250 MB Cartridge for DOS-O/S		STPC0026C	5300	
2.	Segment based miniature standard data products				
	5-1/4" Floppy 1.2 MB for DOS-O/S (512×512×4 bands)			STPC0026A	900
	5-1/4" Floppy 1.2 MB for DOS-O/S (1024×1024×1 bands)			STPC0026A	900
	GEOCODED data products as per SOI topo sheet in Polyconic projection & Cubic Convolution resampling				
1.	Data products corresponding to 15'×15' SOI topo sheet				
1.1	Photo products				
1.1.1	Black & White (B&W) per band scene				
	1000mm Paper print		1:50,000	G3PC01225	1600
1.1.2	False Colour Composite (FCC)				
	1000mm Paper print		1:50,000	G3PC01255	4400
1.2	Digital products with Band Interleaved by Line (BIL) format				
	Computer Compatible Tape 1600 BPI			STPC00266	4000
	Computer Compatible Tape 6250 BPI			STPC00267	4000
	250 MB Cartridge for DOS-O/S			STPC0026C	4000
2.	Segment based miniature standard data products				
	5-1/4" Floppy 1.2 MB for DOS-O/S (512×512×4 bands)			STPC0026A	900
	5-1/4" Floppy 1.2 MB for DOS-O/S (1024×1024×1 band)			STPC0026A	900

Note: 1. All the above standard data products can be supplied with Space Oblique Mercator (SOM) projection--Replace the 3rd character of product code with 'S'.

2. Digital data products can be supplied with Nearest Neighbour (NN) resampling--Replace the 4th character of product code with 'N'.

3. Digital data products can be supplied in BSQ (Band Sequential) format also. Replace the 8th character of product code with 7.

4. Floppies and Cartridges can be supplied to work under XENIX operating system. Replace the 9th character of product code with 'D' for cartridge and 'B' for floppy.
5. In the case of geocoded products, if a given topo sheet is covered in two geocoded products, each of the resulting geocoded products will be priced 0.75 times the unit price.
6. Data can be supplied on 3-1/2" 1.44 MB floppies also. The product code is STPD00268 in BIL format for DOS and G3PD00268 for Geocoded products. Please replace the last character by 9 for data in XENIX operating system. For BSQ format replace the eighth character by 7. The price is the same as that of 5-1/4" floppy products.
7. The cartridge products can also be supplied in 150 MB, for which the ninth character in the product code should be E (for DOS OS) and F (for Xenix OS).

	Product description		Nominal scale LISS-I	Nominal scale LISS-II	Product code	Price in Rs.
PRECISION data products with polyconic projection and cubic convolution resampling						
1.	Full scene based data products					
1.1	Photo products					
1.1.1	Black & White (B&W) per band scene					
	240mm	Film negative	1:1,000,000	1:500,000	STPC01402	3300
	240mm	Film positive	1:1,000,000	1:500,000	STPC01412	3300
	240mm	Paper print	1:1,000,000	1:500,000	STPC01422	3300
	480mm	Paper print	1: 500,000	1:250,000	STPC01423	3500
	960mm	Paper print	1: 250,000	1:125,000	STPC01424	4300
1.1.2	False Colour Composite (FCC)					
	240mm	Film negative	1:1,000,000	1:500,000	STPC01432	6700
	240mm	Film positive	1:1,000,000	1:500,000	STPC01442	6100
	240mm	Paper print	1:1,000,000	1:500,000	STPC01452	6000
	480mm	Paper print	1: 500,000	1:250,000	STPC01453	6100
	960mm	Paper print	1: 250,000	1:125,000	STPC01454	8700
1.2	Digital products with Band Interleaved by Line (BIL) format					
	Computer Compatible Tape 1600 BPI				STPC00466	8100
	Computer Compatible Tape 6250 BPI				STPC00467	8100
	250 MB Cartridge for DOS-O/S				STPC0046C	8100

Note:
1. IRS Precision data products can be supplied with Space Oblique Mercator (SOM) projection-- Replace the 3rd character of product code with 'S'.
2. CCT/Cartridge data products can be supplied with Nearest Neighbour (NN) resampling--Replace the 4th character of product code with 'N'.
3. CCT/Cartridge data products can be supplied in BSQ (Band Sequential) format also. Replace the 8th character of product code with 7.
4. Supply of precision data products is subject to availability of suitable ground control points.
5. The cartridge products can also be supplied in 150 MB, for which the ninth character in the product code should be E (for DOS OS) and F (for Xenix OS).

LANDSAT DATA PRODUCTS

	Product description	Nominal scale	Product code	Price in Rs.
	STANDARD data products based on Space Oblique Mercator (SOM) projection			
	Multi Spectral Scanner (MSS) data products			
1.	Full scene based data products			
1.1	Photo products with Cubic Convolution resampling			
1.1.1	Black & White (B&W) per band scene			
	240mm Film positive	1:1,000,000	STSC01212	2800
	960mm Paper print	1: 250,000	STSC01224	3900
1.1.2	False Colour Composite (FCC)			
	240mm Film positive	1:1,000,000	STSC01242	6100
	960mm Paper print	1: 250,000	STSC01254	8900
1.2	Digital products with Band Interleaved by Line (BIL) format and Nearest Neighbour resampling			
	Computer Compatible Tape 1600 BPI		STSN00266	14400
	Thematic Mapper (TM) data products			
1.	Full scene based data products			
1.1	Photo products with Cubic Convolution resampling			
1.1.1	Black & White (B&W) per band scene			
	240mm Film positive	1:1,000,000	STSC01212	4700
	960mm Paper print	1: 250,000	STSC01224	7700
1.1.2	False Colour Composite (FCC)			
	240mm Film positive	1:1,000,000	STSC01242	9500
	960mm Paper print	1: 250,000	STSC01254	21200
2	Quadrant data products			
2.1	Digital products with Band Interleaved by Line (BIL) format and Nearest Neighbour resampling			
	Computer Compatible Tape 1600 BPI		QUSN00266	22300
	250 MB Cartridge for DOS-O/S		QUSN0026C	22300
3.	Segment based miniature standard data products			
	5-1/4" Floppy 1.2 MB for DOS-O/S (512×512×4 bands)		STSN0026A	1800
	5-1/4" Floppy 1.2 MB for DOS-O/S (1024×1024×1 band)		STSN0026A	1800
4.	Full scene products			
	Computer Compatible Tape 6250 BPI (4 bands) STPDB4267		23200	
	Computer Compatible Tape 6250 BPI (7 bands) STPD00267		43700	
	250 MB Cartridge for DOS-O/S (4 Bands)		STPDB426C	23200
	250 MB Cartridge for DOS-O/S (7 Bands)		STPD0026C	43700

Note: 1. Floppies and Cartridges can be supplied to work under XENIX operating system. Replace the 9th character of product code with 'D' for cartridge and 'B' for floppy.

2. Digital data products can be supplied in BSQ (Band Sequential) format also. Replace the 8th character of product code with 7.

3. MSS and TM CCT/cartridge/floppy data products can be supplied in 6250 BPI also. Replace the 9th character of product code with 7.

4. Data can be supplied on 3-1/2" 1.44 MB floppies also. The product code is STPD00268 in BIL format for DOS and G3PD00268. Please replace the last character by 9 for data in XENIX operating system. The price is the same as that of 5-1/4" floppy products.

5. The cartridge products can also be supplied in 150 MB, for which the ninth character in the product code should be E (for DOS OS) and F (for Xenix OS).

	Product description	Nominal scale	Product code	Price in Rs.
TM GEOCODED data products as per Survey of India topo sheet with Polyconic projection				
1.	Data products corresponding to 15' × 15' SOI topo sheet			
1.1	Photo products			
1.1.1	Black & White (B&W) per band scene			
	1000mm Paper print	1: 50,000	G3PD01225	5400
1.1.2	False Colour Composite (FCC)			
	1000mm Paper print	1: 50,000	G3PD01255	14500
1.2	Digital products with Band Interleaved by Line (BIL) format			
	Computer Compatible Tape 1600 BPI		G3PD00266	15300
	Computer Compatible Tape 6250 BPI		G3PD00267	15300
	250 MB Cartridge for DOS-O/S		G3PD0026C	15300
2	Segment based miniature data products			
	5-1/4" Floppy 1.2 MB for DOS-O/S (512×512×4 bands)		G3PD0026A	1800
	5-1/4" Floppy 1.2 MB for DOS-O/S (1024×1024×1 band)		G3PD0026A	1800

Note: 1. Floppies and Cartridges can be supplied to work under XENIX operating system. Replace the 9th character of product code with 'D' for cartridge and 'B' for floppy.

2. If a given topo sheet is covered in two geocoded products, each of the resulting geocoded products will be priced 0.75 times the unit price.

3. TM CCT/Cartridge data products can be supplied in BSQ (Band Sequential) format also. Replace the 8th character of product code with 7.

4. Data can be supplied on 3-1/2" 1.44 MB floppies also. The product code is G3PD00268 in BIL format in DOS. Please replace the last character by 9 for data in XENIX operating system. The price is the same as that of 5-1/4" floppy products.

5. The cartridge products can also be supplied in 150 MB, for which the ninth character in the product code should be E (for DOS OS) and F (for Xenix OS).

SPOT DATA PRODUCTS

Product description		Nominal scale	Product code	Price in Rs.
STANDARD data products based on polyconic projection				
Multispectral Linear Array (MLA) data products				
1.	Full scene based standard data products.			
1.1	Photo products			
	Black & White (B&W) per band scene			
	240mm Film positive	1:400,000	STPD01212	8100
	960mm Paper print	1:100,000	STPD01224	10500
	False Colour Composite (FCC)			
	240mm Film positive	1:400,000	STPD01242	11800
	960mm Paper print	1:100,000	STPD01254	17800
1.2	Digital products with Band Interleaved by Line (BIL) format			
	Computer Compatible Tape 1600 BPI		STPD00266	23100
	Computer Compatible Tape 6250 BPI		STPD00267	23100
	250 MB Cartridge for DOS-O/S		STPD0026C	23100
2.	Segment based miniature standard data products			
	5-1/4" Floppy 1.2 MB for DOS-O/S (512×512×4 bands)		STPD0026A	1000
	5-1/4" Floppy 1.2 MB for DOS-O/S (1024×1024×1 band)		STPD0026A	1000
Panchromatic Linear Array (PLA) data products				
1.	Data products corresponding to 15'×15' SOI topo sheet			
1.1	Photo products			
	Black & White (B&W) per band scene			
	240mm Film positive	1:400,000	STPD01212	11800
	960mm Paper print	1:100,000	STPD01224	17000
1.2	Digital products			
	Computer Compatible Tape 1600 BPI		STPD00266	23100
	Computer Compatible Tape 6250 BPI		STPD00267	23100
	250 MB Cartridge for DOS-O/S		STPD0026C	23100
2.	Segment based miniature standard data products			
	5-1/4" Floppy 1.2 MB for DOS-O/S (512×512×4 bands)		STPD0026A	1000

Note: 1. MLA CCT/Cartridge/floppy data products can be supplied in BSQ (Band Sequential) format also. Replace the 8th character of product code with 7.

2. Floppies and Cartridges can be supplied to work under XENIX operating system. Replace the 9th character of product code with 'D' for cartridge and 'B' for floppy.

3. To order a stereo pair, replace the first two characters of the product code with 'SR' (ST for standard data, SR for stereo pair).

4. Photo and digital products can be supplied with raw and only radiometric correction. Replace the

7th character of product code with 1.
(0 for raw, 1 for radiometrically corrected, 2 for standard)

5. Data can be supplied on 3-1/2" 1.44 MB floppies also. The product code is STPD00268 in BIL format for DOS. Please replace the last character by 9 for data in XENIX operating system. The price is the same as that of 5-1/4" floppy products.
6. The cartridge products can also be supplied in 150 MB, for which the ninth character in the product code should be E (for DOS OS) and F (for Xenix OS).

	Product description	Nominal scale	Product code	Price in Rs.
SPOT GEOCODED data products as per Survey of India topo sheet with Polyconic projection.				
Multispectral Linear Array (MLA) data products				
1.	Full scene based standard data products.			
1.1	Photo products			
1.1.1	Black & White (B&W) per band scene			
	1000mm Paper print	1:25,000	G4PD01225	7800
1.1.2	False Colour Composite (FCC)			
	1000mm Paper print	1:25,000	G4PD01255	14500
2	Data products corresponding to 15' × 15' SOI topo sheet			
2.1	Photo products			
2.1.1	Black & White (B&W) per band scene			
	1000mm Paper print	1:50,000	G3PC01225	7800
2.1.2	False Colour Composite (FCC)			
	1000mm Paper print	1:50,000	G3PC01225	14500
2.2	Digital products with Band Interleaved by Line (BIL) format			
	Computer Compatible Tape 1600 BPI		G3PD00266	17500
	Computer Compatible Tape 6250 BPI		G3PD00267	17500
	250 MB Cartridge for DOS-O/S		G3PD0026C	17500
3	Segment based miniature data products			
	5-1/4" Floppy 1.2 MB for DOS-O/S (512×512×4 bands)		G3PD0026A	1800
	5-1/4" Floppy 1.2 MB for DOS-O/S (1024×1024×1 band)		G3PD0026A	1800
Panchromatic Linear Array (PLA) data products				
1.	Data products corresponding to 15'×15' SOI topo sheet			
1.1	Photo products			
1.1.1	Black & White (B&W) per band scene			
	1000mm Paper print	1:25,000	G4PD01225	13100
2	Data products corresponding to 15' × 15' SOI topo sheet			
2.1	Photo products			
2.1.1	Black & White (B&W) per band scene			
	1000mm Paper print	1:50,000	G3PD01225	13100

2.2	Digital products		
	Computer Compatible Tape 1600 BPI	G3PD00266	17500
	Computer Compatible Tape 6250 BPI	G3PD00267	17500
	250 MB Cartridge for DOS-O/S	G3PD0026C	17500
3	Segment based miniature data products		
	5-1/4" Floppy 1.2 MB for DOS-O/S (1024×1024×1 band)	G3PD0026A	1000

Note: 1. MLA CCT/Cartridge/floppy data products can be supplied in BSQ (Band Sequential) format also. Replace the 8th character of product code with 7.

2. Floppies and Cartridges can be supplied to work under XENIX operating system. Replace the 9th character of product code with 'D' for cartridge and 'B' for floppy.

3. If a given topo sheet is covered in two geocoded products, each of the resulting geocoded products will be priced 0.75 times the unit price.

4. Data can be supplied on 3-1/2" 1.44 MB floppies also. The product code is STPD00268 in BIL format for DOS. Please replace the last character by 9 for data in XENIX operating system. The price is the same as that of 5-1/4" floppy products.

5. The cartridge products can also be supplied in 150 MB, for which the ninth character in the product code should be E (for DOS OS) and F (for Xenix OS).

6. Digital Geocoded products can be supplied for 7-1/2' × 7-1/2' also, the second character in the product code should be 4.

METSAT DIGITAL DATA PRODUCTS

Product description		Product code	Price in Rs.	
AVHRR	Level 1B CCT	1600 BPI	ST0000086	6800
	Level 1B CCT	6250 BPI	ST0000087	6800
	Lgsowg CCT	1600 BPI	ST0000066	7600
	Lgsowg CCT	6250 BPI	ST0000067	7600
TOVS	Level 1B CCT	1600 BPI	ST0000086	5600
	Level 1B CCT	6250 BPI	ST0000087	5600

Note: Lgsowg CCTs can be supplied in BSQ (Band Sequential) format also. Replace the 8th character of product code with 8. (6 for BIL format and 7 for BSQ format).

ERS-SAR (IMAGE MODE) STANDARD DATA PRODUCTS

Product description		Nominal scale	Product	Price in Rs.
1	Photo products			
1.1	Black & White (B&W) per band scene			
	240mm Film positive	1:500,000	STPC01212	17900
	240mm Paper print	1:500,000	STPC01222	17900
	480mm Paper print	1:250,000	STPC01223	17900
	960mm Paper print	1:125,000	STPC01224	17900
2	Digital data products			
	Computer Compatible Tape 1600 BPI		STPC00266	29000
	Computer Compatible Tape 6250 BPI		STPC00267	29000

Note: The data will be supplied from existing archives. For programming requests, access fees is charged extra.

SPECIFICATIONS OF IRS-P4 PAYLOAD

Ocean colour monitor

IFOV (meters)	VNIR : 250, SWIR : 500	
Swath (km)	1500	
Instrument bands	Wavelength (nm)	NE (%)
1	402-422	0.045
2	433-453	0.039
3	490-510	0.030
4	500-520	0.035
5	555-575	0.030
6	655-675	0.030
7	745-785	0.025
8	845-885	0.060
9	1500-1700	0.200

M.F.T.	>0.2 at Nyquist frequency
Digitization	12 bits
Data rate (Mbps)	17.35
Along track steering	+20, 0, -20°
Payload weight	75 kg
Payload power	1 electro-optics module
No. of packages	4 electronoics packages

Multi-frequency scanning microwave radiometer

Frequency (GHz)	6.6, 10.6, 18, 21
Altitude (km)	743
Antenna dia (cm)	86
Polarization	V&H, V&H, V&h, H
3 dB beam width (deg)	4.2, 2.6, 1.6, 1.4 for respective frequencies
Spatial resolution (km)	120, 75, 45, 40 for respective frequencies
Swath (km)	1500

Pemp. resolution	>1.0° K
Dynamic temp. range	10.330° K
Data rate (kbps)	5
DC power (W)	76
Weight (kg)	6

Source: M.G. Chandrasekhar, V. Jayaraman & Mukund Rao (1966), Future prespectives of remote sensing, *Current Science*, vol. 70, No. 7, pp. 648-53.

BIBLIOGRAPHY

Abdalla, A.E. (1991), A short note on the production of satellite image maps for Sudan, *ITC Journal,* vol. 1991-1, pp. 1-2.

Adearn, S.C. & C. Wee (1991), Data space volumes and classification optimization of SPOT and Landsat TM data, *PERS*, vol. 57 (1), pp. 61-5.

Adeniyi, P.O. (1983), An aerial photographic method for estimating urban population, *PERS,* vol. 49 (4), pp. 545-60.

Adeniyi, Peter O. (1987), Using remotely sensed data for census surveys and population estimation in developing countries: Examples from Nigeria, *Geocarto International*, vol. 4, pp. 11-31.

Allen, Kathleen M., Stanton W. Green & Ezra B.W. Zubrow (1990), *Interpreting Space: GIS and Archaeology*, Taylor & Francis, London.

Anderson, J.R., E.E. Hardy, J.T. Roack & R.E. Witmer (1976), A land use and land cover classification system for use with remote sensing sensor data, Professional Paper, *ASPRS/ACSM Annual Convention,* pp. 276-91

Anson, R.W. & F.J.Ormerling (eds)(1991), *Basic Cartography for Students and Technicians–Exercise Manual*, volumes I and II, ICA & Elsevier Applied Science, London and New York.

Anuta, P.E., L.A. Bartolucci, M.E. Dean, D.F. Lozano, E. Malaret, C.D. McGillem, J.A. Valdes & C.R. Valenzuela (1984), Landsat 4 MSS and thematic mapper data quality and information content analysis, *IEEE-TGRS,* GE-22, pp. 152-9.

Armstrong, M. & P. Densham (1990), Database organisation strategies for spatial decision support systems, *IJGIS*, vol. 4, No. 1, pp.3-20.

August, Peter, Joanne Michaud, Charles Labash & Christopher Smith (1994), GPS for environmental applications : Accuracy and precision of locational data, *PERS*, vol. 60 (1), pp. 41-5.

Barnsley, M.T., J. Sheperd & Y. Sun (1988), Conversion and evaluation of remotely sensed imagery for town planning purposes in J.C. Muller (ed), *Environmental Applications of Digital Mapping*, Proceedings of the Eurocarto Seven, ITC Publication, No. 8, Enschede, pp. 134-43.

Barnsley, M.J., G.J. Sadler & J.S. Shepherd (1989), Integrating remotely-sensed images and digital map data in the context of urban planning, *Proceedings of the 15th Annual Conference of the Remote Sensing Society*, 13-15 September 1989, Bristol, pp. 25-32.

Begin, D.F. (1991), System of integrated acquisition procedure with satellite data SINAPS project, in *Proceedings of the 1991 ACSM/APPRS Annual Convention*, vol. 3, Remote Sensing, Baltimore, pp. 1-5.

Benediktsson, J.A., P.H. Swain & O.K. Esroy (1990), Neural network approaches versus statistical methods in classification of multisource remote sensing data, *IEEE TGRS,* vol. 28(4), pp. 540-52.

Berry, J.K. (1987), Fundamental operations in computer assisted map analysis, *IJGIS*, vol. 1 (2), pp. 119-36.

Black, James (1996), Data conversion—Follows the Sun, *International Journal for Geomatics*, vol. 10, No. 4, pp. 66-7.

Blodget, Herbert W. (1991), Caldera-structure relationships in Kenya : Observations from satellite data, *1991 ACSM/ASPRS Annual Covention*, vol. 3, Baltimore, pp. 16-22.

Bohme, R. (1989), *Inventory of World Topographic Mapping*, ICA & Elseview Applied Sciences, London and New York.

Bolstad, P.V. & T.M. Lillesand (1991), Rapid maximum likelihood classification, *PERS,* vol. 57 (1), pp. 67-74.

Bolstad, Paul & T.M. Lillesand (1991), Automated GIS integration in landcover classification, *1991 ACSM/ ASPRS Technical Papers*, vol. 3, pp. 23-32.

Brannon, David P., Charles L. Hill & Bruce A. Davis (1994), Commercial remote sensing program, *PERS*, vol. 60 (3), pp. 317-30.

Brugioni, D.A. (1983), The census : It can be done more accurately with space-age technology, *PERS*, vol. 49 (9), pp. 1337-9.

Buchman, G.M. & N.K., Hubbard (1986), Remote sensing in land-use planning : An application in west central Scotland using SPOT-simulation data, *IJRS*, vol. 7 (6), pp. 767-77.

Bugayevshiy, L.M. & John P. Synder (1995), *Map Projections: A Reference Manual*, Taylor & Francis, London.

Bullard, R.K. & R.W. Dixon-Gough (1985), *Britain from Space: An Atlas of Landsat Images—In Colour*, Taylor & Francis, London.

Burroughs, P.A. (1986), *Principles of Geographic Information Systems for Land Resources Assessment*, Oxford University Press.

Campbell, Heather & Ian Masser (1995), *GIS in Organizations: How Effective are GIS in Practice ?*, Taylor & Francis, London.

Campbell, J.B. (1981), Spatial correlation effects upon accuracy of supervised classification of land cover, *PERS*, vol. 47, pp. 355-63.

Chandrasekhar, M.G., V. Jayraman & Mukund Rao (1996), Future perspectives of remote sensing, *Current Science*, vol. 70, No. 7, pp. 648-53.

Changchui, He, Robert A. Ryerson & Barry N. Haack (1991), The ESCAP/UNDP Regional remote sensing program for Asia and the Pacific, *1991 ACSM/ASPRS Technical Papers,* Baltimore, vol. 3, pp. 33-41.

Chatterjee, Rajat S., Devashis Banerjee, Jibanananda Roy & Amit K. Bhattacharya

(1994), Landsat TM data processing techniques for identifying and delineating environmental impacts of coal mining, *ITC Journal*, vol. 1994-2, pp. 156-162

Chaveg, P.A. Jr. & B. Baner (1982), An automatic optimum kernel-size selection technique for edge enhancement, *RSE*, vol. 12, pp. 23-38.

Cheves, Marc (1996), The U.S. surveying market-place--1966 : The year of GPS, *International Journal for Geomatics*, vol. 10, No. 4, pp. 25-7.

Chiao, Kuo-mu (1991), A study on age class discrimination of cryptomeria with greenness indicators, *1991 ACSM/ASPRS Technical Papers*, vol. 3, Baltimore, pp. 42-51.

Chiesa, Chistopher C. & Williams A. Tyler (1991), Mosaicking techniques developed for AVHRR data, *1991 ACSM/ASPRS Annual Convention*, vol. 3, Batlimore, pp. 52-66.

Chouhan, T.S. & K.N. Joshi (eds) (1995), *Readings in Remote Sensing Applications*, Scientific Publishers, Jodhpur.

Chuvieco, E. & R. Congalton (1998), Using cluster analysis to improve the selection of training statistics in classifying remotely sensed data, *PERS*, vol. 54 (9), pp. 1275-81.

Civco, Daniel L. (1991), Landsat TM image classification with an artificial neural network, *1991 ACSM/ASPRS Annual Convention*, Baltimore, vol. 3, pp. 67-77.

Civco, D.L. (1989), Knowledge-based land use and land cover mapping, in *Proceedings of the Annual ASPRS Convention*, Baltimore, pp. 176-91.

Colvocoresses, Alden P. (1991), Landsat 7--A challenge to America, *1991 ACSM/ ASPRS Annual Convention*, vol. 3, pp. 78-82.

Colvocoresses, A. (1986), Image mapping with the Thematic Mapper, *PERS*, vol. 52 (9), pp. 1499-505.

Colwell, Stephen (1992), *GIS and GPS: A New Partnership*, Space Technology International, Cornhill Publications.

Conese, C., F. Maselli & L. Bottai (1993), Topographic normalization of TM scenes through the use of an atmospheric correction method and digital terrain models, *PERS*, vol. 59 (12), pp. 1745-53.

Cook, Elizabeth A. & Louis R. Iverion (1991), Inventory and change detection of urban land cover and Illinois using Landsat Thematic Mapper data, in *1991 ACSM/ASPRS Technical Papers*, vol. 3, pp. 83-92.

Cosentino, Brein L. & Thomas M. Lillesand (1991), Towards automated statewise land cover mapping in Wisconsin using satellite remote sensing and GIS Techniques, *1991 ACSM/ASPRS Technical Papers*, vol. 3, pp. 93-102.

Crane, Aaron I., Peter R. Weiler & Frank L. Scarpace (1991), The use of transputers to reduce maximum likelihood classification time, *1991 ACSM/ ASPRS Technical Papers*, vol. 3, pp. 103-10.

Crankell, A.P. *et al.* (eds)(1989), *Computer Programs for Image Processing of Remote Sensing Data*, University of Dundee, Dundee,

Cranknell, A.P. & Paithoonwattanakij (1989), Pixel and sub-pixel accuracy in geometrical correction of AVHRR imagery, *IJRS*, vol. 10, pp. 661-7.

Cranknell, A.P. & L.W.D. Hayes (1991), *Introduction to Remote Sensing*, Taylor and Francis, London.

Cranknell, A.P. & Ladson Hayes (eds) (1990), *Remote Sensing Yearbook*, 1990, Taylor and Francis, London.

Cromp, Robert F. (1991), Automated extraction of meta-data from remotely sensed satellite imagery, *1991 ACSM/ASPRS Technical Paper*, vol. 3, pp. 111-20.

Dale, P.F. & J. McLaughlin (1988), *Land Information Management: An Introduction with Special Reference to Cadastral Problems in the Third World Countries*, Oxford University Press.

Dangermond, J. (1991), Where is GIS technology going ?, *ARC News*, vol. 13, No. 2, pp. 24-5.

Curran, J.P. *et al.* (eds)(1988), *Compendium of Cartographic Techniques*, ICA & Elsevier Applied Sciences, London and New York.

Curran, Paul J. (1985), *Principles of Remote Sensing*, ELBS/Longman, Essex.

Dasgupta, A.R. (1995), IRS-1C data processing, *SAC Courier*, vol. 20, No. 2, pp. 2-6.

Davis, W.A. & F.G. Peet (1977), A method of smoothing digital thematic maps, *RSE*, vol. 6, pp. 45-9.

Deekshatulu, B.L. & Y.S. Rajan (eds) (1984), *Remote Sensing*, Macmillan India Press, Madras.

Deekshatulu, B.L. (1991), Remote sensing of environment--critical issues, *Photonirvachak*, vol. 19(4), pp. 237-44.

Deffies, R.S. & J.R.G. Townshend (1994), NDVI-derived land cover classifications at a global level, *IJRS*, vol. 15, No. 17, pp. 3567-86.

Denegre, J. (ed) (1988), *Thematic Mapping from Satellite Imagery: An International Report*, ICA and Elsevier Applied Sciences, London and New York.

Derenyi, Eugene E. (1994), Simple techniques for thematic mapping, *ISPRS Journal of Photogrametry and Remote Sensing*, vol. 49, No.6, pp. 27-31.

Di Zenzo, S., S.D. DeGloria, R. Bernstein & H.G. Kolsky (1987), Gaussian maximum likelihood and contextual classification algorithms for multicrop classification experiments using thematic mapper and multispectral scanner data, *IEEE-TGRS*, GE-22, vol. 6, pp. 37-44.

Dickenson, H. & H. Calkins (1988), The economic evaluation of implementing GIS, *IJGIS*, vol. 2, No. 4, pp. 307-28.

Dillworth, Mary E. (1991), Geographic windows in remote sensing : Does windows size matter ? *1991 ACSM/ASPRS Technical Papers*, vol. 3, pp. 122-8.

Duggin, M.J., R. Rowntree, M. Emmon, N. Hubbard, A.W. Odell, H. Sakhavat & J. Lindsay (1986), The use of multidate multichannel radiance data in urban future analysis, *RSE*, vol. 20, pp. 95-105.

Durry, S.A. (1990), *A Guide to Remote Sensing : Interpreting Images of the Earth*,

Oxford University Press, Oxford.

Dutta, Sujay, S.A. Sharma & Ajai (1994), Accuracy assessment in cotton acreage estimation using Indian Remote Sensing Satellite data, *ISPRS Journal of Photography and Remote Sensing*, vol. 49, No. 6, pp. 21-6.

Ehlers, Manfred, David Greenlee, Terence Smith & Jeffery Star (1991), Integration of remote sensing and GIS: Data and data access, *PERS*, vol. 57(6), pp. 669-75.

Estes, John E. (1992), Remote sensing and GIS integration : Research needs, status and trends, *ITC Journal*, vol. 1992-1, pp. 2-9.

Eyre, L.A., B. Addphers & M. Amiel (1970), Census analysis and population studies, *Photogrammetric Engineering*, vol. 36 (5), pp. 460-6.

Faust, Nickolas L., William H. Anderson & Jeffery L. Star (1991), Geographic information systems and remote sensing future computing environment, *PERS*, vol. 57 (6), pp. 655-68.

Ferndndez, R. Norberto, D. Fabian Lozano-Garcia, Greg Deeds & Chris J. Johannsen (1991), Accuracy assessment of map coordinate retrieval, *PERS*, vol. 57, No. 11, pp. 1447-52.

Ferrante, R.D., M.H. Carlutto. H. Pomaraede & P.W. Baim (1989), Multispectral image analysis system, *Proceedings of the 1st Conference on Artificial Intelligence Applications*, pp. 357-63.

Fleming, M.D. & R. M. Hoffer (1979), Machine processing of Landsat MSS data and DMA topographic data for forest cover type mapping, *Proceedings of the Machine Processing of Remotely Sensed Data*, Purdue University, pp. 377-90.

Forster, B.C. (1985), An examination of some problems and solutions in monitoring urban areas from satellite platforms, *IJRS,* vol. 6, pp. 139-57.

Forster, B.C. (1980), Urban residential ground cover using Landsat digital data, *PERS*, vol. 46, pp. 547-58.

Fotheringham, A.S. & P. Rogerson (eds) (1994), *Spatial Analysis and GIS*, Taylor & Francis, London.

Frank, Andrew U., Max J. Egenhofer & Werner Kuhn (1991), A perspective on GIS technology in the nineties, *PERS*, vol. 57, No. 11, pp. 1433-6.

Fuller, R.M. & G.B. Groom (1994), The land cover map of Great Britain: An automated classification of Landsat Thematic Mapper data, *PERS*, vol. 60 (5), pp. 553-62.

George, Nich (1996), GIS & information technology, *GIS Asia Pacific*, vol. 2, No. 2, pp. 34-7.

Goldberg, M., D.G. Goodenough, M. Alvo & G.M. Karma (1988), A hierarchical expert system for updating forestry maps with Landsat data, *Proceedings of IEEE*, vol. 73, pp. 1054-63.

Goodchild, Michael F. (ed)(1989), *The Accuracy of Spatial Databases*, Taylor & Francis, London.

Goodchild, Michael & Sucharita Gopal (eds) (1989), *Accuracy of Spatial*

Database, Taylor and Francis, London.

Gopal, Sucharita & Curtis Woodcock (1994), Theory and methods for accuracy assessment of thematic maps using fuzzy sets, *PERS*, vol. 60 (2), pp. 181-8.

Gopalan, A.K.S. (1996), Payload for P4 satellite, *SAC Courier*, vol. 21, No. 1, pp. 7-12.

Gordon, S.J. (1980), Utilizing Landsat imagery to monitor land-use change : A case study of Ohio, *RSE*, vol. 9, pp. 189-96.

Green, David & David Rix (eds)(1993), *Geographic Information: 1994*, Taylor & Francis, London.

Green, Kass (1994), Using remote sensing to detect and monitor land-cover and land-use chnages, *PERS*, vol. 60 (3), pp. 331-7.

Gregory, A.F. & H.D. More (1986), Economical maintenance of a national topographical data base using Landsat images, *PERS*, vol. 52 (4), pp. 519-24.

Gupta, S.C. (1995), India's space launch vehicles, *Photonirvachak*, vol. 23, No. 4, pp. 163-74.

Gurney, Charlotte M. (1981), The use of contextual information to improve land cover classification of digital remote sensing data, *IJRS*, vol. 2(4), pp. 379-88.

Gurney, Charlotte M. & J.R.G. Townshend (1983), The use of contextual information in the classification of remotely sensed data, *PERS*, vol. 49(1), pp. 55-64.

Haack, B., N. Bryant & S. Adams (1987), An assessment of Landsat MSS and TM data for classification, *RSE*, vol. 13, pp. 149-60.

Haack, Barry & Susan Jampoler (1994), Agricultural classification comparisons using Landsat thematic mapper data, *ITC Journal*, vol. 1994-2, pp. 119-28.

Haines-Young, Roy, David Green & Steven Cousins (eds)(1993), *Landscape Ecology and Geographical Information Systems*, Taylor & Francis, London.

Handerson, F.M. (1979), Housing and population analysis, in K. Ford (ed), *Remote Sensing for Planners*, Rutgers Centre for Urban Policy Research, New Brunswick, N.J., pp. 135-54.

Hardy, E.E. (1982), Remote sensing for land and water resource management, in Johnansen and Sanders (eds), *Remote Sensing for Resources Management*, Soil Conservation Society of America, pp. 3-15.

Harris, R. (1987), *Satellite Remote Sensing : An Introduction*, Routledge and Kegan Paul, London & New York.

Harris, R. (1980), Spectral and image processing for remote sensing, *IJRS*, vol. 1, pp. 361-75.

Healey, R.G. & G.B. Desa (1989), Transputer based parallel processing for GIS analysis : Problems and potentials, *Auto-Carto 9 Proceedings*, April, pp. 90-99.

Hebbar, K. Jairam, J. Dakshina Murthy, K.M.M. Rao, B. Laxmi, V. Vittal Reddy, R. Joseph Arokiadas, D. Kaveri Devi, D. Chandrashekaran, G. Padma Rani & V. Raghu (1996), IRS-1C data products generation and dissemination, *Current*

Science, vol. 70, No. 7, pp. 551-61.

Helfert, M.R. & K. Lulla (1989), Monitoring tropical environments with space shuttle photography, *Geocarto International*, vol. 4 (1), pp. 55-69.

Hepner, G.F., T. Logan, N. Rittner & N. Bryant (1990), Artificial neural network classification using a minimal training set: comparison to conventional supervised classification, *PERS*, vol. 56(4), pp. 469-73.

Hixson, M., D. Scholtz & N. Fuhs (1980), Evaluation of several schemes for classification of remotely sensed data, *PERS*, vol. 46 (12), pp. 1547-53.

Hodgson, Michael E. (1991), Characteristics of the window for neighbourhood analysis of nominal data, *1991 ACSM/ASPRS Technical Papers*, vol. 3, pp. 206-14.

Hopkin, P.J., A.L. Maclan & T.M. Lillesand (1988), Assessment of Thematic Mapper imagery for forestry applications under Lakes States conditions, *PERS*, vol. 54(1), pp. 61-70.

Horler, D.N.H. & F.J. Ahern (1986), Forestry information content of Thematic Mapper data, *IJRS*, vol. 7(1), pp. 405-28

Hudson, W.D. (1987), Evaluating Landsat classification accuracy from forest cover-type maps, *CJRS*, vol. 13(1), pp. 39-42.

Hunter, Gary (1996), Accuracy and GIS--Why worry about it ?, *GIS Asia Pacific*, vol. 2, No. 2, pp. 16-7.

Hutchinson, C.F. (1982), Techniques for combining Landsat and ancillary data for digital classification improvement, *PERS*, vol. 48(1), pp. 123-30.

Ilsaka, J. & E. Hagedus (1982), Population estimates from Landsat imagery, *RSE*, vol. 12, pp. 259-72.

Indian Society of Geomatics (1996), *Tutorial Notes on Geographic Information System: GIS--The Basics*, Space Applications Centre, Ahmedabad.

James, M. (1986), *Classification Algorithms*, John Wiley and Sons, New York.

Jansen, J.R., E.W. Ramsey, J.M. Holmes, J.E. Michel, B. Savitsky & B.A. Davis (1990), Environmental sensitivity Index (ESI) mapping for oil spills using remote sensing and geographical information system technology, *IJGIS*, vol. 4(2), pp. 181-201.

Jansen, Lucas L.F. (1994), Accuracy of satellite derived land-cover data: A review, *PERS*, vol. 60 (4), pp. 419-26.

Jansen, J.R., M.E. Hodgson, E.J.Christensen, H.E. Mackey, L.R. Timey & R.R. Sharitz (1986), Remote sensing inland wetlands: A multispectral approach, *PERS*, vol. 52(1), pp. 87-100.

Jansen, J.R., E.J. Christensen & R.R. Sharitz (1984), Nontidal wetland mapping in South Carolina using airborne multispectral scanner data, *RSE*, vol. 16, pp. 1-12.

Jansen, John R. (1986), *Introductory Digital Image Processing--A Remote Sensing Perspective*, Prentice-Hall, New York.

Jayraman, V. & B. Manikiam (1993), *Remote Sensing in India : Facilities &*

Infrastructure, National Natural Resources Management System, Department of Space, Technical Report, DOS-NNRMS-TR-97-93, Bangalore.

Jensen, John R., David J. Cowen & Joanne Halls (1994), Improved urban infrastructure mapping and forecasting for Bellsouth using remote sensing and GIS technology, *PERS*, vol. 60 (3), pp. 339-46.

Jensen, J.K. & D.L. Toll (1982), Detecting residential land use development at the urban fringe, *PERS*, vol. 48 (4), pp. 629-43.

Jenson, S.K. & J.O. Domingue (1988), Extracting topographical structure from digital elevation data for geographic information system analysis, *PERS*, vol. 54, pp. 1593-1600.

Jere, Ajit V. (1996), Computer aided siting analysis--A GIS approach, *GIS India*, vol. 5, No. 1111, pp. 25-8.

Johannsen, C.T. & J.L. Sanders (eds) (1982), *Remote Sensing for Resources Management*, Soil Conservation Society of America, Ankery (Iowa).

Johnsson, Katarina (1994), Segment-based land-use classification from SPOT satellite data, *PERS*, vol. 60 (1), pp. 47-53.

Justice, C.O. & J.R. Townshend (1994), Data sets for global remote sensing : Lessons learnt, *IJRS*, vol. 15, No. 17, pp. 3621-39.

Kalyanaraman, S. & R.K. Rajangam (1996), Advanced features and specifications of IRS-1C spacecraft, *Current Science*, vol. 70, No. 7, pp. 501-9.

Kasturirangan, K., R. Aravamudan, B.L. Deekshatulu, George Joseph & M.G. Chandrasekhar (1996), Indian remote sensing satellite (IRS)-1C--The beginning of a new era, *Current Science*, vol. 70, No. 7, pp. 495-500.

Kasturirangan, K. (1994), *Sustainable Development and Role of Geomatics (Remote Sensing & Conventional Data Integration)*, Indian Space Research Organisation, Special Report ISRO-SP-68-94, Bangalore.

Kasturirangan, K. (1995), Remote sensing in India--Present scenario and future thrusts, *Photonirvachak*, vol. 23 (1), pp. 1-11.

Kelly, Glenn & Joy Hood (1991), AVHRR conterminous United States reference data set, *1991 ACSM/ASPRS Technical Papers*, vol. 3, pp. 232-9.

Kontones, C., G.G. Wilkinson, A. Burrill, S. Goffredo & J. Megier (1993), An experimental system for the integration of GIS data in knowledge-based image analysis for remote sensing of agriculture, *IJGIS*, vol. 7, pp. 247-262 and vol. 8, p. 489.

Kudrat, M., S.K. Saha & A.K. Tiwari (1992), Soil resource mapping using IRS 1A LISS II digital data--A case study of Kandi area adjacent to Chandigarh, India, *IJRS*, vol. 4, No. 2, pp. 129-33.

Kudrat, M. & S.K. Saha (1993), Land productivity assessment and mapping through integration of satellite and terrain slope data, *Photonirvachak*, vol. 21, No. 3, pp. 157-66.

Kudrat, M., S.K. Saha & A.K. Tiwari (1990), Potential use of IRS-LISS II digital data in soil and landuse mapping and productivity assessment, *Proceedings of the National Symposium on Agricultural Applications of Remote Sensing*, New Delhi, pp. 25-34.

Kudrat, M., S.K. Saha & A.K. Tiwari (1990), The potential of IRS-LISS II digital data for soil mapping and land evolution studies, *Asian Pacific Remote Sensing Journal*, vol. 2, No. 2, pp. 73-8.

Kumar, Anil & T.C. Sarma (1996), Data reception system for IRS-1C, *Current Science*, vol. 70, No. 7, pp. 534-42.

Lauer, Donald T., John E. Estes, John R. Jensen & David D. Greenlee (1991), Institutional issues affecting the integration and use of remotely sensed data and geographic information system, *PERS*, vol. 57 (6), pp. 647-54.

Lee, J., R.C. Weger, S.K. Sengupta & R.M. Welch (1990), A neural network approach to cloud classification, *IEEE Transaction on Geosciences and Remote Sensing*, vol. 28(5), pp. 846-55.

Lillesand, T.M. (1990), Satellite remote sensing : Its evolution and synergism with GIS technology, *Government Information Quarterly*, vol. 7, No. 3, pp. 307-27.

Lo, C.P. & F.F. Chan (1980), Rural population estimation from aerial photographs, *PERS*, vol. 46, pp. 337-45.

Lo, C.P. (1986), Accuracy of population estimation from medium-scale aerial photography, *PERS*, vol. 52, pp. 1859-69.

Lock, Gary (ed) (1995), *Archaeology and Geographic Information Systems*, Taylor & Francis, London.

Loveland, Thomas R., James W. Merchant, Donald O. Ohlen & Jesslyn F. Brown (1991), Development of land-cover characteristics database for the conterminous U.S., *PERS*, vol. 57, No. 11, pp. 1453-63.

Lunetta, Ros S., Russel G. Congalton, Lynn K. Fenstemaker, John R. Jensen, Kenneth C. McGwire & Larry R. Tinney (1991), Remote sensing and geographic information system data integration : Error sources and research issues, *PERS*, vol. 57 (6), pp. 677-87.

Ma, Z. & C.E. Olson (1989), A measurement of spectral overlap among cover types, *PERS*, vol. 55, No. 10, pp. 1441-4.

Maguire, D.J., M.F. Goodchild & D.W. Rhind (1991), *Geographical Information Systems : Principles and Applications*, Longman Scientific and Technical Publications, London.

Majumdar, K.L. (1996), IRS-1C data products, *SAC Courier*, vol. 21, No. 1, pp. 3-4.

Malingreen, J.P., C. Tucker & N. LaPorte (1989), AVHRR for monitoring global tropical deforestation, *IJRS*, vol. 10 (4), pp. 855-67.

Marble, D.E. & D.J. Peuquet (1983), Geographic information systems and remote sensing, in R.N. Colwell (ed), *Manual of Remote Sensing*, American Society of Photogrammetry, Falls Church, Virginia, Chapter 22, pp. 923-58.

Marrow, P.J. & R.H. (1988), The design and implementation of low-land image processing algorithms on a transputer network, in I. Page (ed), *Parallel Architecture and Computer Vision*, Oxford Press, pp. 243-59.

Mason, D.C., D.G. Corr, A. Cross, D.C. Hogg, D.H. Lawrence, M. Petron & A.M. Tailor (1988), The use of digital map data in the segmentation and classification of remotely-sensed images, *IJGIS*, vol. 2(3), pp. 195-215.

Mather, Paul M. (ed)(1992), *Terra-1: Understanding the Terrestrial Environment: The Role of Earth Observations from Space*, Taylor & Francis, London.

Max, R. (1990), The TIGER system : Yesterday, today and tomorrow, *Cartography and Geographic Information Systems*, vol. 17, no. 1, pp. 89-97.

McDonald, Kenneth R. & Deborah J. Blake (1991), Information management challenges of the EOS data and information system, *1991 ACSM/ASPRS Technical Papers*, vol. 3, pp. 258-67.

Mckeown, D.M. (1987), The role of artificial intelligence in the integration of remotely sensed data with geographic information system, *IEEE-TGRS*, vol. GE-25(3), pp. 330-49.

Mead, R.A., T.L. Sharik *et al.* (1981), A computerized spatial analysis system for assessing wildlife habitat from vegetation maps, *CJRS*, vol. 7(1), pp. 34-40.

Mehldan, G, & R.A. Schowengerdt (1990), A C-extension for rule-based image classification systems, *PERS*, vol. 56(6), pp. 887-92.

Merson, R.H. (1989), An AVHRR mosaic image of Antarctica, *IJRS,* vol. 10 (4 & 5), pp. 669-74.

Michener, W.K. & J.W. Brunt (eds) (1994), *Environmental Information Management and Analysis*, Taylor & Francis, London.

Milazzo, V.A. & R.H. DeAngelie (1984), Application of simulated SPOT data to mapping land cover patterns and changes in an urban fringe environment, *Proceedings of the 1984 SPOT Simulation Symposium*, American Society of Photogrammetry, Falls Church, pp. 177-86.

Moellering, H. (ed) (1991), *Spatial Database Transfer Standards : Current International Status*, ICA & Elsevier Applied Sciences, London and New York.

Mohie El Deen, F.A. (1991), The use of GIS, GPS and satellite remote sensing to map woody vegetation in Kazagail area, Sudan, *ITC Journal*, vol. 1991-1, pp. 3-10.

Molenaar, Martian (1989), Towards a geographic information theory, *ITC Journal*, vol. 1989-1, pp. 5-11.

Morrow-Jones, H.A. (1984), Remote sensing technology and the U.S. Census, *PERS*, vol. 50 (1), pp. 229-12.

Mounsey, Helen & Roger Tomlinson (eds)(1988), *Building Databases for Global Sciences*, Taylor and Francis, London.

Muller, Jean-Claude (1991), The cartographic agenda of the 1990s: Updates and prospects, *ITC Journal*, vol. 1991-2, pp. 55-62.

Muller, Jean-Claude (1993), Latest developments in GIS/LIS, *IJGIS*, vol. 7, No. 4, pp. 293-303.

Muller, J.C. (ed) (1991), *Advances in Cartography*, ICA & Elsevier Applied Sciences, London and New York.

Muller, Jan-Peter (ed) (1988), *Digital Image Processing in Remote Sensing*, Taylor and Francis, London.

Nag, P. (ed) (1984), *Census Mapping Survey,* Concept Publishing Company, New Delhi.

Nag, P. (1987), A proposed base for a geographical information system for India, *IJGIS*, vol. 1(2), pp. 181-7.

Nag, P. (ed) (1992), *Thematic Cartography and Remote Sensing*, Concept Publishing Company, New Delhi.

Navalgund, R.R., J.S. Parihar, L. Venkataratnam, M.V. Krishna Rao, S. Panigrahy, M.C. Chakraborty, K.R. Hebbar, M.P. Oza, S.A. Sharma, N. Bhagia & V.K. Dadhwal (1996), Early results from crop studies using IRS-1C data, *Current Science*, vol. 70, No. 7, pp. 568-74.

Nayak, Shailesh, Prakash Chauhan, H.B. Chauhan, Anjali Bahuguna & A. Narendra Nath (1996), IRS-1C applications for coastal zone management, *Current Science*, vol. 70, No. 7, pp. 614-8.

Nelson, R.F., R.S. Latty & G. Mott (1984), Classifying northern forests with Thematic Mapper simulator data, *PERS,* vol. 50(5), pp. 607-17.

NRSA (1988), *Landsat Thematic Mapper Data Applications--Illustrated Examples*, National Remote Sensing Agency, Hyderabad.

NRSAA (1995), *Integrated Mission for Sustainable Development--Technical Guidelines*, National Remote Sensing Agency, Hyderabad.

Ogata, Al (1992), Making the most of today's data conversion methods, *GIS World*, April, pp. 82-6.

Olorunfemi, J.F. (1984), Land use and population : A linking model, *PERS*, vol. 50 (2), pp. 221-7.

Oza, M.P. (1989), Selection of band combination of IRS data, *Photonirvachak*, vol. 17, No. 2, pp. 23-6.

Paine, David P. (1981), *Aerial Photography and Image Interpretation for Resource Management*, John Wiley & Sons, New York.

Parker, H. Dennison (1991), The role of the private sector in GIS, *PERS*, vol. 57, No. 11, pp. 1465-6.

Paul, C.K. & A.F. Mascarenhas (1981), Remote sensing in development. *Science*, vol. 214 (4517), pp. 139-45.

Petrie, Gordon & Abdelazim E.H. El Niweiri (1994), Comparative testing of space images for small-scale topographic mapping of Sudan, *ITC Journal*, vol. 1994-2, pp. 113-8.

Peuquet, Donna J & Todd Bacastow (1991), Organisational issues in the development of geographical information systems: A case study of U.S. Army topographical infor-mation automation, *IJGIS,* vol. 5(3), pp. 303-19.

Peuquet, Donna J. & Duane F. Marble (eds)(1990), *Introductory Readings in Geographical Information System*, Taylor and Francis, London.

Polle, V.F.L. (1984), Population estimation from aerial photos for non-homogeneous urban residential areas, *ITC Journal*, vol. 1984-2, pp. 116-22.

Poller, N.G. & J.C. Colwell (1986), Coarse-resolution satellite data for ecological surveys, *Biosciences*, vol. 36, pp. 468-75.

Price, Martin & Ian Heywood (eds) (1994), *Mountain Environments and Geographic Information Systems*, Taylor & Francis, London.

Radhakrishnan, K., S. Adiga, Geeta Vardhan & P.G. Diwakar (1966), Enhanced geographic information system application using IRS-1C data--Potential for urban utility mapping and modelling, *Current Science*, vol. 70, No. 7, pp. 629-37.

Raghavswamy, V., S.K. Pathan, P. Ram Mohan, R.J. Bhandari & Padma Priya (1996), IRS-1C applications for urban planning and development, *Current Science*, vol. 70, No. 7, pp. 582-8.

Rao, Mukund, V. V. Jayaraman, K.R. Sridhar Murthy, N. Sampath & M.G. Chandrasekhar (1996), Commercialization of remote sensing--Issues and perspectives, *Current Science*, vol. 70, No 7, pp. 642-7.

Rao, Mukund, V, Jayaraman & M.G. Chandrasekhar (1994), *Organising Spatial Information Systems Around a GIS Core*, Indian Space Research Organisation, Special Publication, ISRO-NNRMS-SP-70-94.

Rao, U.R. (1991), Remote sensing for sustainable development, *Photonirvachak*, vol. 19(4), pp. 217-35.

Rao, U.R. (1989), *The Next 40 Years in Space,* Indian Space Research Organisation, Special Report ISRO-SP-46-89, Bangalore.

Rao, D.P., N.C. Gautam, R. Nagaraja & P. Ram Mohan (1996), IRS-1C applications in landuse mapping and planning, *Current Science*, vol. 70, No. 7, pp. 575-81.

Rao, D.P., A. Bhattacharya & P.R. Reddy (1996), Use of IRS-1C data for geological and geomorphological studies, *Current Science*, vol. 70, No. 7, pp. 619-23.

Rao, D.P., R.R. Navalgund & Y.V.N. Krishnamurthy (1996), Cadastral applications using IRS-1C data--Some case studies, *Current Science*, vol. 70, No. 7, pp. 624-8.

Rao, Mukund & V. Jayaraman (1995), *Guidelines for GIS Standardisation*, Indian Space Research Organisation, Technical Report ISRO-NNRMS-TR-105-95, Bangalore.

Raper, Jonathan (ed)(1989), *Three Dimensional Applications in GIS*, Taylor and Francis, London.

Rhind, D.W. (ed) (1989), *Cartography: Past, Present and Future*, ICA and Elsevier Applied Sciences, London and New York.

Rhind, D.W. & Helen Mounsey (1990), *Understanding Geographic Information*

System, Taylor and Francis, London.

Richards, J.A. (1986), *Remote Sensing Digital Image Analysis*, Springer-Verlag.

Roger. A., L. Charbonneau & F. Bonn (1988), Urbanization and Landsat MSS albedo change in the Windsor-Quebec corridor since 1972, *IJRS*, vol. 9, pp. 555-66.

Roy, P.S., C.B.S. Dutt, R.N. Yadav, B.K. Rangnath, M.R.S. Murthy, B. Gharai, V. Udaya Lakshmi, A.K. Kandya & P.S. Thakker (1996), IRS-1C data utilization for forestry applications, *Current Science*, vol. 70, No. 7, pp. 606-13.

Sabins, F.F. (1978), *Remote Sensing: Principles and Interpretation*, W.H. Freeman & G. San Franciso.

Sadler, G.J. & M.J. Barnsley (1990), Use of population density data to improve classification accuracies in remotely-sensed images of urban areas, *Proceedings of the First European Conference on GIS*, Amsterdam, April 10-13, pp. 968-77.

Sadowski, F.G., J.A. Sturdevant & R.A. Rowtree (1987), Testing the consistency for mapping urban vegetation with high-altitude aerial photographs, and Landsat MSS data, *RSE*, vol. 21, pp. 129-41.

Sahai, Baldev (ed) (1988), *Remote Sensing in Agriculture*, Indian Society for Remote Sensing, Ahmedabad Chapter, Ahmedabad.

Sanyal, A.K. (1991), *Map Projection*, Survey Training Institute, Hyderabad (mimeo.).

Schwarz, K.P., M.A. Chapman, M.W. Cannon & P. Gong (1993), An integrated INS/GPS approach to the georeferencing of remotely sensed data, *PERS*, vol. 59 (11), pp. 1667-74.

Sebastian, M., V. Jayaraman & M.G. Chandrasekhar (1995), *Space Technology Applications for Sustainable Development of Watersheds*, Indian Space Research Organisation, Technical Report ISRO-HQ-TR-104-95, Bangalore.

Sheldon, R.A. (1990), Satellite image analysis using neural networks, in *Proceedings of the 1990 Goddard Conference on Space Applications of Artificial Intelligence,* NASA Goddard Space Flight Center, Greenbelt, MD, pp. 349-55.

Shivakumar, S.K., K.S. Sarma, N. Nagarajan, M.G. Raykar, H.R. Rao, K.V.S.R. Prabhu & N. Rajan Paramananthan (1996), IRS-1C mission planning, analysis and operations, *Current Science*, vol. 70, No. 7, pp. 516-23

Short, N.M. (1989), Accuracy of image mapping from space, *1989 ACSM/ASPRS Technical Papers*, vol. 3, pp. 376-88.

Siegal, B.S. & A.R. Gillespie (1979), *Remote Sensing in Geology*, John Wiley & Sons, New York.

Sinclair, Susan (1996), The expanding role of GIS satellites in Asia Pacific, *GIS Asia Pacific*, vol. 2, No. 3, June, pp. 27-32.

Singh, K.D. (1990), Design of a global tropical forest resources assessment, *PERS*, vol. 56(10), pp. 1353-54.

Snyder, J.P. (1982), *Map Projection Used by the U.S. Geological Survey*, Geological Survey Bulletin 1532, U.S. Government Printing Office, Washington D.C,

Srivastava, P.K., B. Gopala Krishna & K.L. Majumdar (1996), Cartography and terrain mapping using IRS-1C data, *Current Science*, vol. 70, No. 7, pp. 562-7.

Stadelmann, M. & G.D. Lodwick (1993), A rule-based system for the extraction of cartographic features from Landsat TM imagery, *Cartography*, vol. 22, No. 2, pp. 1-8.

Stahler, A.H. (1981), Stratification of natural vegetation for forest and rangeland inventory using Landsat digital imagery and collateral data, *IJRS*, vol. 2, pp. 15-41.

Star, Jeffrey & John Estes (1990), *Geographical Information System : An Introduction*, Prentice-Hall, New Jersey.

Story, M. & R. Congalton (1986), Accuracy assessment : A user's perspective, *PERS*, vol. 52 (3), pp. 397-9.

Swain, Philip H. & Shirley M. Davis (1978), *Remote Sensing--The Quantitative Approach*, McGraw-Hill Inc., New York.

Swain, P.H., H.J. Siegel & B.W. Smith (1980), Contextual classification of multispectral remote sensing data using a multi-processor system, *IEEE--TGRS*, GE-18, No. 2, pp. 197-203.

Thiruvengadachari, S., S. Jonna, K.A. Hakeem, P.V. Raju, V. Bhanumurthy, G. Srinivasa Rao, P.R. Paul, E. Siva Sankar & A.T. Jeyaseelen (1996), Improved water management : The IRS-1C contribution, *Current Science*, vol. 70, No. 7, pp. 589-99.

Thomas, I.L. (1980), Spatial post processing of spectrally classified Landsat data, *PERS*, vol. 46(a), pp. 1201-26.

Toll, D.L. (1985), Analysis of digital Landsat MSS and SEASAT SAR data for use in discriminating land cover at the urban fringe of Denver, Colorado, *IJRS*, vol. 6 (7), pp. 1209-29.

Toll, D.L. (1985), Effect of Landsat Thematic Mapper sensor parameters on land cover classification, *RSE*, vol. 17, pp. 129-40.

Toll, D.L. (1984), An evaluation of simulated Thematic Mapper data and Landsat MSS data for discriminating suburban and regional land use and land cover, *PERS*, vol. 50 (12), pp. 1713-24.

Townshend, F.E. (1986), The enhancement of computer classifications by logical smoothing, *PERS*, vol. 52, pp. 213-21.

Vanghan, R.A. & R.P. Kirby (eds)(1989), *Geographical Information Systems and Remote Sensing for Local Resource Planning*, University of Dundee, Dundee.

Venkatachary, K.V., P. Soma, V. Jalaramaiah, A. Karuppaiyan, N. Sengupta, O. Chiranjeevi & Y.K. Singhal (1996), TTC network and spacecraft control centre for IRS-1C, *Current Science*, vol. 70, No. 7, pp. 524-33.

Victorov, Serge (1995), *Regional Satellite Oceanography*, Taylor & Francis, London.

Wall, Stephen D. & Kenneth W. Ledbetter (1991), *Design of Mission Operations Systems for Scientific Remote Sensing*, Taylor & Francis, London.

Wang, F. and R. Newkirk (1987), Design and implementation of a knowledge based system for remotely sensed change detection, *Journal of Imaging Technology*, vol. 13 (4), pp. 116-22.

Watkins, J.F. & H.A. Marrow-Jones (1985), Small area population estimates using aerial photography, *PERS*, vol. 51(12), pp.1933-5.

Waugh, Tom & Richard Healey (eds)(1994), *Advances in GIS Research,* Taylor & Francis, London.

Wayumba, G.O. & W.R. Philipson (1985), Remote sensing of shifting cultivation and grazing patterns in Kenya's semi-arid regions, *ITC Journal,* vol. 4, 1985, pp. 261-7.

Weiler, P.R., & F.L. Scarpace (1990), Using the architecture of the Intel 80287/80387 to optimize maximum likelihood computations, *1990 ACSM/ASPRS Technical Papers*, vol. 4, pp. 475-82.

Wharton, S.W. (1987), A spectral-knowledge-based approach for urban land cover discrimination, *IEEE--TGRS*, GE-25(1), pp. 272-82.

Woldai, Tsehale (1995), Lithologic and structural mapping in a vegetated low-relief terrain using multiple-source remotely sensed data : A case study of the Calanas area in southwest Spain, *ITC Journal*, vol. 1995-2, pp. 95-114.

Worboys, Michael F. (ed) (1994), *Innovations in GIS*, Taylor and Francis, London and Bristol.

Worboys, M.F. (ed) (1995), *GIS: A Computer Science Perspective*, Taylor & Francis, London.

Worrall, Les (1994), Justifying investment in GIS : A local government perspective, *IJGIS*, vol. 8(6), pp. 545-65.

Wright, J.W. (1967), Three types of reconnaissance mapping, *Cartographic Journal*, vol. 2 (1), pp. 16-22.

Zobrist, Albert L., Nevin A. Bryant, Ronald G. McLeod (1983), Technology for large digital mosaics of Landsat data, *PERS*, vol. 49, No. 9, pp. 1325-35.

Zubrow, E., K. Allen & S. Green (eds) (1990), *Interpreting Space : GIS in Archaeology and Anthropology*, Taylor and Francis, London.

Abbreviations used in Bibliography

ACSM	:	Americian Congress of Surveying and Mapping
CJRS	:	Canadian Journal of Remote Sensing
GIS	:	Geographical Information System
IEEE-TGRS	:	IEEE Transactions of Geosciences and Remote Sensing
IJGIS	:	International Journal on Geographical Information System
IJRS	:	International Journal on Remote Sensing
PERS	:	Photogrammetry Engineering and Remote Sensing
RSE	:	Remote Sensing of Environment

INDEX